Peptide-Based Drug Design

METHODS IN MOLECULAR BIOLOGY™

For other titles published in this series, go to
www.springer.com
select the subdiscipline
search for your title

METHODS IN MOLECULAR BIOLOGY™

Peptide-Based Drug Design

Edited by

Laszlo Otvos

Editor
Laszlo Otvos, PhD
Center for Biotechnology
Temple University
Philadelphia, PA 19122

Series Editor
John M. Walker
School of Life Sciences
University of Hertfordshire
Hatfield, Hertfordshire Al10 9 AB
UK

ISBN: 978-1-58829-990-1   e-ISBN: 978-1-59745-419-3
DOI: 10.1007/978-1-59745-419-3

Library of Congress Control Number: 2008930838

Printed on acid-free paper

9 8 7 6 5 4 3 2 1

springer.com

Preface

Natural products chemistry books usually open with a chapter on amino acids, peptides, and proteins. Peptides represent the interaction site between bioactive proteins. Together with peptide hormones, these molecules promise to be the starting points of drug development. Indeed, with the fast advances of peptide synthetic techniques, injectable peptide drugs obtained regulartory approval. Currently, peptide/protein drugs contstitute more than 10% of the ethical drug market. However, they are orally not available and expensive to mass produce. Thus, in the late 1980s, pharmaceutical companies quickly abandoned peptide research and focused on small molecules with more advantageous in vivo stability and pharmacokinetic properties. At the turn of the new century, the unexpected toxicity and cross-reactivity of small molecule drugs turned investors back to peptides, with their high specificity and, in most cases, low toxicity profile. The 15 years in exile was not completely useless; we all learned how to modify peptides to be competitive with small molecules in the drug development arena. Peptide synthesis technology further developed both as a research tool to produce biopolymers containing more than 100 natural or non-natural amino acid residues and a variable industrial production opportunity. Fast computers and improved algorithms brought structure-based peptide drug design to our desks. Understanding peptide metabolism, delivery, and clearance provided tools not only to increase the half-life in biological media, but also targeting to tissues and cell compartments unavailable earlier.

My colleagues frequently come to me and ask: "I have this exciting peptide sequence (real or virtual). It produces miraculous biological results in vitro (or so I hope). What should I do with it? What can I do to make it work in vivo, attrack investor's interest, or just grant funding more readily available for developmental projects than basic science?" These questions prompted us to assemble a handbook that offers a selection of research and production tools suitable for transforming a promising protein fragment or stand-alone bioactive peptide to a pharmaceutically acceptable composition. The chapter subjects include peptide-based drug development, from the early stages of computer-aided design and target identification through sequence modifications to satisfy pharmacologists until the actual production of difficult sequences and in vivo imaging. As with all contemporary and somewhat complicated chemistry projects, the devil is in the details. The format of the *Methods in Molecular Biology* series allows the contributing authors to provide sufficiently detailed information to reproduce the

0227032711

protocols or make the technology available for other projects calling for similar solutions.

The current buzzwords, chemical biology, somewhat degrade peptide chemistry to a service of biology. With the current book, we demonstrate that peptide-based drug design is an independent science that is well and on the rise. Our ultimate satisfaction will arrive when the first molecules, helped to conception or manufacture from the ideas set forth in this volume, gain regulatory approval and market share as stand-alone or combination medicines.

Laszlo Otvos

Contents

Contributors

NIKOLINKA ANTCHEVA • *Department of Life Sciences, University of Trieste, Trieste, Italy*

STEIN IVAR ASPMO • *Recovery Development, Novozymes AS, Bagsvaerd, Denmark*

JOSEPH M. BACKER • *SibTech, Inc., Brookfield, CT*

MARINA V. BACKER • *SibTech, Inc., Brookfield, CT*

KEVIN J. BARNHAM • *Department of Pathology, University of Melbourne, Victoria, Australia, and The Mental Health Research Institute of Victoria, Parkville, Victoria, Australia*

MONICA BENINCASA • *Department of Life Sciences, University of Trieste, Trieste, Italy*

FRANCIS G. BLANKENBERG • *Department of Pediatrics & Division of Nuclear Medicine/Department of Radiology & MIPS (Molecular Imaging Program at Stanford,) Stanford University, Stanford, CA*

PHILIPPE BULET • TIMC-IMAG, UMR5525 UFF CNRS, Archamps, France

GAYLE D. BURSTEIN • *Department of Chemistry and Biochemistry, Florida Atlantic University, Boca Raton, FL*

KEYLA PEREZ CAMACARO • *Department of Pathology, University of Melbourne, Victoria, Australia*

WILLIAM D. CAMPBELL • *Biomime Solutions, Inc., Surrey, British Columbia, Canada*

ARTEM CHERKASOV • *Division of Infectious Diseases, Faculty of Medicine, University of British Columbia Vancouver, British Columbia, Canada*

BRENDON Y. CHUA • *Department of Microbiology and Immunology, University of Melbourne, Parkville, Victoria, Australia*

JEFFREY COPPS • *Department of Biomedical Sciences, Creighton University, Omaha, NE*

DAVID J. CRAIK *Institute for Molecular Bioscience, University of Queensland, Brisbane, Queensland, Australia*

MARE CUDIC • *Department of Chemistry and Biochemistry, Florida Atlantic University, Boca Raton, FL*

PREDRAG CUDIC • *Department of Chemistry and Biochemistry, Florida Atlantic University, Boca Raton, FL*

CHRISTOPHER D. FJELL • *Division of Infectious Diseases, Department of Medicine, Faculty of Medicine, University of British Columbia, Vancouver, British Columbia, Canada*

RENATO GENNARO • *Department of Life Sciences, University of Trieste, Trieste, Italy*

ELENI GIANNAKIS • *Howard Florey Institute, University of Melbourne, Parkville, Victoria, Australia; and Bio-Rad Laboratories, Hercules, CA*

KAI HILPERT • *Centre for Microbial Diseases and Immunity Research, University of British Columbia, Vancouver, British Columbia, Canada*

RALF HOFFMANN • *Institute of Bioanalytical Chemistry, Center for Biotechnology and Biomedicine, Faculty of Chemistry and Mineralogy, University of Leipzig, Leipzig, Germany*

LIN-WAI HUNG • *Howard Florey Institute and Department of Pathology, University of Melbourne, Parkville, Victoria, Australia*

DAVID C. JACKSON • *Department of Microbiology and Immunology, University of Melbourne, Parkville, Victoria, Australia*

HÅVARD JENSSEN *Centre for Microbial Diseases and Immunity Research, University of British Columbia, Vancouver, British Columbia, Canada*

ZOIA LEVASHOVA • *Department of Pediatrics & Division of Nuclear Medicine/Department of Radiology & MIPS (Molecular Imaging Program at Stanford), Stanford University, Stanford, CA*

RICHARD LEVENSON • *Biomedical Systems, CRI, Inc., Woburn, MA*

SÁNDOR LOVAS • *Department of Biomedical Sciences, Creighton University, Omaha, NE*

MAURA MATTIUZZO • *Department of Life Sciences, University of Trieste, Trieste, Italy*

RICHARD F. MURPHY • *Department of Biomedical Sciences, Creighton University, Omaha, NE*

LASZLO OTVOS, JR. • *Sbarro Institute for Cancer Research and Molecular Medicine, Temple University, Philadelphia, PA*

MARCO SCOCCHI • *Department of Life Sciences, University of Trieste, Trieste, Italy*

DAVID SINGER • *Institute of Bioanalytical Chemistry, Center for Biotechnology and Biomedicine, Faculty of Chemistry and Mineralogy, University of Leipzig, Leipzig, Germany*

DAVID P. SMITH • *Department of Pathology, University of Melbourne, Parkville, Victoria, Australia; and Mental Health Research Institute of Victoria, Parkville, Victoria, Australia*

MACIEJ STAWIKOWSKI • *Department of Chemistry and Biochemistry, Florida Atlantic University, Boca Raton, FL*

CHRISTIN STEGEMANN • *Institute of Bioanalytical Chemistry, Center for Biotechnology and Biomedicine, Faculty of Chemistry and Mineralogy, University of Leipzig, Leipzig, Germany*

ALESSANDRO TOSSI • *Department of Life Sciences, University of Trieste, Trieste, Italy*

JOHN D. WADE • *Howard Florey Institute and School of Chemistry, University of Melbourne, Parkville, Victoria, Australia*

JAN-CHRISTOPH WESTERMANN • *Institute for Molecular Bioscience, University of Queensland, Brisbane, Queensland, Australia*

DIRK F.H.WINKLER • *Peptide Array Facility of the Brain Research Centre, University of British Columbia, Vancouver, British Columbia, Canada*

WEIGUANG ZENG • *Department of Microbiology and Immunology, University of Melbourne, Parkville, Victoria, Australia*

1

Peptide-Based Drug Design: Here and Now

Laszlo Otvos, Jr.

Summary

After many years of stagnation, peptide therapeutics once again became the focus of innovative drug development efforts backed up by venture funds and biotechnology companies. Designer peptide drugs overcome the unattractive pharmacological properties of native peptides and protein fragments and frequently feature nonnatural amino acid or backbone replacements, cyclic or multimeric structures, or peptidic or nonpeptidic delivery modules. With their high specificity and low toxicity profile, biologicals offer viable alternatives to small molecule therapeutics. The development of peptide drugs requires specific considerations of this family of biopolymers. Historically, peptide vaccines to viral infections and antibacterial peptides led the way in clinical development, but recently many other diseases have been targeted, including the big sellers AIDS, cancer, and Alzheimer's disease. This book gives practical advice to the most important steps in peptide-based drug development such as isolation, purification, characterization, interaction with targets, structural analysis, stability studies, assessment of biodistribution and pharmacological parameters, sequence modifications, and high throughput screening. This brief overview provides historical background for each of the listed techniques and diseases.

Key Words: Backbone; design; nonnatural amino acids; pharmacology; recognition; stability; toxicity.

The vast majority of current designer drugs target biopolymers or biopolymer interactions. There is no surprise here: the more we know about complex biochemical pathways, the larger the number of potential target proteins, nucleic acids, or lipidic structures that surface and offer themselves for agonist or antagonist development. In many cases a protein fragment, either part of the unprocessed protein or a cleaved piece, can be identified that serves as a ligand or the target itself. Naturally occurring peptide drugs are just the beginning. With the

From: *Methods in Molecular Biology, vol. 494: Peptide-Based Drug Design*
Edited by: L. Otvos, DOI: 10.1007/978-1-59745-419-3_1, © Humana Press, New York, NY

advance of computer power and molecular interaction databases, ligands can be designed for those protein fragments where nature has not yet succeeded but the models show validated structural features. With their high specificity and low toxicity profile, peptide-based drugs should be the first choice as contemporary therapeutics.

Unfortunately, until very recently, large pharmaceutical companies and biotech investors saw a more promising business opportunity in making one cent per dose of a small molecule—frequently generic—drug than a hundred dollars per dose of a groundbreaking biological therapeutic. This trend was the consequence of large amounts of money invested in the 1980s in peptide drugs only to learn that natural peptides lack pharmacological properties, such as serum resistance, tissue penetration, oral bioavailability, and delayed elimination—all needed for a fast and painless drug-development process. In addition, large-scale peptide manufacturing was considered prohibitively expensive, and only peptide hormones, given in low doses, could attract commercial interest. However, the rising number of and publicity about side effects seen for small molecule blockbusters (e.g., cancer chemotherapeutics or COX-2 inhibitors) together with innovative synthetic strategies and low monomer prices suddenly opened closed doors and liberated closed minds for peptide-based drug development.

Since the turn of the millennium, many biotech companies have discovered new peptides with interesting pharmacological properties, and the solid-phase peptide synthesis was optimized, allowing routine synthesis of large polypeptides and small proteins *(1)*. In 2000 the total ethical pharmaceutical market was worth about $265 billion, with peptides and proteins, excluding vaccines, accounting for more than 10%. The number of new chemical entities (NCEs) has been almost stable for about 10 years, with around 35–40 each year, but the number of peptide and protein NCEs has been continuously increasing during recent years. The most prominent recent peptide drug is T-20 (Enfuvirtide), which blocks HIV entry into cellular CD4 *(2)* and can be considered the turning point of investor attitude to biotech in general and to peptide drugs in particular. This book gives practical advice for researchers who have already decided to venture into peptide-based therapeutics. The short review below takes bits and pieces of the history of how we got here and points out the theoretical considerations and practical solutions offered by the individual chapter authors.

The poster child of modern peptide drugs T-20 inhibits gp41 hexamer formation *(3)*. With a 3.0 nM IC_{50} value in the HIV-mediated cell–cell fusion assay and serum half-life of 3.8 h, it has everything drug developers ask for. T-20 is a 36-mer peptide with no nonnatural amino acids except an acetylated amino-terminus. The drug is a clear success even if a yearly supply costs approximately $20,000. Another HIV drug candidate, the CCR5 transmembrane receptor ligand RANTES (68 residues), is an example of the current trends in peptide modification. N-terminal derivatization, first with aminooxypentane, later with nonanoic acid, together with nonnatural amino acid incorporation into positions 1–3 resulted in significant increase of potency *(4,5)* and likely protease resistance. The N-terminus of peptides is a favored position for medicinal chemistry modifications, as the automated solid-phase peptide synthesis can be finished off by manual addition of various N-capping amino acid residues or other organic acids. While acetylation is the most common technique used to prevent aminopeptidase cleavage, incorporation of modified and difficult-to-cleave residues, such as glycoamino acids, can be equally effective *(6)*. Chapter 11 details how to introduce sugars and glycoamino acids into synthetic peptides.

Peptides designed to work in the central nervous system (CNS) have to cross the blood-brain barrier (BBB). Glycosylation of enkephalin analogs improves both the stability of peptide drugs and their penetration across the BBB, just as increased hydrophobicity does *(7,8)*. In addition to active transport, peptides and cationized proteins enter the brain with the help passive transport mechanisms. The major sequence features of peptides and proteins that generally penetrate through biological membranes are a concentration of positive charges (arginines, lysines) interspersed with hydrophobic residues *(9)*. These structures can interact with both the hydrophobic protective barriers and the negatively charged cell surface lipids and can deliver peptides through cell layers, including the BBB model endothelial cell-astrocyte model *(10)*. Similar structural features promote entry into prokaryotic and eukaryotic cells. Currently only a few peptide drugs can be administered orally, but this gap will very be soon be overcome by innovative delivery technologies *(11)*. The peptides are either coupled to active or passive penetration enhancers or simply mixed with formulations able to pass cell and tissue layers.

This leads us to antimicrobial peptides, a land of promise not yet realized. As effectors of the innate immune system, antimicrobial peptides naturally represent the first line of defense against bacterial infections *(12)* and as thera-peutic agents indeed made through human clinical trials *(13)*. Their isolation from natural sources is relatively easy, but the poor pharmacological parameters of peptides are magnified for antimicrobial peptides. The low systemic stability of proline- and arginine-containing peptides, common sequence features of bacterial cell-penetrating biopolymers, requires detailed stability studies in serum. Even if the stability parameters are acceptable (as is the case with antimicrobial peptide dimers containing nonnatural amino acids in terminal positions) *(14)*, the fast renal clearance of peptides suggests that targeting urinary tract infections offers more hope than systemic applications *(14)*. Never-theless, *in vivo* active antimicrobial peptides are possible to develop, and as these products are multifunctional boosters of the immune system *(15)*, we should see derivatives active in bacteremia models in the foreseeable future. Chapters 2 and 3 look at the intricacies of the identification and characterization of antimicrobial peptides, and Chapter 8 presents a computer-based optimization strategy to finally break into the drug marketplace. Chapter 9 investigates the mode of action of proline-rich antimicrobial peptides by using genetic approaches. Recognizing the vital importance of systemic stability studies of peptide-based drugs, Chapter 10 presents the most frequently used *in vitro* and *in vivo* stability assessment methodologies.

Chemical synthesis, solution and solid-phase alike, is steadily gaining importance in manufacturing peptide drugs. In 2000 more than 40 different peptide drugs were made by chemical synthesis, with a sequence limit of 75 linear residues *(16)*. However, with improvements in coupling reagents and fluidic transport machinery, we can expect the ceiling to soon reach 100 amino acids, the arbitrary border between peptides and proteins. The production price of peptide drugs made in large scale under good manufacturing practice (GMP) regulations hovers around $50–75 per gram, although the inclusion of nonnatural amino acids for which the synthetic monomers are not easily available can take it up to $1,000 per gram (for leuprolide made for a controlled release application). Indeed, the synthetic difficulties, sometimes precipitated in the price, or delivery difficulties limit the volume of peptide-based drug sales. At the time of its regulatory approval in late 2004, Elan's ziconotide, a synthetic conopeptide for the treatment of chronic pain, was expected to reach a sales peak at $250 million per year, well below a small molecule painkiller similarly working by calcium channel blockage, approved around the same time *(17)*.

In addition to T-20, currently angiotensin-converting enzyme (ACE) inhibitors (zestril, enalapril) as well as peptide hormones are the major sellers, including insulin and calcitonin. However, many more peptide drugs are in the development pipeline. Among the potential drug families, clinically viable peptide-based subunit vaccines or epitope-based diagnostics were long considered possible to create. Monoclonal antibodies recognize a relatively extended stretch of protein fragments (6–20 residues), impossible to cover by small molecules *(18)*. Likewise, binding to the major histocompatibility complex proteins for T-cell vaccines requires a 9- to 12-amino-acid stretch of continuous peptide. Peptide vaccines based on the human papillomavirus against cervical cancer and cancer of the head and the neck reached the clinical trial stage *(19)*. As with almost all other peptide-based dugs, modern peptide vaccines are multimeric constructs often carrying additional boosters of the immune system. Chapters 14 and 15 show examples how complex peptide vaccines can be designed and manufactured. Due to their small size compared to proteins, peptides are preferred carriers of labels used for *in vivo* imaging where it is essential that the amount of label be sufficient per probe per receptor for optimal signal-to-noise ratio *(20)*.

In Chapter 16, Joseph Backer (the inventor of site-specific labels by using Cys-tags) *(21)* presents imaging alternatives for peptide and protein drugs. In addition, Chapter 16 provides useful strategies to deliver polyamide-based drugs into biological systems to improve the diagnostic and therapeutic potential of these hydrophilic substances.

Alzheimer's disease (AD) is characterized by the brain accumulation of peptide and protein aggregates, i.e., the 42-mer Aβ peptide and the hyperphosphorylated τ protein *(22)*. Synthetic peptides were essential to identify short τ fragments that are specifically recognized by AD-derived monoclonal antibodies *(23)*. Due to the severity of the disease and the easy synthetic access to the culprit protein aggregates and their fragments, aggregation inhibitors to AD were always in the forefront of peptide drug studies. Peptidic β-sheet breakers were developed, but medicinal chemistry manipulations were needed to improve the *in vitro* and *in vivo* stability parameters *(24)*. In general, frequently used peptide modifications include the introduction not only of nonnatural amino acid residues (or preassembly side-chain variations, if you will), but also of alterations of the peptide backbone as well as postassembly sequence modifications *(25)*. These possibilities are too diverse to list; companies specialized in amino acid and peptide sales are well equipped to provide mimics of all 20 proteinogenic amino acid residues, monomers or oligopeptide carriers of structural determinants (β-turns, for example), or altered amide bonds. Chapter 13 presents a panel of potential synthetic routes to apply the most commonly used medicinal chemistry changes to the peptide backbone. Chapters 5 and 12 show how Aβ or τ phosphopeptides, respectively, can help the development of diagnostic or therapeutic tools in AD. It needs to be mentioned that the last two techniques are widely applicable to either additional peptide–protein interactions or non-AD phosphopeptides, for example, those dominating signal transduction research.

With the completion of the human genome sequence identification together with developments in the proteomics field, peptides are emerging as important molecules for many other small- or high-volume drug applications, including the most emotionally attractive cancer therapy *(26)*. Several peptides with exciting preclinical results have now entered into clinical trials for the treatment of human cancers as inhibitors of oncogenic signaling pathways *(27)*. However, the complexity of these signaling processes and the large number of potential targets involved require a rational approach for drug development, or at least the availability of a high-throughput screening technology. SOM230, a result of rational drug design based on somatostatin–receptor interactions, is a cyclic hexapeptide mimetic containing nonnatural Lys, Phe, Tyr, and Trp analogs. SOM230 shows improved receptor-binding profile and pharmacological properties compared to natural somatostatin *(28)* and is highly effective against pituitary adenomas in mice *(29)*. Another example is related to a similar cyclic peptidomimetic that was designed based on ligand-binding models of the Grb2 SH2 protein domain *(30)*. As a third example, paper-bound peptide arrays identified nanomolar antagonists to E-selectin *in vitro* with ensuing inhibition of lung metastasis at a 30 mg/kg single dose *in vivo (31)*. Chapter 6 presents the state of the art in the nuclear magnetic resonance (NMR) structural analysis of peptides and their interactions with target molecules. If the conformations involved are known, we can switch to computer-assisted de novo design of potent agonists or antagonists, as described in Chapter 7. Finally, if the computer modeling does not provide peptides with the expected activities in the molecular assays, high-throughput screening on peptide macroarrays, as outlined in Chapter 4, can come to rescue. Actually, I truly believe that the concomitant use of all these designing and screening techniques together with the rest of the methodologies presented in this book will offer the greatest hope for the successful development of many peptide-based drugs to save or improve our lives and those of our companion animals.

References

1. Loffet, A. (2002) Peptides as drugs: Is there a market? *J. Pept. Sci.* **8**, 1–7.
2. Ryser, H.J., and Fluckiger, R. (2005) Progress in targeting HIV-1 entry. *Drug Discov. Today* **10**, 1085–1094.
3. Kazmierski, W.M., Kenakin, T.P., and Gudmundsson, K.S. (2006) Peptide, peptidomimetic and small molecule drug discovery targeting HIV-1 host-cell attachment and entry through gp120, gp41, CCR5 and CXCR4. *Chem. Biol. Drug Des.* **67**, 13–26.
4. Mack, M., Luckow, B., Nelson, P.J., et al. (1998) Aminooxypentane-RANTES induces CCR5 internalization but inhibits recycling: a novel inhibitory mechanism of HIV infectivity. *J. Exp. Med.* **187**, 1215–1224.
5. Hartley, O., Gaertner, H., Wilken, J., et al. (2004) Medicinal chemistry applied to a synthetic protein: Development of highly potent HIV entry inhibitors. *Proc. Natl. Acad. Sci. USA* **101**, 16460–16465.
6. Powell, M.F., Stewart, T., Otvos, L., Jr., et al. (1993) Peptide stability in drug development. II. Effect of single amino acid substitution and glycosylation on peptide reactivity in human serum. *Pharmacol. Res.* **10**, 1268–1273.
7. Polt, R., Porreca, F., Szabo, L Z., et al. (1994). Glycopeptide enkephalin analogues produce analgesia in mice; evidence for penetration of the blood-brain barrier. *Proc. Natl. Acad. Sci. USA* **91**, 7114–7118.
8. Egleton, R.D., Mitchell, S.A., Huber, J.D., et al. (2000) Improved bioavailability to the brain of glycosylated Met-enkephalin analogs. *Brain Res.* **881**, 37–46.
9. Otvos, L., Jr., Cudic, M., Chua, B.Y., Deliyannis, G., and Jackson, D.C. (2004) An insect antibacterial peptide-based drug delivery system. *Mol. Pharmaceut.* **1**, 220–232.

10. Josserand, V., Pelerin, H., de Bruin, B., et al. (2006) Evaluation of drug penetration into the brain: a double study by *in vivo* imaging with positron emission tomography and using an *in vitro* model of the human blood-brain barrier. *J. Pharmacol. Exp. Ther.* **316**, 79–86.

11. Werle, M. (2006) Innovations in oral peptide delivery. *Future Drug Deliver.* 39–40.

12. Brogden, K.A. (2005) Antimicrobial peptides: pore formers or metabolic inhibitors in bacteria? *Nat. Rev. Microbiol.* **3**, 238–250.

13. Zasloff, M. (2002) Antimicrobial peptides of multicellular organisms. *Nature* **415**, 389–395.

14. Cudic, M., Lockatell, C.V., Johnson, D.E., and Otvos, L., Jr. (2003) *In vitro* and *in vivo* activity of antibacterial peptide analogs against uropathogens. *Peptides* **24**, 807–820.

15. Otvos, L., Jr. (2005) Antibacterial peptides and proteins with multiple cellular targets. *J. Pept. Sci.* **11**, 697–706.

16. Bray, B.L. (2003) Large-scale manufacture of peptide therapeutics by chemical synthesis. *Nat. Rev. Drug Discovery* **2**, 587–593.

17. Garber, K. (2005) Peptide leads new class of chronic pain drugs. *Nat. Biotechnol.* **23**, 399.

18. Meloen, R., Timmerman, P., and Langedijk, H. (2004) Bioactive peptides based on diversity libraries, supramolecular chemistry and rational design; A class of peptide drugs. Introduction. *Mol. Diversity* **8**, 57–59.

19. Muderspach, L., Wilczynski, S., Roman, L., et al. (2000) A phase I trial of a human papillomavirus (HPV) peptide vaccine for women with high-grade cervical and vulvar intraepithelial neoplasia who are HPV 16 positive. *Clin. Cancer Res.* **6**, 3406–3416.

20. Agdeppa, E.D. (2004) Rational design for peptide drugs. *J. Nucl. Med.* **47**, 22N–24N.

21. Backer, M.V., Levashova, Z., Patel, V., et al. (2007) Molecular imaging of VEGF receptors in angiogenic vasculature with single-chain VEGF-based probes. *Nat. Med.* **13**, 504–509.

22. Arai, H., Lee, V.M.-Y., Otvos, L., Jr., et al. (1990) Defined neurofilament, tau and beta-amyloid precursor protein epitopes distinguish Alzheimer from non-Alzheimer senile plaques. *Proc. Natl. Acad. Sci. USA* **87**, 2249–2253.

23. Hoffmann, R., Lee, V.M.Y., Leight, S., Varga, I., and Otvos, L., Jr. (1997) Unique Alzheimer's disease paired helical filament specific epitopes involve double phosphorylation at specific sites. *Biochemistry* **36**, 8114–8124.

24. Adessi, C., Frossard, M-J., Boissard, C., et al. (2003). Pharmacological profiles of peptide drug candidates for the treatment of Alzheimer's disease. *J. Biol. Chem.* **278**, 13905–13991.

25. Kotha, S., and Lahiri, K. (2005) Post-treatment assembly modifications by chemical methods. *Curr. Med. Chem.* **12**, 849–875.

26. Sehgal, A. (2002) Recent development in peptide-based cancer therapeutics. *Curr. Opin. Drug Discov. Devel.* **5**, 245–250.

27. Borghouts, C., Kunz, C., and Groner, B. (2005) Current strategies for the development of peptide-based anti-cancer therapeutics. *J. Pept. Sci.* **11**, 713–726.
28. Bruns, C., Lewis, I., Briner, U., Meno-Tetang, G., and Weckbecker, G. (2002) SOM230: a novel somatostatin petidomimetic with broad somatotropin release inhibiting factor (SRIF) receptor binding and a unique antisecretory profile. *Eur. J. Endocrin.* **146**, 707–716.
29. Fedele, M., DeMartino, I., Pivonello, R., et al. (2007) SOM230, a new somatostatin analogue, is highly effective in the therapy of growth hormone/prolactin-secreting pituitary adenomas. *Clin. Cancer Res.* **13**, 2738–2744.
30. Phan, J., Shi, Z-D., Burke, T.R., Jr., and Waugh, D.S. (2005) Crsytal structures of a high-affinity macrocyclic peptide mimetic complex with the Grb2 SH2 domain. *J. Mol. Biol.* **353**, 104–115.
31. O, I., Otvos, L., Jr., Kieber-Emmons, T., and Blaszczyk-Thurin, M. (2002) Role of SA-Le[a] and E-selectin in metastasis assessed with peptide antagonist. *Peptides* **23**, 999–1010.

2

Strategies for the Discovery, Isolation, and Characterization of Natural Bioactive Peptides from the Immune System of Invertebrates

Philippe Bulet

Summary

Intensive research efforts for developing new anti-infectious drugs for human health rely mostly on technological advancements in high-throughput screening of combinatorial chemical libraries and/or natural libraries generated from animal/plant extracts. However, nature has done a fascinating job engineering its own mutational program through evolution. This results in an incredible diversity of natural bioactive molecules that may represent a starting matrix for developing new generations of therapeutics of commercial promise to control infectious diseases. Among the natural bioactive molecules, peptides are opening promising perspectives. The search for novel bioactive peptides for therapeutic development relies mainly on a conventional approach driven by a desired biological activity followed by the purification and structural characterization of the bioactive molecule. Nevertheless, this strategy requires large quantities of biological material for activity screening and is thus restrained to animal species of large size or that are widely distributed.

During the past 10 years, thanks to the technological improvements of mass spectrometry (MS) and liquid chromatography, highly sensitive approaches have been developed and integrated into the drug-discovery process. We have used several of these sensitive biochemical technologies to isolate and characterize defense/immune peptides from tiny invertebrates (essentially arthropods) and to limit investigations on a restricted number of individuals. These defense/immune peptides, which are mostly cationic molecules with a molecular mass often below 10 kDa, are the natural armamentarium of the living organisms, and they represent good starting matrices for optimization prior their development as future anti-infectious therapeutics.

From: *Methods in Molecular Biology, vol. 494: Peptide-Based Drug Design*
Edited by: L. Otvos, DOI: 10.1007/978-1-59745-419-3_2, © Humana Press, New York, NY

Key Words: Invertebrate immunity; antimicrobial peptides; immune effectors; mass spectrometry; drug discovery; *Drosophila*; arthropods; peptide purification; molecular mass fingerprints; bioactive peptides.

1. Introduction

To date, three major classes of molecules have been developed for the treatment of human diseases. These types of molecules are small molecules, protein drugs, and peptides. Currently, most of the therapeutic molecules developed and marketed are small molecules. However, with the recent advancements in several technologies in the areas of (1) peptide manufacturing (chemical synthesis, recombinant expression), (2) screening, (3) stability, (4) modifications, and (5) improvements in peptide drug delivery systems, peptides are now considered as lead molecules for therapeutics (http://pubs.acs.org/cen/business/83/i11/8311bus1.html). Nature is an unlimited pipeline of bioactive peptides that are active regulators, information/signaling factors, and key components of the defense reactions, properties that make them interesting for drug discovery and development. Among the peptides, the bioactive peptides from the defense mechanisms of living organisms have such a therapeutic potential *(1–4)*. Host-defense peptides are widely distributed in nature. Bacteria produce peptide antibiotics to mediate their microbial competitions *(5,6)*. In multicellular organisms from both the vegetal and animal kingdoms, host-defense peptides such as antimicrobial peptides (AMPs) form a first line of host defense against pathogens *(7–11)*. AMPs are critical effectors of the innate immunity of many species and are playing an essential role in terms of resistance to infection and survival. In living organisms, AMPs were found to be either constitutively stored or produced upon infection (natural or experimental). In invertebrates, for example, AMPs were found to be stored in immune competent blood cells (hemocytes) and in epithelial cells of a range of tissues. Interestingly, the presence of AMPs and other immune defense molecules in the hemolymph (the blood of invertebrates) or at the surface of the epithelia is only present if an infection is detected *(12–17)*. AMPs were also reported to be constituents of the venom of arthropods *(18)*. In addition to AMPs, a variety of other immune-induced molecules are produced to clear microbial infections; this was illustrated in the fruit fly model, *Drosophila melanogaster (19–21)*. Therefore, together with other effectors of invertebrate immunity, they are considered potential candidates for their development as therapeutic drugs or as markers for diagnosis with applications in disease control. Such a distribution within body fluids (hemolymph but also venom) or immunocompetent cells, which is dependant of an immune state, is particularly important to consider when collecting natural samples for the isolation of bioactive peptides playing a function in immune defenses.

The purpose of this chapter is to provide a handbook for (1) induction of an immune state, (2) extraction, (3) purification, and (4) characterization (primary structure and biological properties) of immune-defense peptides (with a special reference to AMPs) from invertebrates. The methods detailed in this contribution are adapted to the extraction of immune factors from hemocytes, epithelial tissues, and hemolymph including small individuals such as *Drosophila*. Methods are largely based on those we used for the isolation and characterization of *Drosophila* immune-induced effectors *(19,20)*. Methods reported in this contribution are also derived/updated/optimized from those reported for the identification and characterization of AMPs from noninsect arthropods and other invertebrates *(22)*. Purification methods (liquid chromatography) and characterization (MALDI- and ESI-MS, MS/MS [tandem MS or MS^n], Edman sequencing, enzymatic digestions) strategies are based on the common physicochemical properties of defense peptides from the immune system, in particular on their relatively high hydrophobicity, cationic character at physiological pH, and short length (mostly below 10 kDa). Finally, structure determination techniques appropriate for the full characterization of immune factors are given together with a brief description of well-recognized antimicrobial assays (antibacterial and antifungal) when the immune effectors are members of the AMP armamentarium.

2. Materials
2.1. Animals
2.1.1. Unchallenged Animals

Wild or bred laboratory individuals or animals from breeding farms at the same developmental stage (larvae or adults) with a similar size.

2.1.2. Experimentally Infected Individuals

1. Living or heat-killed bacteria: Gram-positive and Gram-negative ATCC strains.
2. Spores from a filamentous fungus (Pasteur Collection or private collections) or a yeast strain.
3. Lipopolysaccharides or peptidoglycans (Sigma, St. Louis, MO).
4. Infection with parasites: e.g., *Crithidia* species *(23)*, *Leishmania major (24)*.
5. Standard nutrient media for bacterial (Luria Broth Miller, Muller-Hinton Broth) and fungal growth (Sabouraud, yeast culture media).

6. 5 μL or higher volume Hamilton syringe with adapted needles or ultrafine stainless steel needles.
7. 2.5-mL Eppendorf tubes to prepare the inoculum.

2.2. Sample Collection

2.2.1. Entire Bodies

1. Liquid nitrogen or dry ice for fast freezing.
2. Mortar to reduce the individuals in powder.

2.2.2. Hemolymph

1. Ice-cold water bath.
2. Refrigerated centrifuge (500–1000g).
3. 0.1–1 mL Hamilton syringes or nanoinjector (Nanoject, Drummond Scientific, Broomall) used as a glass capillary holder for either hemolymph collection or experimental infection.
4. Phenylthiourea (PTU, Sigma), stock solution 20 mM in ethanol.
5. 2.5-mL low-protein absorption polypropylene tubes—either Eppendorf or Mini-sorp NUNC Immuno tubes.
6. Anticoagulant adapted to the animal model investigated (citrate buffer, carbonate buffer; modified Alsever solution, etc.).
7. Protease inhibitors: aprotinin or/and phenylmethanesulfonylfluoride (PMSF) as serine protease inhibitor.

2.2.3. Salivary Glands, Gut Tissue, and Any Other Epithelial Tissue

1. Micro-operative dissection equipment (forceps and microscissors) and a binocular or an optical microscope.
2. 10 mM phosphate-buffered saline (PBS) pH 7.4.
3. Liquid nitrogen.

2.3. Extraction

1. Phenylthiourea (PTU), stock solution 20 mM in ethanol.
2. Protease inhibitors: aprotinin or/and PMSF as serine protease inhibitor.
3. Trifluoroacetic acid (TFA) high-purity (sequencing) grade (Pierce), acetic acid.
4. Ultrapure water (MilliQ™).
5. Filter units with 0.8-μm membranes (Millex unit, Millipore).
6. Beakers, ice-cold water bath, a stir bar, and a magnetic stirrer.
7. Refrigerated centrifuge (high speed) with rotors (up to 12,000g) adapted to Eppendorf and Falcon (15–50 mL) tubes.
8. Ultrasonicator specific for cell lysis and Dounce (maximum 152 μm, minimum 76 μm) and Potter apparatus.

2.4. Purification of the Immune-Defense Peptides

1. Solid-phase extraction (SPE) on prepackaged, disposable cartridges containing chromatographic reversed-phase (C_{18} being the most frequently used). Sorbent weight ranges from 0.1 to 10 g. This step can be performed either individually with a syringe or with a multiposition vacuum manifold or with specific 96-well plates.
2. Disposable Ultrafree-CL (Millipore) ultrafiltration cartridge units (10 kDa nominal molecular weight cutoff).
3. Methanol, acetonitrile, TFA are HPLC grade. Water is Ultrapure (MilliQ™ or of HPLC quality). A proposed system for elution, solution A: acidified water (0.05% TFA) and solution B: acetonitrile in acidified water.
4. Reversed-phase high-performance liquid chromatography (RP-HPLC) columns (C_8, C_{18}) from semipreparative (7 mm internal ∅) down to capillary (<1 mm). Phase porosity of 300 Å and granulometry of 7 μm are preferred.
5. Size exclusion (also referred to as gel filtration or gel permeation, SEC) HPLC columns (7.5 × 300 mm) equipped with a precolumn (7.5 × 75 mm).
6. HPLC system: gradient controller, pump (one or two) optimized for low flow rates (frequently used flow rate below 1 mL/min), and a photodiode array detector or a variable wavelength detector (preferred wavelength 214 or 225 nm). An oven might be used for temperature control in the column and of the solvents delivered. An analogic recorder to directly follow the optical density of the hand-collected fractions. A fraction collector may be useful but not necessary.
7. Centrifuge vacuum drier or vacuum lyophilizator.

2.5. Primary Structure Elucidation

2.5.1. Edman Sequencing

Pure peptides (native or reduced and alkylated) are sequenced by automatic Edman degradation on a pulse liquid automatic sequenator. Reagent and solvents are purchased from the manufacturer (Applied Biosystem Division).

2.5.2. Mass Spectrometry

Purity control, molecular mass determination and molecular mass fingerprints can be performed either by electrospray ionization (ESI) or matrix-assisted laser desorption/ionization time of flight (MALDI-TOF) MS. ESI is preferred for sequence determination by MS^n and MALDI-TOF for molecular mass fingerprints. Both technologies are appropriate for molecular mass determination and purity control. We have edited a critical review on the use of MS strategies for discovery and peptide sequencing of bioactive peptide *(25)*.

2.5.2.1. MALDI-TOF-MS

1. MALDI-TOF-MS may be operated in either linear or reflector mode. Due to the cationic character of the peptides investigated, detection is performed in a positive mode.
2. Appropriate standard peptides for calibration within the mass range 500–5000 in reflector mode and 2000 to 20–30 kDa for the linear mode.
3. A matrix adapted for <10 kDa peptides such as α-cyano-4-hydroxycinnamic acid and the adapted solvents (4HCCA, 40 mg/mL in acetone) for preparing the different solutions for sample preparation: acetone, TFA, acetonitrile, and water (all HPLC grade). To select the best methodologies for sample preparation, some tricks are discussed in *(26)*.

2.5.2.2. ESI-MS AND LIQUID CHROMATOGRAPHY-MS^n (LC-ESI-MS^n)

1. An ESI mass spectrometer equipped with a standard or a nanospray source and with the option for MS^n may be used (Q-TOF, Ion Trap, etc.) as well as appropriate software. Ion Trap mass spectrometer (Esquire-LC, Bruker) equipped with a nanospray source was used.
2. Gold/palladium-coated nanospray capillaries (Protana) or a PicoTip^{TM} emitter (New Objective).
3. Appropriate standard peptides for calibration with monoisotopic molecular masses covering, for example, the mass range 500–3000 Da *(27)*. Glu-fibrinopeptide-B is also used as external calibrant for MS^n experiments *(28)*.
4. A solution of water/acetonitrile/formic acid (composition 49.9/49.9/0.2, v/v/v).

2.5.3. Enzymatic Treatments

Enzymatic cleavage is often requested when sequence information is too limited for cDNA cloning experiments (if this strategy is appropriate to the animal model), database search or when no N-terminal sequence is obtained by Edman chemistry, and for cysteine pairing elucidation.

1. Thermostatted water bath or temperature-controlled chamber (30–50°C).
2. 1.5–2 mL low-protein absorption polypropylene Eppendorf tubes.
3. Proteases of sequencing grade and reaction buffers of high purity:

   a. Lysyl endoproteinase (*Achromobacter* protease I) (10–25 mM Tris-HCl buffer pH 8, 0.01% Tween 20 with or without 1 mM EDTA) that is specific for Lys-X cleavage.
   b. Arginyl endoproteinase (10 mM Tris-HCl buffer, pH 8, 0.01% Tween 20) that is specific for Arg-X cleavage.
   c. Trypsin (10 mM Tris-HCl buffer, pH 8, 0.01% Tween 20, 10 mM CaCl$_2$) that is specific for Lys/Arg-X cleavage.
   d. α-Chymotrypsin (100 mM Tris-HCl pH 7.5, 10 mM CaCl$_2$) that cleaves at the C-terminus of aromatic residues.

e. *Staphylococcus aureus* V8 proteinase (50 m*M* ammonium carbonate, pH 7.8, 0.01% Tween 20) that is specific for Glu-X cleavage.

f. Asparaginyl endopeptidase (20 m*M* sodium acetate buffer, pH 5, 0.01%, Tween 20, 1 m*M* dithiothreitol (DTT), 1m*M* EDTA) that is specific for Asn-X cleavage.

g. Endoproteinase Asp-N (20 m*M* sodium phosphate buffer, pH 8.0, 0.01% Tween 20) that is specific for X-Asp cleavage.

h. Pyroglutamate aminopeptidase (100 m*M* sodium phosphate buffer, pH 8, 10 m*M* EDTA, 5% glycerol, 5 m*M* DTT) for removal of pyroglutamic acid (N-terminal cyclic residue that prevents Edman degradation).

i. Carboxypeptidase Y and P (50 m*M* sodium citrate buffer, pH 6 [for Y] and pH 4 [for P]) for C-terminal amino acid removal.

j. Thermolysin (100 m*M* MES [2-(*N*-morpholine) ethane sulfonic acid, pH 6.5, 2 m*M* CaCl$_2$]) for disulfide bridge assignment or for generating a large series of peptidic fragments.

2.5.4. Chemical Treatment: Reduction and Alkylation

To facilitate sequencing analysis and enzymatic cleavage, reduction and alkylation of the peptides to open their potential disulfide bridges is recommended. Several alkyl groups can be used: acrylamide, iodoacetamide, and 4-vinylpyridine (4-VP). S-pyridylethylation is preferred over the other alkyl groups when Edman degradation (PTH-4-pyridylethylated-Cys is commercially available) and MS (addition of 57 Da per alkylated cysteine residue) have to be performed.

1. Reaction buffer: 0.5 *M* Tris-HCl, 2 m*M* ethylenediaminetetraacetic acid (EDTA), pH 7.5 containing 6 *M* guanidinium hydrochloride.

2. 2.2 *M* DTT; tris(2-carboxyethyl)phosphine hydrochloride (TCEP) may be an alternative.

3. Distilled 4-VP (*see* **Note 1**). Keep the solution under nitrogen atmosphere in a deep freezer. Iodoacetamide is also frequently used. An addition of 107 or 57 Da per alkylated cysteine residue is observed with vinylpyridine and iodoacetamide, respectively. Cysteine residues derivatized with *N*-methyl iodoacetamide can be also analyzed by the Edman sequencing (no PTH standard commercially available).

4. Stop solution: acidic solution, 1% TFA or 1% acetic acid or less than 3% formic acid.

5. Desalting by reversed-phase HPLC (*see* **Subheading 2.4.**) or SEC or micro-SPE onto C$_{18}$ Zip-Tip (used the conditions recommended by the manufacturer Millipore). The recently developed magnetic beads from Bruker may represent a fast and economical alternative.

6. Nitrogen or argon gas tank.

7. Small-volume low-protein absorption polypropylene tubes—either Eppendorf or Minisorp NUNC Immuno tubes.
8. Water bath or dry incubator adjusted at 45°C.

2.6. Antimicrobial Assays

Bacteria, yeast, and fungi should be purchased from certified institutions such as ATCC or Pasteur Institute or from private collections. For antimicrobial activity screening during the course of purification, sensitive strains are recommended: *Micrococcus luteus* (Gram-positive), *Escherichia coli* D31 (Gram-negative, *see* **Note 2**), and *Neurospora crassa* (filamentous fungus) are good test organisms. Pathogenic strains for the animal model can also be used *(29)*. The liquid growth inhibition assays (antibacterial and antifungal) are preferred to the radial diffusion assays for high-throughput screening (HTS). In addition, the liquid growth inhibition assays allow evaluating within a same experiment growth inhibition and colony-forming unit (CFU) counting, a method frequently used for discriminating between bacteriostatic and bacteriolytic activities.

1. Adapted media for the microbial strains: potato dextrose broth (PDB, Difco) used at half-strength (12 g/L for filamentous fungi supplemented with antibiotics) and Poor broth (PB, 1% bactotryptone, 0.9% NaCl w/v, pH 7.4) if bacterial strains for activity screening during isolation.
2. Luria Bertani broth (LB, 1% bactotryptone, 0.5% yeast extract, 0.9% NaCl w/v, pH 7.4) for CFU counting.
3. Antibiotics: 10 mg/mL tetracyclin (Sigma) in dimethylsulfoxide (DMSO) and 100 μg/mL cefotaxim (Sigma) in water.
4. 96-well microtitration plates for cell culture (Nunc) and microplate reader and a spectrophotometer equipped with a filter between 580 and 620 nm.

3. Methods

The following protocols can be used for the isolation and structural characterization of any natural bioactive peptides from the immune system of invertebrates. The different procedures that will be detailed below refer to the identification and primary structure determination of the *Drosophila* immune-induced peptides *(19,20,23,27,30)* and of bioactive peptides from the immune system of other Diptera *(17,21,24,31)*. These approaches were also successfully used for the discovery of bioactive peptides from crustaceans, arachnids, and mollusks. These methods should be considered as a guideline and not as the exact procedure to follow (*see* **Note 3**). The suggested procedures will be reported following the normal order of execution, (1) induction of the immune response by an experimental infection, (2) collection of the immuno-competent cells (hemocytes), tissues (epithelia, trachea, salivary glands, etc.)

and hemolymph, (3) isolation of the immune-induced peptides following *in vitro* assays or molecular mass fingerprints by differential display mass spectrometry analysis, and (4) structural elucidation.

3.1. Experimental Infection: Induction of the Immune Response

1. An appropriate number of individuals at the same developmental stage (e.g., third instar larvae or 3-d old *Drosophila*, newly emerged [teneral] *Glossina*).
2. Prepare a known inducer for the immune response of invertebrates (*see* **Note 4**). An efficient inducer is a cocktail of overnight cultures of a Gram-positive bacteria (*Micrococcus luteus*) and a Gram-negative bacteria (wild-type *Escherichia coli*).
3. Inject the cocktail of the living bacteria selected into the body cavity or, for small-size invertebrates, optimize the induction by pricking the individuals with a fine needle soaked with a concentrated moist pellet of bacteria.
4. Maintain the experimentally infected individuals and an equivalent number of unchallenged individuals (*see* **Note 5**) for 24–48 h in appropriate conditions for keeping the animals.

3.2. Extraction

3.2.1. Sample Collection for Isolation and Purification

1. Withdraw the hemolymph from one to several hundred individuals in order to get a sample volume of at least 0.1–100 μL in prechilled polypropylene tubes containing an anticoagulant buffer supplemented with PTU (1 μg/mL) and a protease inhibitor (e.g., aprotinin at a final concentration of 40 μg/mL; PMSF is also an alternative).
2. Separate the cell-free hemolymph from hemocytes (*see* **Note 6**) by centrifugation at low speed (500–700*g*) at 4°C for a brief period (10–15 min). Collect and wash the hemocytes with 10 m*M* PBS at pH 7.4 or with an anticoagulant buffer.
3. Dissect the organs (e.g., midguts or salivary glands).
4. Midguts are collected either in 200 m*M* sodium acetate at pH 4.5 supplemented with 154 m*M* sodium chloride (*Stomoxys calcitrans (17)*) or in PBS (*Anopheles gambiae (32)*). Salivary glands from termites are collected in a 10 m*M* PBS at pH 7.4 *(33)*.
5. Keep the cell-free hemolymph, hemocytes, and organs in ice (if freshly used) or at −80°C until use.

3.2.2. Extraction from Cell-Free Hemolymph

1. Thaw the cell-free hemolymph in a large volume of aqueous TFA (0.1%) to get a final pH below 3–4.
2. Extract the peptides overnight at 4°C under gentle stirring in 0.1% TFA supplemented with a protease inhibitor (e.g., aprotinin) and PTU.

3. Centrifuge between 8000 and 12,000g (30 min, 4°C) to remove lipids (upper phase), denaturized proteins, and debris. If necessary, the supernatant can be clarified by an additional step with filter units (0.8-μm membranes, Millipore).

3.2.3. Extraction from Hemocytes, Epithelia, and Salivary Glands

1. Thaw the sample (hemocytes, epithelia) on ice if not freshly prepared.
2. The salivary glands are snap-frozen in liquid nitrogen.
3. Homogenize (*see* **Note 7**) using sonication (3 × 30 s) at medium power in the presence of 2 *M* acetic acid (in an ice-cold water bath, a solution of 0.2 *M* acetic acid is also an option to decrease the protein content of the extracted material). Extract overnight at 4°C under gentle stirring in 2 *M* acetic acid supplemented with a protease inhibitor (e.g., aprotinin).
4. Centrifuge between 8000 and 12,000g (30 min, 4°C) to separate the supernatant (cytosolic fraction) from the organelle-rich fraction or tissue debris (pellet). If necessary, the supernatant can be clarified by additional step with filter units (0.8-μm membranes, Millipore).

3.3. Purification of the Natural Bioactive Peptides

Because of their well-recognized physicochemical properties (relatively high hydrophobicity, cationic character, and short length) reversed-phase, size-exclusion, and cation exchange chromatographies are particularly appropriate to purify bioactive peptides from the immune system of invertebrates. The sensitivity of HPLC, MS, Edman degradation, and liquid growth inhibition assays allow one to use from narrow (e.g., 2.1-mm internal diameter) down to micro- or nano-columns.

3.3.1. Prepurification

To limit the co-precipitation of the peptides of interest with remaining proteins, the first purification step is performed either by SPE on open-columns or cartridges that supports large volumes of diluted extracts or using disposable Ultrafree-CL (Millipore) ultrafiltration (UF) cartridge units (both strategies can be combined, UF prior SPE or SPE prior UF; *see* **ref. 28**). Open columns can be bought manufactured (e.g., Sep-Pak cartridges, Waters™) or handmade (bulk phase available, e.g., from Waters™).

3.3.1.1. SOLID-PHASE EXTRACTION

This procedure has been already reported in **ref. 22**. Briefly:

1. Solvate the cartridge with methanol and equilibrate with acidified water (0.05% TFA).
2. Load the acidified extract (pH < 4).

3. Elute in a stepwise fashion with increasing percentages of acetonitrile prepared in acidified water: 40 and 80% (*see* **Note 8**) with 5–10 times the hold-up volume of the column (see manufacturer instructions).
4. Remove the organic solvent by lyophilization/freeze-drying or centrifugation under vacuum.
5. Reconstitute the fractions in MilliQ water (Millipore) and keep them at 4°C for immediate biological activity screening (*see* **Note 9**) or in the freezer until use.

3.3.1.2. ULTRAFILTRATION (UF PRE-SPE)

1. Dilute the extract in a large volume of 0.1% TFA.
2. Filter the diluted extract according to the manufacturer's recommendations.
3. Load the filtrate onto the equilibrated SPE cartridge and pursue the SPE procedure.

3.3.1.3. ULTRAFILTRATION (UF POST-SPE)

1. Filter the reconstitute fractions eluted during SPE (*see* **Subheading 3.3.1.1.**).
2. Dry under vacuum and store at -20°C until use.

3.3.2. Purification by HPLC

In general, the first step of purification of the fractions eluted from SPE, SPE-UF, or UF-SPE is subjected to RP-HPLC followed either by a second RP-HPLC or by size exclusion chromatography (SEC-HPLC). The last step is often a highly sensitive RP-HPLC using highly selected elution conditions (gradient with a soft slope, low flow rate, and high sensitivity detection). However, before finalizing purification, several steps may be requested (at least three steps are needed to get a sufficient amount [1 nmol or 100 μg] of highly pure [>98%] material for structural characterization either by Edman chemistry or MS strategies (enzymatic digestion, on-line or off-line micro-RP-HPLC mass fingerprints, MS^n sequencing).

3.3.2.1. STEP 1: REVERSED-PHASE HPLC (SEE NOTE 10)

1. Load the sample on an appropriate RP column (internal diameter of the column chosen according to the starting biological material available, 300 Å for granulometry, internal diameter from 3.8 to 7 mm, a C_{18} or C_8 column is recommended for **step 1**). It is recommended to equilibrate the column with 2% acetonitrile in acidified water (0.05 % TFA) to remove the hydrophilic molecules and to properly stack the interesting material.
2. Elute the peptides with a linear gradient of acetonitrile in acidified water at a flow rate between 0.8 and 1.3 mL/min depending of the internal diameter of the column. Flow rate can be 0.8–1 mL/min if the column has 4.6 mm of internal diameter and increased to 1.3 mL/min for a semi-preparative column (7 mm of

internal diameter). Increasing the flow rate will give a better resolution but will also generate higher elution volumes.

3. Fractions are collected manually according to the absorbance measured either at 214 nm (increased sensitivity, higher background) or at 225 nm (best signal/solvent ratio). Using this procedure, each fraction corresponds to an individual absorbance peak. This minimizes the number of steps to get pure material for structural analysis.

4. Fractions are analyzed by MS or subjected to *in vitro* assays. For MS analysis, the collected fractions can be directly analyzed following dilution in the appropriate solvent. For activity screening, fractions are dried in a vacuum-centrifuge, reconstituted either in water or in a suitable buffer adapted to the activity monitoring.

3.3.2.2. STEP 2: RP-HPLC OR SIZE EXCLUSION HPLC

Depending of the results of MS, **step 2** can be performed either by RP-HPLC or by SEC. For RP-HPLC, the same column as in **step 1** can be used, but the fractions are eluted using linear biphasic gradients of acetonitrile in acidified water at a same flow rate and at the same temperature (an increase in the temperature would result in a better resolution). The gradient is chosen according to the following strategy: from 2% to ([C]-2)% acetonitrile in 10 min and from ([C]-2)% to ([C]+15)% acetonitrile at an increase concentration of 0.2–0.3%, where [C] corresponds to the exact concentration of acetonitrile for the elution of the compound of interest (*see* **Note 11**). SEC is preferred when MS measurements reveal the presence of molecules with molecular masses rather distant (e.g., >10 kDa and 2–4 kDa).

1. Inject the fraction of interest (biologically active, immune induced) on two serially linked size exclusion columns (equivalent to the former Beckman SEC 3000 and SEC 2000 columns, 300 mm × 7.5 mm or any other brand). Use of a precolumn is recommended to protect expensive columns. The injected volume should not exceed 100 µL for adequate resolution.

2. Elute with 30% acetonitrile in acidified water at a flow rate below 0.5 mL/min. Fractions are collected manually according to the absorbance at 214 nm. Using SEC, detection at 214 nm is not a problem as isocratic conditions are used.

3. For the procedure, *see* **Subheading 3.3.2.1., step 4**.

3.3.2.3. FINAL PURIFICATION STEP: RP-HPLC

The final purification step is a linear biphasic gradient of acidified acetonitrile at a flow rate between 0.08 and 0.25 mL/min (0.08 mL for the microbore column, 0.1–0.25 mL for the narrowbore column). The following two-step gradient for **step 2** is used. However, to increase resolution, a column with a smaller internal diameter is needed and column temperature is increased to 37°C. For fraction handling, use the procedure according to **Subheading 3.3.2.1., part 4**.

3.4. Reduction and Alkylation

1. Dissolve or dilute the peptide (few to 100 pmol) in an Eppendorf tube using the reaction buffer.
2. Add 2 μL of the reducing agent in a large excess (e.g., 2.2 M DTT or TCEP).
3. Flush the tube containing the sample with N_2 to prevent oxidation and incubate at 45°C for 1 h.
4. Add 2 μL of the alkylated agent: pure colorless 4-VP or 2.5% (w/v) iodoacetamide for alkylation of the peptide and of the residual reducing agent. Flush again the Eppendorf tube with N_2 and incubate for 10 min in the dark.
5. Stop the reaction by desalting the reaction mixture by SPE onto RP-cartridges, by RP-HPLC or SEC, or by functionalized magnetic beads (*see* **Note 12**).

3.5. Enzymatic Cleavages

1. Reconstitute the dried peptide in the digestion buffer corresponding to the enzyme of interest in the conditions recommended by the manufacturer.
2. Incubate at the appropriate temperature (37°C is often recommended) for 1 to 16 h depending on the time-course properties of the enzyme. For the C-terminal sequencing by carboxypeptidase P digestion, a time course digestion is suitable (e.g., 30 s, 2 min, 5 min, 10 min, etc. up to 48 h).
3. Stop the reaction by addition of 1 volume of an acidic solution (e.g., 0.1% TFA) and control the pH using pH paper (pH < 4). For direct molecular mass fingerprint *see* **Note 13**.
4. Perform immediately the purification of the digest mixture on a reversed-phase column or by fast desalting by micro SPE using a C_{18} Zip-Tip (Millipore).

3.6. Microsequencing by Edman Chemistry

The following procedure may be used for native and reduced/alkylated peptides as well as for enzymatic peptide fragments.

1. Reconstitute the peptide in less than 10–15 μL with 30% acetonitrile in 0.1% TFA. Adapt the quantity to sequence according to the sequencer, to the molecular mass of the sample to analyze and to the number of amino acids to sequence (e.g., 50–100 pmol for a Procise System, <10 pmol for a Procise cLC).
2. Load the peptide in 5-μL aliquots on an appropriate sequencing membrane (pre treated). Dry carefully the sample for at least 10–15 min.

3.7. Mass Spectrometry

The different types of mass spectrometers provide a wealth of information going from simple molecular masses of intact components to an inference of the amino acid composition, sequence order, substitution site and nature of posttranslational modifications, and tissue distribution of the molecule of interest. As no instrument offers all capabilities simultaneously, complementary mass spectrometers (ESI and MALDI) are often required for peptidomics and proteomics studies. *See* **Note 14** for additional comments on MS.

3.7.1. MALDI-TOF

For MALDI-TOF MS several sample preparations are available with different matrices *(26)*. The choice of the matrix and of the sample preparation should be adapted to the molecular mass of the compounds and to the complexity of the samples to analyze. α-Cyano-4-hydroxycinnamic acid (4HCCA) is preferred for peptides between 1 and 15 kDa, and the sandwich sample preparation can be universally used for molecular mass determination of pure peptides and enzymatic fragments or complex mixtures (e.g., crude hemolymph, enzymatic digests). The procedures reported below are the ones used for the discovery of the *Drosophila* immune-induced peptides *(19,27,30)*.

1. 0.5 μL of a saturated solution of 4HCCA in acetone is placed on the MALDI sample plate.
2. When this is dried, deposit 0.5 μL of 1% TFA on the crystallized matrix bed.
3. For molecular mass determination, dissolve/dilute the purified peptide in acidified water (0.5% TFA) and deposit 0.5 μL of the acidified peptide solution.
4. Immediately add a droplet of 0.2 μL of a saturated solution of matrix 4HCCA in 50% acetonitrile in acidified water.
5. Dry the target under gentle vacuum and wash (also used for desalting if crude hemolymph or digestion mixture) the sample preparation twice with 1 μL of 1% TFA. Remove the washing solution after a few seconds using forced air, and dry the sample under vacuum.
6. Insert the sample probe in the MALDI-TOF mass spectrometer. Calibration is performed in external mode with peptides covering the mass range of 500 Da to 5 kDa (*see* **Note 15**).

3.7.2. ESI-MS & -MS^n

1. Dissolve the fraction aliquot to analyze in a solution of water/acetonitrile/formic acid (composition 49.9/49.9/0.2, v/v/v) (*see* **Note 16**).
2. Inject, using a Hamilton syringe (total volume of the syringe 5 μL), a fivefold dilution first for evaluating the intensity of the mass signals. If necessary, inject the rest of the sample.
3. For nano-ESI-MS & MS^n, the sample is loaded either in a previously rinsed gold/palladium-coated nanospray capillary (Promega) or in a PicoTip™ emitter (New Objective) and analyzed in a nanospray source.
4. Calibrate the mass analyzer with the multicharged ions of at least three to five standard peptides or following **Note 15**.

3.7.3. On-Line LC-ESI-MS^n

1. Deliver the sample automatically using an autosampler in a reversed-phase capillary precolumn (e.g., internal diameter 100 μm).
2. Wash the sample on the precolumn with a solution of 5% acetonitrile in 0.1% formic acid.
3. Elution of the sample is performed using a linear gradient of acetonitrile in formic acid (5–60% over 120 min) on a capillary RP column (e.g., internal diameter 75 μm) connected to the ionization source.
4. The eluted material is delivered on-line to the mass spectrometer at a flow-rate of 300 nL/min, for example.

3.8. Liquid Antimicrobial Assays

This section is intentionally restricted to antimicrobial assays as these assays can be routinely managed in research laboratories. Bioassays are used either to follow the antibacterial/antifungal peptides during the course of their purification or to define their biological properties. As such, they should be easy to perform, sensitive, and reliable. To meet these requirements, liquid growth inhibition assays are preferred to the more conventional radial diffusion assays. However, assays performed in liquid media may give slightly different results than when assays are performed using solid media. The liquid growth inhibition assay used to follow antimicrobial (antibacterial, antifungal) activities in HPLC fractions can also be used to determine the minimal inhibitory concentration (MIC) value of a pure compound. Fractions are replaced by serial dilutions of the peptide of interest. For details *see* **ref. 22**.

3.8.1. Antibacterial Assay

1. Inoculate 5–10 mL of fresh culture broth with a single colony of the bacterial strain of interest and incubate in a 37°C dry incubator or a temperature-controlled chamber.
2. When absorbance at 600 nm indicates an exponential growth phase culture, the cultured bacteria are diluted in Poor Broth to a $A_{600} = 0.001$.
3. To 90 μL of the exponential phase culture at $A_{600} = 0.001$, add 10 μL of test fraction (HPLC fraction or pure peptide reconstituted in water or in an appropriate culture medium). In control wells, water is replacing test fractions.
4. Incubate plates with gentle shaking for 16 h at 25°C (37°C is also possible but will accelerate the growth) in an environmental incubator.
5. Measure bacterial growth between 580 and 620 nm using a microplate reader.

3.8.2. Antifungal Assay

1. Dispense 10 μL of the fractions to be tested with 10 μL of water into wells of a 96-well microtiter plate. For control wells, 20 μL of water are used.
2. Add 80 μL of a spore suspension (10^4 fungal spores).

3. Incubate the plates in the dark for 48 h in a moist chamber at 25°C.
4. Measure fungal growth by detecting culture absorbency in a microplate reader (600 nm). A control of the growth can be performed after 24 h using a photonic microscope.

4. Notes

1. If the solution is yellowish discard it.
2. To kill *E. coli* D31, ciprofloxacin, which belong to the group of fluroquinolones, may be used as a positive control antibiotic.
3. Experimental infections, tissue collection, extraction conditions, purification procedures, and strategies for determining the primary structure of the natural bioactive peptides should be adapted to the animal model (size, rarity, possibility to conduct molecular biology investigations, availability of EST, and genomic databases, etc.) considered and to the complexity of the bioactive peptide.
4. Different microorganisms or parasites are used as inducers of the immune response of invertebrates. Constituents of the bacterial membrane may also be considered as successful inducers. Nevertheless, they may not develop the full immune response, limiting the possibility to recover all the armamentarium build by the invertebrate to fight off an infection *(34)*.
5. A relevant set of not experimentally infected individuals (control) is required for differential analysis by MALDI-TOF-MS or RP-HPLC in order to discriminate between immune-induced molecules and constitutively present substances *(19)*. This is a prerequisite when no *in vitro* assays are used to select the bioactive peptides from the immune system of the model invertebrate investigated.
6. Working on an animal model with a limited volume of hemolymph (<1 μL), it is recommended to escape the centrifugation step and to transfer the withdrawn hemolymph (*see* **Subheading 2.2.2., step 3**) directly in an aqueous solution of 0.1% TFA.
7. If large quantities of tissues or hemocytes are used, a preliminary homogenization with a Dounce apparatus may be suitable (e.g., 25 strokes in ice-cold Tris buffer at pH 8.7 containing 50 mM NaCl and 40 μg/mL aprotinin; *see* **ref. 29**).
8. Other percentages can be selected, but a first elution with less than 10% acetonitrile is suitable to remove the hydrophilic compounds. After this first "cleaning" step, two successive more hydrophobic solutions are often used to have first the bulk of bioactive peptides and then the highly hydrophobic molecules in a different fraction. This step also allows lipid removal from the sample by an irreversible binding of the lipids to the single-use cartridge.
9. The equivalent of extract to be tested should be calculated at this point of the purification procedure and should be kept until the final purification step. Increasing the extract equivalent during the course of the purification may be needed as the recovery after each step of purification is often between 50–75%.
10. For the purification of the *Drosophila* immune-induced peptides *(19,27,30)*, the hemolymph extract (equivalent to 10 μL of hemolymph collected from

120 experimentally infected flies) was loaded onto a microbore RP column (Aquapore RP 300 C_8, 1×100 mm, Brownlee). Elution was performed either with a linear gradient of 0–80% acetonitrile in acidified water over 80 min *(19,27)* or with a linear gradient of 2–62% acetonitrile in acidified water over 60 min *(30)*, both at a flow rate of 80 μL/min at 35°C.

11. If the molecule of interest is eluting during the first HPLC run with the front, the following purification step should be performed in an isocratic condition with a minimum of acetonitrile (0.5%) or directly in acidified water without acetonitrile.

12. Using RP-HPLC, the alkylated peptide is separated from reagents and the unreacted peptide with a gradient of acidified acetonitrile from 2 to 80% in 120 min. The large quantity of reagents is eliminated owing to long washing period (at least 30 min) with 2% acidified acetonitrile prior starting the gradient. Using the 4-vinylpyridine, an increase in the absorbance is observed on the alkylated peptide compared to the native molecule. When using magnetic beads, the chemically modified peptide is recovered by elution with 60% acetonitrile.

13. For molecular mass fingerprint of an enzymatic digestion, the acidity of the matrix and of the 1% TFA droplet is sufficient to quench any further digestion.

14. For determining the molecular mass of a polypeptide, a single-stage mass spectrometer is appropriate, whereas for analysis of structural features tandem MS (MS^n) is required. In this latter case, after molecular mass measurement, specific ions are selected and then subjected to fragmentation through collision in a specific chamber (collision cell) supplied with an appropriate gas (e.g., argon). Instrument performance (resolution, sensitivity, mass accuracy) depends on the instrument type, the ionization method, and the scanning capabilities.

15. For the analysis of a trypsin digest by MS, an internal calibration is performed with fragments of autolysed trypsin (e.g., $[M-H]^+$ at m/z 842.510, 1045.546, and 2211.104).

16. To avoid the inconvenience of the TFA (decrease sensitivity) during MS, if molecular mass measurements are performed on fractions that may contain traces of TFA, the sample has to be desalted and concentrated onto a C_{18} Zip-Tip. Elution of the peptide(s) will then be performed by 50% aqueous acetonitrile supplemented with 1% formic acid.

5. Conclusion

Intensive research efforts for developing new anti-infectious and antitumoral drugs for human health rely on technological advancements in screening of combinatorial chemical libraries and/or natural extracts generated from tissues and fluids from animals and plants. Presently, the use of natural extracts and the random HTS of libraries of synthetic or natural molecules have successively allowed the discovery of new therapeutic molecules. The race to discover the new drugs of tomorrow has begun. In the area of antibiotics, facing the increased resistance of bacteria and fungi to conventional antibiotics, the natural bioactive peptides from the innate immune defenses have considerably attracted the interest of pharmaceutical/biotechnology discovery companies. The isolation and structural characterization of natural bioactive peptides from the immune systems of invertebrates can be considered a promising step to discovering new innovative candidates for development as therapeutics. Although the AMPs exert relatively lower activities against susceptible microbial strains compared to conventional low molecular mass antibiotics, they have several advantages such as a fast target killing, a large activity spectrum, limited toxicity, and a limited tendency for developing resistance in the target microorganisms *(35)*. The constant and rising need for innovative therapeutic approaches combined with the potential of peptides, as active pharmaceutical component for effective drug formulation, is an important factor in the rapid development of the peptide market. Throughout our organism, peptides are active regulators and information brokers that make them interesting for drug development. Peptides have several virtues compared to the classical small molecules. They show higher specificity, have few toxicology problems, can be more potent, and do not accumulate in organs or face drug–drug interactions. Nevertheless, they have some drawbacks. Peptides are bigger and therefore more expensive to manufacture, and they may be less stable than chemical drugs, requiring an improved formulation. However, with recent manufacturing improvements (transgenic expression, recombinant or synthetic methods) and the development of highly powerful delivery protocols and techniques to improve stability, the global market for peptide-based therapeutics is expected to considerably expand. The various methodologies described here were successful for the characterization of natural bioactive immune-induced peptides from invertebrates. Depending on the objectives of the research and of the biological model investigated, the complete panel of approaches reported in this chapter should be considered together with their specific limitations.

Acknowledgments

I would like to thank Professors Laurence Ehret-Sabatier (Institut Pluridisciplinaire Hubert Curien, Strasbourg, France) and Sirlei Daffre (São Paulo University, Brazil) for critical reading of the manuscript.

References

1. Zhang, L., and Falla, T.J. (2006) Antimicrobial peptides: therapeutic potential. *Expert Opin. Pharmacother.* **7(6)**, 653–663.
2. Hancock, R.E., and Sahl, H.G. (2006) Antimicrobial and host-defense peptides as new anti-infective therapeutic strategies. *Nat. Biotechnol.* **24(12)**, 1551–1557.

3. Pereira, H.A. (2006) Novel therapies based on cationic antimicrobial peptides. *Curr. Pharm. Biotechnol.* **7(4)**, 229–234.
4. McPhee, J.B., and Hancock, R.E. (2005) Function and therapeutic potential of host defence peptides. *J. Pept. Sci.* **11(11)**, 677–687.
5. Riley, M.A., and Wertz, J.E. (2002) Bacteriocin diversity: ecological and evolutionary perspectives. *Biochimie* **84(5–6)**, 357–364.
6. Guder, A., Wiedemann, I., and Sahl, H.G. (2000) Posttranslationally modified bacteriocins—the lantibiotics. *Biopolymers* **55**, 62–73.
7. Jenssen, H., Hamill, P., and Hancock, R.E. (2006) Peptide antimicrobial agents. *Clin. Microbiol. Rev.* **19(3)**, 491–511.
8. Castro, M.S., and Fontes, W. (2005) Plant defense and antimicrobial peptides. Protein Pept. Lett. **12(1)**, 13–18.
9. Bulet, P., Stöcklin, R., and Menin, L. (2004) Anti-microbial peptides: from invertebrates to vertebrates. *Immunol. Rev.* **198**, 169–184.
10. Sitaram, N., and Nagaraj, R. (2002) Host-defense antimicrobial peptides: importance of structure and activity. *Curr. Pharm. Des.* **8(9)**, 727–742.
11. Beisswenger, C., and Bals, R. (2005) Functions of antimicrobial peptides in host defense and immunity. *Curr. Protein Pept. Sci.* **6(3)**, 255–264.
12. Hancock, R.E., Brown, K.L., and Mookherjee, N. (2006) Host defence peptides from invertebrates—emerging antimicrobial strategies. *Immunobiology* **211(4)**, 315–322.
13. Bachère, E., Gueguen, Y., Gonzalez, M., de Lorgeril, J., Garnier, J., and Romestand, B. (2004) Insights into the anti-microbial defense of marine invertebrates: the penaeid shrimps and the oyster *Crassostrea gigas*. *Immunol. Rev.* **198**, 149–168.
14. Imler, J.L., and Bulet, P. (2005) Antimicrobial peptides in *Drosophila*: structures, activities and gene regulation. *Chem. Immunol. Allergy* **86**, 1–21.
15. Bulet, P., and Stöcklin, R.(2005) Insect antimicrobial peptides: structures, properties and gene regulation. *Protein Pept. Lett.* **12(1)**, 3–11.
16. Lemaitre, B., and Hoffmann, J. (2007) The host defense of *Drosophila melanogaster*. *Annu. Rev. Immunol.* **25**, 697–743.
17. Boulanger, N., Munks, R.J., Hamilton, J.V., et al. (2002) Epithelial innate immunity. A novel antimicrobial peptide with antiparasitic activity in the blood-sucking insect *Stomoxys calcitrans*. *J. Biol. Chem.* **277(51)**, 49921–49926.
18. Kuhn-Nentwig, L. (2003) Antimicrobial and cytolytic peptides of venomous arthropods. *Cell Mol. Life Sci.* **60(12)**, 2651–2668.
19. Uttenweiler-Joseph, S., Moniatte, M., Lagueux, M., Van Dorsselaer, A., Hoffmann, J.A., and Bulet, P. (1998) Differential display of peptides induced during the immune response of *Drosophila*: a matrix-assisted laser desorption ionization time-of-flight mass spectrometry study. *Proc. Natl. Acad. Sci. USA* **95(19)**, 11342–11347.
20. Levy, F., Rabel, D., Charlet, M., Bulet, P., Hoffmann, J.A., and Ehret-Sabatier, L. (2004) Peptidomic and proteomic analyses of the systemic immune response of *Drosophila*. *Biochimie* **86(9–10)**, 607–616.

21. Chernysh, S., Kim, S.I., Bekker, G., et al. (2002) Antiviral and antitumoral peptides from insects. *Proc. Natl. Acad. Sci. USA* **99(20)**, 12628–12632.
22. Hetru, C., Bulet, P. (1997) Strategies for the isolation and characterization of antimicrobial peptides of invertebrates, in *Antimicrobial Peptide Protocols* (Shafer, W.M., ed.), Humana Press, Totowa, NJ, pp. 35–49.
23. Boulanger, N., Ehret-Sabatier, L., Brun, R., Zachary, D., Bulet, P., and Imler, J.L. (2001) Immune response of *Drosophila melanogaster* to infection with the flagellate parasite *Crithidia* spp. *Insect Biochem. Mol. Biol.* **31**, 129–137.
24. Boulanger, N., Lowenberger, C., Volf, P., et al. (2004) Characterization of a defensin from the sand fly *Phlebotomus duboscqi* induced by challenge with bacteria or the protozoan parasite *Leishmania major. Infect. Immun.* **72(12)**, 7140–7146.
25. Favreau, P., Menin, L., Michalet, S., et al. (2006) Mass spectrometry strategies for venom mapping and peptide sequencing from crude venoms: case applications with single arthropod specimen. *Toxicon* **47**, 676–687.
26. Bulet, P., and Uttenweiler-Joseph, S. (2000) A MALDI-TOF mass spectrometry approach to investigate the defense reactions in *Drosophila melanogaster*, an insect model for the study of innate immunity, in RM Kamp, D Kyriakidis, Th Choli-Papadopoulos, ed. Springer *Proteome and Protein Analysis* (Kamp, R.M., Kyriakidis, D., and Choli-Papadopoulos, T., eds.), Springer-Verlag, Berlin, pp. 157–174.
27. Carte, N., Cavusoglu, N., Leize, E., Van Dorsselaer, A., Charlet, M., and Bulet, P. (2001) *De novo* sequencing by nano-electrospray multiple-stage tandem mass spectrometry of an immune-induced peptide of *Drosophila melanogaster. Eur. J. Mass Spectrom.* **7(4)**, 399–408.
28. Favreau, P., Cheneval, O., Menin, L., et al. (2007) The venom of the snake genus *Atheris* contains a new class of peptides with clusters of histidine and gylcine residues. *Rapid Commun. Mass Spectrom.* **21(3)**, 406–412.
29. Destoumieux, D., Munoz, M., Cosseau, C., et al. (2000) Penaeidins, antimicrobial peptides with chitin-binding activity, are produced and stored in shrimp granulocytes and released after microbial challenge. *J. Cell Sci.* **113**, 461–469.
30. Rabel, D., Charlet, M., Ehret-Sabatier, L., et al. (2004) Primary structure and *in vitro* antibacterial properties of the *Drosophila melanogaster* attacin C pro-domain. *J. Biol. Chem.* **279(15)**, 14853–14859.
31. Boulanger, N., Munks, R.J.L., Hamilton, J.V., et al. (2002) Epithelial innate immunity. A novel antimicrobial peptide with antiparasitic activity in the blood-sucking insect *Stomoxys calcitrans. J. Biol. Chem.* **32(4)**, 369–375.
32. Vizioli, J., Richman, A.M., Uttenweiler-Joseph, S., Blass, C., and Bulet, P. (2001) The defensin of the malaria vector mosquito *Anopheles gambiae*: antimicrobial activities and expression in adult mosquitoes. *Insect Biochem. Mol. Biol.* **31**, 241–248.
33. Lamberty, M., Zachary, D., Lanot, R., et al. (2001) Insect immunity. Constitutive expression of a cysteine-rich antifungal and a linear antibacterial peptide in a termite insect. *J. Biol. Chem.* **276(6)**, 4085–4092.

34. Royet, J., Reichhart, J.M., and Hoffmann, J.A. (2005) Sensing and signaling during infection in *Drosophila. Curr. Opin. Immunol.* **17,** 11–17.
35. Marr, A.K., Gooderham, W.J., and Hancock, R.E. (2006) Antibacterial peptides for therapeutic use: obstacles and realistic outlook. *Curr. Opin. Pharmacol.* **6(5),** 468–472.

3

Sequence Analysis of Antimicrobial Peptides by Tandem Mass Spectrometry

Christin Stegemann and Ralf Hoffmann

Summary

The emergence of new diseases as well as the increasing resistance of bacteria against antibiotics over the past decades has become a growing threat for humans. This has driven a sustained search for new agents that possess antibacterial activities against bacteria being resistant against conventional antibiotics and prompted an interest in short to medium-sized peptides called antimicrobial peptides (AMPs). Such peptides were isolated from different species, including mammals, insects, birds, fish, and plants. As these peptides circulate only at low concentrations in body fluids, a multidimensional purification strategy is obligatory to obtain pure peptides. The resulting low peptide amounts require highly sensitive analytical techniques to sequence the peptides, such as Edman degradation or mass spectrometry. Here we describe the protocols used routinely in our laboratory to identify peptides with antimicrobial activities by mass spectrometry including de novo sequence analysis.

Key Words: Antibacterial peptide; collision-induced dissociation (CID); electrospray ionization (ESI); matrix-assisted laser desorption/ionization (MALDI); peptide derivatization.

1. Introduction

Despite remarkable advances in antibiotic therapy over the last century, the incidence of serious bacterial and fungal infections in humans is increasing due to the rapid emergence of drug-resistant Gram-positive and Gram-negative bacterial strains *(1–3)*. Thus one of the most urgent topics is the search for new compound classes which exert their antimicrobial activity by molecular mechanisms different from the currently used antibiotics. One of the most

From: *Methods in Molecular Biology, vol. 494: Peptide-Based Drug Design*
Edited by: L. Otvos, DOI: 10.1007/978-1-59745-419-3_3, © Humana Press, New York, NY

promising candidates are gene-encoded antimicrobial peptides isolated from a wide variety of different species, such as insects, birds, reptiles, amphibics, fishes, and mammals *(4,5)*. These naturally occurring AMPs form a first line of host defense against pathogens and are involved in innate immunity. They are divided into several subgroups based on their amino acid composition and structure *(6)*. Among the several thousand sequences possessing antimicrobial activities, only a few peptides have been thoroughly investigated using a broad spectrum of bacteria or fungi. Whereas all these peptides appear to attach first to the membrane, they kill the bacteria by different mechanisms, such as forming pores in the membrane or targeting specific intracellular bacterial targets.

The search for new AMPs starts from body fluids, cells, organs, or tissues, for example, to isolate active peptides and small proteins to homogeneity combining different orthogonal separation techniques, such as liquid chromatography (size exclusion, ion-exchange, reversed-phase chromatography) or gel electrophoresis (SDS-PAGE). Fractions or bands showing antimicrobial activities against a selected panel of microbes are further separated until a single active peptide is finally obtained, typically at the low μg or even ng scale, judged by matrix-assisted laser desorption/ionization (MALDI) or electrospray ionization (ESI) mass spectrometry (MS). Besides the mass information of the intact peptide, these techniques can also be used to determine their sequences using tandem mass spectrometry. These instruments rely on various combinations of mass analyzers, such as ion traps (IT), triple quadrupole (QqQ), hybrid quadrupole/time-of-flight (QqTOF), or time-of-flight/time-of-flight (TOF/TOF) instruments. Hybrid instruments provide especially good access to the peptide sequences due to their high sensitivity and mass accuracy.

The product ions produced from the selected precursor ion contain all information to deduce the peptide sequence from the b- or y-ion series *(7,8)* resulting from cleavage of the peptide bonds by collision-induced dissociation (CID), also called collision-activated dissociation (CAD). In principle, the peptide can be identified from a single MS/MS spectrum automatically if the corresponding protein is found in a data base *(9)*. If not, the complete sequence has to be retrieved from the mass spectrum de novo. However, this requires quite a lot of experience and is typically limited to peptide lengths from approximately 10 to 25 residues. Both longer and shorter peptides are difficult to sequence. Thus, it is often obligatory to digest the sample and sequence the shorter peptides, which are combined afterwards to the complete sequence. Moreover, spectral interpretations are often ambiguous due to overlapping ion series, isobaric amino acids, or a low mass accuracy. Therefore, we always confirm the postulated sequence by comparing the MS/MS spectra of the isolated peptide and the synthetic peptide.

2. Materials

2.1. Solid Phase Extraction

1. ZipTip™ (Millipore Corporation, Billerica, MS) containing 0.6 µL C_{18}-phase.
2. Wetting solution: 50% aqueous acetonitrile (CH_3CN, LC-MS grade; Biosolve BV, Valkenswaard, Netherlands). Store at room temperature (*see* **Note 1**).
3. Wash solution: 0.1% aqueous formic acid (HCOOH) (puriss. p.a., for mass spectroscopy; Fluka Chemie GmbH, Buchs, Switzerland). Store at 4°C.
4. Elution solution: 50% aqueous acetonitrile containing 0.1% formic acid.

2.2. MALDI Mass Spectrometry

1. MALDI-TOF/TOF-MS (4700 Proteomics Analyzer, Applied Biosystems Applera Deutschland GmbH, Darmstadt, Germany).
2. MALDI matrices
3. CHCA matrix solution:

   a. Weigh approximately 3.5 mg α-cyano 4-hydroxy-cinnamic acid (CHCA; Bruker Daltonics GmbH, Bremen, Germany; store at 4°C) in a 1.5-mL polypropylene vial.
   b. Add 250 µL of 0.1% aqueous trifluoroacetic acid (TFA) (for UV-spectroscopy, ≥99.0% (GC); Fluka Chemie GmbH) (*see* **Note 2**) or adjust this volume according to the correct CHCA weight.
   c. Vortex the solution for 10 s.
   d. Add 250 µL acetonitrile (MS grade).
   e. Vortex well until the matrix is completely dissolved.
   f. This solution containing CHCA at a concentration of 7 mg/mL can be stored at −20°C for one month. This matrix solution is only for the derivatization with 2-sulfobenzoic acid cyclic anhydride.
   g. For all other measurements dilute the matrix with 50% aqueous acetonitrile containing 0.1% TFA to 4 mg/mL.

4. DHB matrix solution:

   a. Weigh approximately 5 mg of 2,5-dihydroxybenzoic acid (DHB; Bruker Daltonics GmbH; store at 4°C) in a 1.5-mL polypropylene vial.
   b. Dissolve the 5 mg DHB in 450 µL water or correct the volume according to exact weight.
   c. Add 50 µL methanol (Biosolve BV, Valkenswaard, Netherlands, absolute, HPLC grade).

   d. Vortex well until DHB is completely dissolved.

   e. This DHB solution (10 mg/mL) can be stored at −20°C for one month.

2.3. ESI Mass Spectrometry

1. QSTAR Pulsar I (Applied Biosystems GmbH), which is a hybrid mass spectrometer consisting of a quadrupol and an orthogonal time-of-flight mass analyzer (QqTOF-MS). For data processing use the Analyst QS (version 1.0) and the Bioanalyst QS (version 1.0) software packages provided with the instrument.

2. For static nanoESI-MS connect the static nanospray source (Protana, Odense, Denmark) to the mass spectrometer and use for each sample to be analyzed a new borosilicate spray capillary (ES380 "Medium", PROXEON Biosystems AIS, Odense, Denmark; *see* **Note 3**).

3. Spray solution: 60% aqueous acetonitrile containing 1% formic acid.

4. To couple a gradient HPLC system to the mass spectrometer, replace the needle adapter for static nanospray by a tip adapter and connect the column to Distal Coated SilicaTips™ with an ID of 10 μm (New Objective, Woburn, MA).

5. For highest sensitivity use a nanoHPLC system, such as an Agilent 1100 series nanoHPLC system consisting of an isocratic pump (G1310A), an autosampler (G1377A), two pumps (G2226A) to form the gradient controlled by ChemStation for LC (Rev. A.10.02).

6. Sample plate: 384-well microtiter plate (Greiner Bio-One GmbH, Frickenhausen, Germany).

7. Trap column: Zorbax 300SB-C18 (300 μm internal diameter, 5-mm length, particle size 5 μm; Agilent Technologies GmbH).

8. Eluent A: 3% aqueous acetonitrile containing 0.1% formic acid.

9. Separation column: Zorbax 300SB-C18 (internal diameter of 100 μm, column length 150 mm, particle size 3.5 μm; Agilent Technologies GmbH).

10. Eluent B: Acetonitrile containing 0.1% formic acid.

2.4. Reduction and Alkylation of Disulfide Bonds

1. ABC buffer: 50 mM aqueous ammonium bicarbonate (NH_4HCO_3) (BioChemika Ultra, ≥99.5%, Fluka Chemie GmbH; store at room temperature). Check that pH is between 7.2 and 7.8. Store at room temperature.

2. DTT solution: Dissolve 15.42 mg 1,4-dithiothreitol (≥99% p.a., Carl Roth GmbH & Co. KG, Karlsruhe, Germany; store at 4°C) in 100 μL water. Prepare fresh for each reaction. Do not store.

3. Acrylamide solution: Serva Acrylamide 4X solution (Serva Electrophoresis GmbH, Heidelberg, Germany) corresponding to a 40% (w/v) acrylamide solution. Store at 4°C.

4. Iodoacetamide solution: Dissolve 27.5 mg iodoacetamide (purum ≥98.0%, Fluka Chemie GmbH; store at 4°C) in 100 μL water. Store at −20°C (*see* **Note 4**).

2.5. Enzymatic Digests

1. Three millimolar ABC-buffer: 3 m*M* aqueous ammonium bicarbonate buffer (BioChemika Ultra, ≥99.5%, Fluka Chemie GmbH). Check that pH is between 7.2 and 7.8. Store at room temperature.
2. Trypsin solution: Dissolve trypsin (Promega Sequencing Grade Modified Trypsin Porcine, activity: 5000 U/mg; store at −20°C) in the 3 m*M* ABC-buffer to obtain a final concentration of 20 ng/μL. Store in aliquots at −20°C (*see* **Note 5**).
3. Ten millimolar ABC-buffer: 10 m*M* aqueous ammonium bicarbonate buffer (BioChemika Ultra, ≥99.5%, Fluka Chemie GmbH). Check that pH is between 7.2 and 7.8. Store at room temperature.
4. Lys-C solution: Dissolve endoprotease Lys-C from *Lysobacter enzymogenes* (sequencing grade, Roche Diagnostics GmbH, Roche Applied Science, Mannheim, Germany; store at −20°C) in the 10 m*M* ABC-buffer to a final concentration of 10 ng/μL. Store in aliquots at −20°C (*see* **Note 5**).

2.6. Derivatization of Amino Groups

2.6.1. Guanidation of Lysine (10)

1. OME-reagent: Dissolve 50 mg O-methylisourea hemisulfate (94.0%, Acros Organics, Geel, Belgium; store at room temperature) in 51 μL water. Prepare always a fresh solution and use immediately.
2. Seven molar aqueous ammonium hydroxide solution. Store at room temperature.
3. 0.1% aqueous TFA.

2.6.2. N-Terminal Peptide Derivatization Using 4-Sulfophenyl Isothiocyanate (11)

1. Tris-buffer: 50 m*M* Tris-HCl buffer (pH 8.2). Store at room temperature.
2. SPITC-reagent: Dissolve 2.55 mg 4-sulfophenyl isothiocyanate sodium salt monohydrate (SPITC, technical grade, Sigma Aldrich, St. Louis, MO; store at 4°C) in Tris-buffer at a concentration of 2.55 mg/mL. Prepare always a fresh solution and use immediately.
3. Eighty percent aqueous acetonitrile containing 0.5% trifluoroacetic acid. Store at 4°C.

2.6.3. N-Terminal Peptide Derivatization Using 2-Sulfobenzoic Acid Cyclic Anhydride (12)

1. Reaction buffer: Mixture of acetonitrile and 12.5 m*M* aqueous ammonium bicarbonate buffer (1:1 by vol.) (BioChemika Ultra, ≥99.5%, Fluka Chemie GmbH; store at 4°C).

2. SACA reagent: 2-Sulfobenzoic acid cyclic anhydride (SACA) (technical, ≥95.0%, Fluka Chemie GmbH; store at room temperature) at a concentration of 2 mg/mL in dry tetrahydrofurane (THF, extra dry with molecular sieve, water < 50 ppm, stabilized; Acros Organics, Geel, Belgium). Prepare always a fresh solution and use immediately.

3. Methods

3.1. Solid Phase Extraction on a Reversed Phase

1. Add 5 μL elution solution to the bottom of a 0.6-mL polypropylene vial (*see* **Note 6**) and close the vial. Use 10-μL pipette tips for this and all following steps.
2. Dissolve the peptide in 10 μL wash solution.
3. Wash the ZipTip™ first by aspirating and dispensing 10 μL wetting solution to the waste. Repeat this washing procedure at least twice (*see* **Note 7**).
4. Equilibrate the stationary phase by drawing in wash solution and dispensing it to the waste three times.
5. Put the tip of the ZipTip™ into the peptide solution to be purified and aspirate and dispense the solution slowly at least ten times. Thus the peptide is completely bound to the stationary phase.
6. Remove all polar impurities and salts from the stationary phase by washing the packing material with 10-μL aliquots of wash solution. Repeat this procedure seven times.
7. Slowly aspirate the 5 μL of elution solution stored in the 0.6-mL polypropylene vial (**step 1**) and dispense it again in the tube. Repeat at least five times without getting any air into the packing material.
8. Aspirate 5 μL of fresh elution solution and dispense once into the 0.6-mL polypropylene vial yielding 10 μL peptide solution to be analyzed either by MALDI-MS or static nanoESI-MS (*see* **Note 8**).

3.2. MALDI Mass Spectrometry

3.2.1. Sample Preparation with CHCA as Matrix

1. Spot 0.5 μL of the CHCA matrix solution on the MALDI target (*see* **Note 6**).
2. Wait until the spot is dried (*see* **Note 9**) and add 1 μL of the sample solution in the middle of the crystallized matrix spot (*see* **Note 10**).
3. Wait until the spot is dried again (about 5 min) (*see* **Note 9**). Check with a microscope that light yellow crystals were formed (*see* **Note 11**).

3.2.2. Sample Preparation with DHB as Matrix

1. Spot 0.5 μL of the DHB matrix solution on the MALDI target.
2. Add 1 μL of the sample solution in the middle of the dried matrix spot (*see* **Note 10**).
3. Wait until the spot is dried (*see* **Note 9**).

4. Add again 0.5 µL of the DHB matrix solution. Check with a microscope that white needle-like crystals were formed as an outer circle of the spot (*see* **Note 11**).

3.2.3. Recording MALDI Spectra

1. Calibrate the mass spectrometer with a peptide mixture spanning a similar mass range to be used afterwards for the analysis, such as a mixture of des-Arg bradykinin (M = 903.4603 g/mol), angiotensin I (M = 1295.6775 g/mol), Glu-fibrino-peptide B (M = 1569.6696 g/mol), ACTH 1–17 (M = 2092.0789 g/mol), ACTH 18–39 (M = 2464.1910 g/mol), and oxidized insulin B chain (M = 3493.6435 g/mol) (4700 Proteomics Analyzer Calibration Mixture, Applied Biosystems GmbH; store at −20°C).
2. As the layer technique with CHCA produces a homogeneous peptide distribution within the matrix, record the spectra from all parts of the sample spot. For samples prepared with DHB, use only the outer edge of the matrix spot, as the center part produces mostly salt clusters and sodiated peptide ions.

Record the spectra in reflectron mode using an acceleration voltage of 20 kV, 70% grid voltage and a delay of 1.277 ns. Generate MS spectra by accumulating 2000 laser shots (**Fig. 1**). Analyze the mass spectra with the Data Explorer® Software Version 4.6 (Applied Biosystems GmbH) (*see* **Note 12**).

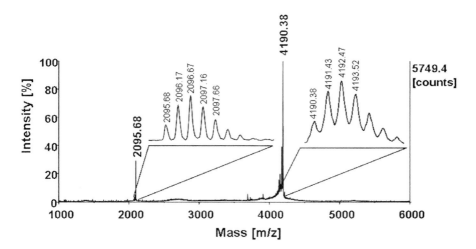

Fig. 1. MALDI-TOF mass spectrum of an antibacterial peptide isolated from dog *Canis familiaris* using the CHCA matrix. The dominant signals at m/z 4190.38 and 2095.68 correspond to the [M+H]$^+$ and [M+2H]$^{2+}$ ions of the same peptide. The doubly charged ion, which is usually not obtained for peptides in MALDI-MS, is probably detected due to the high content of basic lysine and arginine residues.

3.3. ESI Mass Spectrometry

3.3.1. Static Mode

1. Dissolve 5–15 pmol of the peptide in 3 μL spray solution.
2. Transfer the peptide solution with GELoader®Tips (Eppendorf AG, Hamburg, Germany) into the nanospray needle and centrifuge it down to the tip of the needle. Be careful not to touch the metal coating and the tip (*see* **Note 3**).
3. Record the mass spectra in positive ion mode from the m/z 400 to 2000 range using a spray voltage of 1100 V. Set instrumental values for the ion source gas 1 to 6 and for the curtain gas to 25. Analyze the product ion spectra with the Analyst QS software (Applied Biosystems GmbH) (*see* **Notes 12 and 13**).

3.3.2. LC-MS Coupling

1. Dissolve 0.5 to 2 pmol peptide in 20 μL of eluent A and load it isocratically onto the trap column using a flow rate of 30 μL/min.
2. Switch the valve after 10 min (*see* **Note 14**) to connect the trap column to the separation column (Zorbax SB-C18) and start the gradient to elute the peptide at a flow rate of 300 nL/min (*see* **Note 15**).
3. Record the spectra in the "data-dependent acquisition" (DDA) mode (*see* **Note 16**) consisting of one TOF scan followed by two MS/MS scans for automatically selected precursor ions. In both scan modes use positive ion mode, spray voltage of 1800 V, curtain gas of 18 and CAD gas of 5. For the TOF spectrum select an m/z range from 400 to 1500 and record the data for 1 or 2 s. Record the product ion spectra for the two most intense ions with charge states from 2 to 5 in the m/z range from 70 to 1700 for 3 to 5 s. Allow the software to calculate the fragmentation energy individually for each selected parent ion based on its charge state and molecular mass.

3.4. Reduction and Alkylation of Disulfide Bonds

1. Mix 10 μL sample containing 40 pmol peptide, add 9 μL ABC buffer and vortex well.
2. Add 1 μL freshly prepared DTT solution.
3. Incubate at 95°C in the thermomixer at 800 rpm (revolutions per minute) for 5 min (*see* **Scheme 1**).
4. Allow the samples to cool to room temperature.
5. Divide the solution in two parts of equal volume in 0.6-mL polypropylene vials and add 1 μL acrylamide solution to the first part and 1 μL fresh iodoacetamide solution to the second part (*see* **Note 4** and **Scheme 1**).
6. Shake both reaction mixtures at 800 rpm at room temperature for 30 min (*see* **Note 17**).
7. Dry both samples in a SpeedVac and dissolve them again with 10 μL of 0.1% aqueous formic acid.
8. Desalt the alkylated peptides with C18-ZipTips™ (*see* **Subheading 3.1.**)

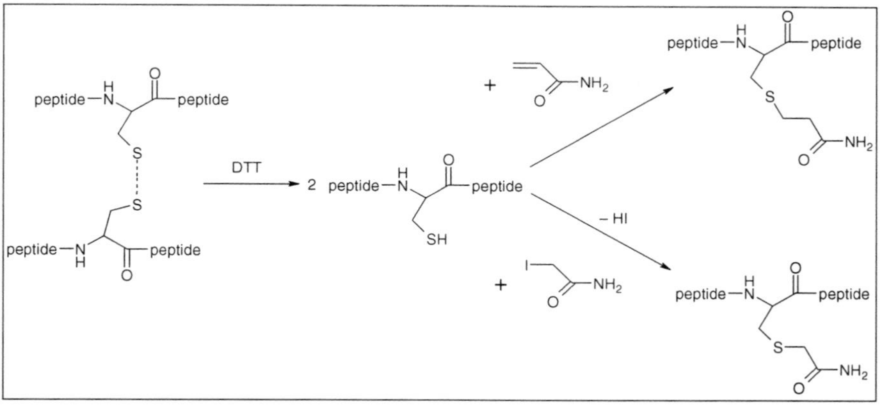

Scheme 1. Reduction of homodimer peptide linked via a disulfide bridge with dithiothreitol (DTT) and alkylation of the resulting thiol groups of both cysteines with either acrylamide (upper reaction) or iodoacetamide (lower reaction).

3.5. Counting Cysteine Residues

1. Determine the molecular masses of the alkylated peptide samples by nanoESI-MS in static mode using 2-µL sample or by MALDI-MS using either CHCA matrix or DHB matrix (*see* **Subheadings 3.2.1.** and **3.2.2.**).
2. Calculate the differences of the original peptide mass and the recorded masses after alkylation with iodoacetamide and acrylamide (e.g., the peptide mass was shifted from 4540.35 to 4972.84 g/mol and 4888.42 after reduction and alkylation with acrylamide and iodoacetamide, respectively, corresponding to mass differences of 432.49 and 348.07 g/mol.).
3. Determine the number of cysteine residues by dividing the mass difference calculated for the acrylamide sample by 72.08 and 58.05 g/mol for the iodoacetamide sample (e.g., the mass shift of 432.49 g/mol for acylamide and 348.07 g/mol for iodoacetamide correspond to six cysteine residues in both cases, i.e., three disulfide bonds in the original sample).

3.6. Tryptic Digest

1. If the peptide to be digested contains cysteine residues, reduce and alkylate them prior to any enzymatic digest (*see* **Subheading 3.4.**) using only 1 µL of either of the two alkylation reagents for 40 pmol peptide.
2. Dry the peptide solution (40 pmol) in the SpeedVac and add 8 µL 3 mM ABC buffer.
3. Add 2 µL trypsin solution and mix carefully by aspirating and dispensing 2 µL of the solution several times with the tip.
4. Split the sample in two parts of equal volume to 0.6-mL polypropylene vials (*see* **Note 6**).

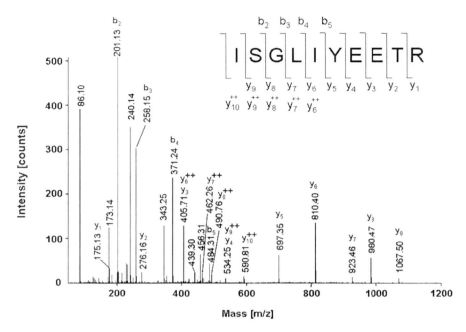

Fig. 2. Manually acquired nanoESI-QqTOF MS/MS spectrum of the doubly charged peptide ion at m/z 590.81 of a tryptic digest of antibacterial peptide with mass 4189.38 (*see* **Fig. 1**). The peptide sequence ISGLIEETR was deduced from the singly and doubly charged y-ions starting at m/z 175.13 corresponding to a C-terminal arginine.

5. Incubate at 37°C and shake continuously at 800 rpm in the thermomixer.
6. After 1 h stop the first digest by addition of 1 μL concentrated formic acid.
7. Stop the second sample after a total digestion time of 4 h by adding 1 μL concentrated formic acid.
8. Analyze the sample without desalting by MALDI-MS or after solid phase extraction by static nanoESI-MS (**Fig. 2**) (*see* **Note 18**).

3.7. Lys-C Digest

1. Cysteine-containing peptides should be reduced and alkylated before the digest.
2. Pipet an aliquot of 10 μL of the solution containing 40 pmol peptide in a 0.6-mL polypropylene vial and dry it in a SpeedVac.
3. Dissolve the peptide in 10 μL of 10 mM ABC buffer.
4. Add 5 μL Lys-C solution.
5. Split the sample in two parts and incubate both at 37°C in the thermomixer (800 rpm).
6. After 1 h stop the first digest by addition of 2 μL concentrated formic acid.
7. Stop the second digest by addition of 2 μL concentrated formic acid after 4 h.

8. Dry both digests in the SpeedVac.
9. Dissolve in 10 μL of 60% aqueous acetonitrile containing 0.1% formic acid.
10. Use these solutions directly for MALDI-MS and nanoLC-ESI-MS or desalt them by solid phase extraction before nanoESI-MS.

3.8. Peptide Derivatization

3.8.1. Guanidation of Lysine

1. Reduce, alkylate, and digest the peptides as described above.
2. Evaporate the solvent completely in a SpeedVac and dissolve the sample in 2 μL of 0.1% aqueous TFA.
3. Add 1.5 μL OME solution, 3 μL water, and 5.5 μL ammonium hydroxide solution (*see* **Note 19** and **Scheme 2**).
4. Incubate in the thermomixer (800 rpm) at 65°C for 10 min.
5. Dry the solution in the SpeedVac.
6. Dissolve the sample in 10 μL of 0.1% aqueous formic acid.
7. Desalt the sample with a C_{18}-ZipTip™ (*see* **Subheading 3.1.**)
8. Analyze the sample by MALDI- or ESI-MS. Each derivatized lysine will increase the peptide mass by 42 u (unified atomic mass unit).
9. To derivatize the sample with SPITC or SACA on the N-terminus, dry it in the SpeedVac.

3.8.2. N-Terminal Derivatization with 4-Sulfophenyl Isothiocyanate

1. Dissolve the dried peptide sample (alkylated, guanidated) in 10 μL of 0.1% aqueous TFA.
2. Load the peptide on a C_{18}-ZipTip™ (**steps 1–6** in **Subheading 3.1.**)

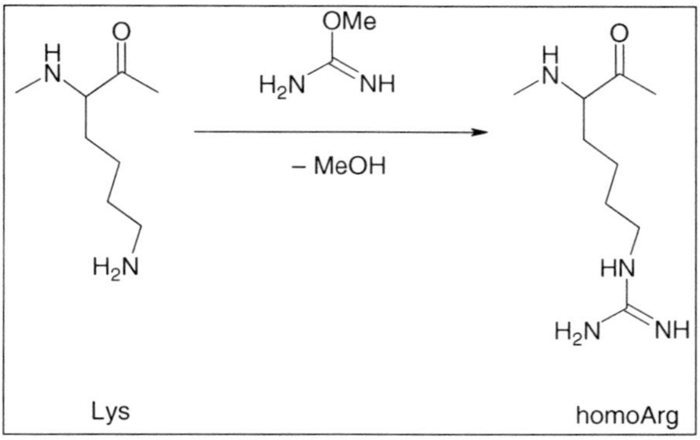

Scheme 2. Reaction of O-methylisourea with the ε-amino group of a lysine. The formed guanidino group of homoarginine has a higher proton affinity than the original amino group, resulting in better protonation efficiencies in the MALDI and ESI process.

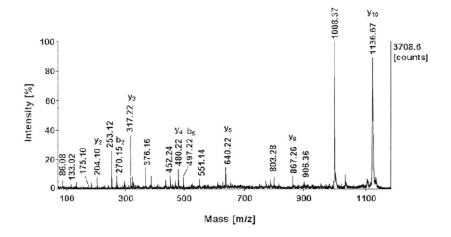

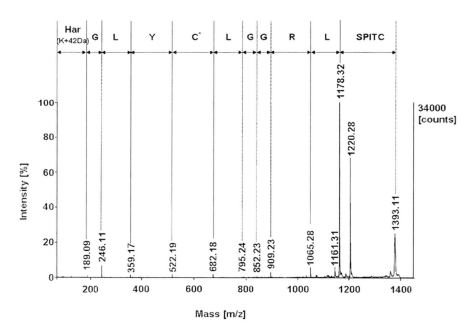

Fig. 3. Automatically acquired MALDI-TOF/TOF MS/MS spectra of a tryptic peptide with m/z 1136.67 (upper panel) and of the same peptide after derivatization with SPITC at m/z 1393.11 (lower panel) using CHCA matrix. Whereas the unmodified peptide displays a rather complex spectrum with intense b- and y-ions in the same mass range, the lower spectrum shows a complete and dominant y-series allowing an easy manual sequence interpretation from m/z 189.09 to m/z 1178.32. The peptide sequence LRGGLC*YLGK was deduced from the y-series with guanidated Lys (homoarginine = Har) and N-terminal derivatization with SPITC. The cysteine residue was reduced and alkylated with DTT and iodoacetamide (M = 160.03 g/mol).

3. Add 10 μL freshly prepared SPITC solution in a clean vial and aspirate and dispense this solution three times through the ZipTip™ loaded with the peptide. Store the ZipTip™ with the SPITC solution on top of the packing material in a 1.5-mL polypropylene vial.
4. Incubate the reaction mixture in the closed vial at 50°C for 1 h without shaking.
5. Put the ZipTip™ carefully on the pipette again (*see* **Note 19**) and dispense the SPITC solution to the waste.
6. Aspirate 10 μL wash solution into the ZipTip™ and dispense to waste. Repeat this procedure seven times without getting air into the packed material.
7. Add 2 μL elution solution to a clean 0.6-mL polypropylene vial and aspirate and dispense it five times through the ZipTip™ without getting air into the stationary phase.
8. Use the sample for MALDI-MS with CHCA and record the spectra in positive and negative ion mode.
9. The increment mass of the N-terminal modification is 214.97 u.
10. For sequence analysis record the MS/MS spectra (**Fig. 3**) for all modified peptides in positive ion mode (*see* **Note 20**).

3.8.3. N-Terminal Derivatization with 2-Sulfobenzoic Acid Cyclic Anhydride

1. Dissolve the dried peptide sample in 2 μL reaction buffer.
2. Add 2 μL freshly prepared SACA reagent (*see* **Note 21**) and mix.
3. Incubate at 24°C for 5 min in the thermomixer (800 rpm).
4. Prepare the sample for MALDI-MS with CHCA and record the spectra in positive and negative ion mode (*see* **Note 20**).
5. The increment mass of the N-terminal modification is 184.17 u.
6. For sequence analysis record the MS/MS spectra for all modified peptides in positive ion mode (*see* **Note 20**).

4. Notes

1. If a reversed-phase HPLC is coupled on-line to a mass spectrometer equipped with an electrospray ionization source, it is important to use acetonitrile with "MS grade" purity instead of "HPLC grade" or "UV grade" to reduce the background noise in the m/z range above 200. Otherwise several strong signals are obtained that may disturb data processing. The eluents should be replaced weekly. We use only "MS grade" acetonitrile to prepare buffers and reagents for MS analysis.
2. The addition of TFA or other volatile acids to the matrix improves the peptide ionization, increasing the signal intensities. Alternatively, the acid can be added to the peptide sample. If the sample already contains a high acid content it is better to add pure matrix without TFA to prevent cluster formation increasing the background noise.

3. The surface of the nanospray capillary tips is coated with a thin metal film to apply the voltage. Do not scratch this coating or touch it, as this may reduce the spray performance. Be careful not to break the thin tip of the needle when mounting it in the tip holder.

4. The fresh iodoacetamide solution can be divided in aliquots and stored at -20°C. Use each aliquot only once and do not freeze again.

5. Prepare aliquots of 25 μL in 0.6-mL polypropylene vials and store these at -20°C.

6. Polypropylene vials, tubes, or tips may contain soluble polymers or low-mass compounds that can interfere with the ionization process. We have often obtained strong signals in both MALDI- and ESI-MS caused by such contaminants, which are typically not removed by solid phase extraction. As these contaminants often suppress peptide signals, we recommend testing new charges of polypropylene equipment first with standard proteins for such contaminants. We have always obtained good results with tubes and tips from Eppendorf AG (Hamburg, Germany), although some lots released some contaminants suppressing the peptide signals. If such problems occur, check also all vials used to store reagents and buffers.

7. Pull in and remove the solutions slowly and be careful that the stationary phase stays wet and that no air bubbles are drawn in. Thus, do not press all liquid out of the ZipTip™ but keep a thin film of liquid on top of the packing material.

8. If the peptide is not analyzed immediately, remove the acetonitrile before prolonged storage. Thus, dry the solution in the SpeedVac, dissolve in 10 μL water or 0.01% aqueous acetic acid (Rotipuran™, 100% p.a., Carl Roth GmbH & Co. KG), and store at 4°C.

9. Alternatively, use the warm air from a hair dryer to dry the matrix from a distance of about 30 cm. Thus, the sample dries faster forming smaller more homogeneous crystals. This improves the quality of the spectra on some instruments.

10. Vortex the sample solution again briefly before you take a 1-μL aliquot to spot.

11. Use a magnification of 10 to check the crystals. Homogeneous opaque layers often indicate a high content of salts or detergents, which disturb the MALDI process. In this case wash the surface of the spot for 2 s with cold water or purify the sample by solid phase extraction and spot it again. Another explanation could be that the peptide amount was too high. In this case use a higher dilution of the sample and spot it again.

12. These values correspond to the standard parameter set provided with the instrument. Use the corresponding standard conditions recommended for peptide analyses on your instrument.

13. In ESI-MS most antibacterial peptides are detected only at high charge states (typically above four) in the low m/z-range due to their high content of basic residues. This complicates the data interpretation of the corresponding product ion spectra, as the charge series overlap. Therefore, it is often useful to decrease the acid concentration in the spray solution and to increase the pressure of the collision gas cell to lower the charge state of the product ions.

14. To achieve a faster separation it is also possible to start the gradient earlier to compensate the dead volume between the gradient mixer and the trap column. On our instrument we have a dead volume of about 1.3 μL. At a flow rate of 300 nL/min we start the gradient 4 min before the valve switches.

15. Due to the high sensitivity it is favorably to couple a nanoHPLC to an ESI-source. As mass spectrometers are concentration dependent detectors, the sensitivity of an instrumental setup is mostly determined by the peptide concentration of the eluate but not by the peptide amount. Thus a nanocolumn with a flow rate of 300 nL/min provides an about thousand times higher sensitivity than a microbore column with a flow rate of 300 μL/min. As an alternative to buying a nanoHPLC system it is also possible to use a relatively inexpensive flow splitter after the pump and before the injection valve and the column. Thereby the flow rate can be reduced to use a capillary column (flow rate 4 μL/min) on an analytical HPLC system or a nanocolumn on a capillary HPLC system. Instead of a flow-splitter it is preferred to couple a nanoHPLC to an ESI-source. Thereby, the flow rate is split according to the column backpressure, i.e., mostly the column volumes if the same packing materials are used. However, these low-cost setups are less reliable than a nanoHPLC and the reproducibility is worse.

16. Alternative to DDA, the term "information-dependent acquisition" (IDA) is used. Both DDA and IDA describe an automatic mode where the MS/MS parameters are collected automatically by calculating the parameter set (e.g., collision energy) based on the mass and charge state of the selected precursor ion.

17. The described conditions alkylate cysteine residues almost quantitatively, even for defensins, which contain three disulfide bonds. Higher temperatures ($>37°C$) and longer reaction times (> 1 h) do not further increase the cysteine alkylation but may produce further byproducts, such as carboxymethyllysine (CML).

18. Despite the high lysine and arginine content of most antimicrobial peptides, tryptic digests typically yield medium-sized peptides, as often a proline residue follows these basic residues eliminating the cleavage site.

19. Prepare a fresh OME solution for every reaction. Storage of this solution reduces the guanidination degree of the lysine residues significantly yielding heterogeneous samples. To process several samples in parallel prepare a larger volume of the solvent (i.e., 30 μL water, 20 μL 0.1% TFA, and 55 μL ammonium hydroxide solution) and dissolve the O-methylisourea hemisulfate (2.55 mg/51 mL in Tris-HCl buffer) just before the guanidation reaction is started, add 15 μL to the solvent solution. Add 12 μL to each dried peptide.

20. Do not press the tip of a ZipTip™ on the bottom of a polypropylene vial or any other surface, as the packing material might break and block the tip. Also remove it carefully with a pair of tweezers and mount it on the pipette by touching the ZipTip™ only on the top.

21. Even if a modified peptide was only detected in negative ion mode it is still possible to record a tandem mass spectrum with reasonable signal intensities in

positive ion-mode. Note that in negative ion mode the [M-H]⁻ is detected and that in positive ion mode the [M+H]⁺ signal at a 2 u higher mass is fragmented.

22. When several samples are processed in parallel it is not possible to mix the SACA reagent with the reaction buffer to add the mixture to the dried peptide. In this case add first 2 μL reaction buffer to each sample to dissolve the peptide and add afterward 2 μL SACA reagent to each sample.

Acknowledgment

We thank Prof. Dr. Vladimir Kokryakov, Alexander Kolobov, and Ekaterina Korableva for providing peptide fractions with antimicrobial activities for sequence analysis. Financial support by the Deutsche Forschungsgemeinschaft (DFG) and the European Regional Development Fund (EFRE, European Union, and Free State Saxonia) is gratefully acknowledged.

References

1. Cohen, M.L. (2000) Changing patterns of infectious disease. *Nature* **406**, 762–767.
2. Toke, O. (2005) Antimicrobial peptides: new candidates to fight against bacterial infections. *Biopolymers (Peptide Science)* **80**, 717–735.
3. Hand, W.L. (2000) Current challenges in antibiotic resistance. *Adolesc. Med.* **11**, 427–438.
4. Otvos L Jr. (2005) Antibacterial peptides and proteins with multiple cellular targets. *J. Pept. Sci.* **11**, 697–706.
5. Bulet, P., Stocklin, R., and Menin, L. (2004) Anti-microbial peptides: from invertebrates to vertebrates. *Immunol. Rev.* **198**, 169–184.
6. Brogden, K.A. (2005) Antimicorbial peptides: pore formers or metabolic inhibitors in bacteria? *Nature Rev.* **3**, 238–250.
7. Steen, H. and Mann, M. (2004) The ABC's (and XYZ's) of peptide sequencing. *Nature Rev. Mol. Cell Biol.* **5**, 699–711.
8. Yergey, A. L, Coorssen, J. R., Backlund, P. S. Jr., Blank, P. S., Humphrey, and G. A., Zimmerberg, J. (2002) De novo sequencing of peptides using MALDI/TOF-TOF. *J Am Soc Mass Spectrom* **13**, 784–791.
9. Perkins, D.N., Pappin, D.J., Creasy, D.M., and Cottrell, J.S. (1999) Probability-based protein identification by searching sequence databases using mass spectrometry data. *Electrophoresis* **20**, 3551–3567.
10. Beardsley R.L. and Reilly J.P. (2002) Optimization of guanidination procedures for MALDI mass mapping. *Anal Chem.* **74**, 1884–1890.
11. Chen, P., Nie, S., Mi, W., Wang, X.-C., and Liang S.-P. (2004) *De novo* sequencing of tryptic peptides sulfonated by 4-sulfophenyl isothiocyanate for unambiguous protein identification using post-source decay matrix-assisted laser desorption/ionization mass spectrometry. *Rapid Commun. Mass Spectrom.* **18**, 191–198.
12. Samyn, B., Debyser, G., Sergeant, K., Devreese, B., and Van Beumen, J. (2004) A case study of de novo sequence analysis of N-sulfonated peptides by MALDI TOF/TOF mass spectrometry. *J. Am. Soc. Mass Spectrom.* **15**, 1838–1852.

4

The Spot Technique: Synthesis and Screening of Peptide Macroarrays on Cellulose Membranes

Dirk F.H. Winkler and William D. Campbell

Summary

Peptide arrays are a widely used tool for drug development. For peptide-based drug design it is necessary to screen a large number of peptides. However, there are often difficulties with this approach. Most common peptide synthesis techniques are able to simultaneously synthesize only up to a few hundred single peptides. Spot synthesis is a positionally addressable, multiple synthesis technique offering the possibility of synthesizing and screening up to 10,000 peptides or peptide mixtures on cellulose or other membrane surfaces. In this chapter we present the basic procedures and screening methods related to spot synthesis and outline protocols for easy-to-use detection methods on these peptide arrays.

Key Words: Spot synthesis; peptide array; screening; detection; probing; HRP; POD; biotin-labeling; peptide-protein interaction.

1. Introduction

Peptide arrays are widely used for the investigation of peptide–protein and protein–protein interactions. Screening peptides for potentially active compounds using peptide arrays is a very convenient tool in the development of drug candidates. For peptide-based drug design, using peptide microarrays it is possible to screen a high number of peptides on a small chip. Due to the miniscule amounts of peptides synthesized directly on chips, this approach has proven to be very difficult and often solution phase assays and analysis are unreliable. These disadvantages can be overcome using peptide macroarrays. The mostly widely used macroarray is produced by synthesis utilizing the spot

From: *Methods in Molecular Biology, vol. 494: Peptide-Based Drug Design*
Edited by: L. Otvos, DOI: 10.1007/978-1-59745-419-3_4, © Humana Press, New York, NY

Table 1
Recommended Literature

Description	Ref.
First description of the spot technique in detail	*(1)*
Protocols for spot synthesis using active esters	*(2)*
Automated spot synthesis using HOBt/DIC activation	*(3)*
Comparison of OPfp and HOBt/DIC activation	*(4)*
Investigations on the quality of spot syntheses	*(5)*
Spot synthesis and applications	*(6)*
Overview over spot synthesis of small organic molecules	*(7)*
Characterization of some useful types of filter paper and description of some linkers	*(8)*
Modifications of cellulose and polypropylene membranes	*(9)*
Short review of applications	*(10)*
Review of applications	*(11)*
Detailed review regarding synthesis and application of spot synthesis	*(12)*
Detection of weak binding by electrochemical blotting to PVDF membranes	*(13)*
Spot synthesis and screening on living cells	*(14)*

method on cellulose membranes *(1)*. One reason for this is that cellulose-based filter paper is available in virtually every laboratory, inexpensive, and easy to handle. In this publication we offer a short introduction of this synthesis and screening technique and describe some common useful probing methods. Due to the broad range of possible applications for probing, the described protocols are only recommendations (*see* **Table 1**) *(15)*.

2. Materials

2.1. Spot Synthesis of Macroarrays and Screening Techniques

1. Solvents: *N,N′*-dimethylformamide (DMF; VWR), methanol or ethanol (VWR), *N*-methylpyrrolidone (NMP; Fluka), diethyl ether (VWR), dichloromethane (methylene chloride, DCM; VWR). Solvents for washing steps should be of at least ACS quality, whereas solvents for dissolving reagents must be amine and water free. Organic solvents (except for methanol and ethanol) should be stored in the dark.
2. Homemade membranes are prepared from filter paper Whatman 50, Whatman 540, or Chr1 (Whatman, Maidstone, UK) *(16,17)*. Ready-to-use membranes are available from AIMS Scientific (Braunschweig, Germany).

3. For amine functionalization of the filter paper the following reagents are needed: diisopropylcarbodiimide (Fluka), *N*-methylimidazole (NMI; Sigma), and Fmoc-β-alanine (EMD Biosciences).

4. Coupling reagents: diisopropylcarbodiimide (DIPC, DIC; Fluka) and *N*-hydroxybenzotriazole (HOBt; EMD Biosciences). Coupling reagents are only necessary if preactivated amino acid derivatives are not used (*see* **Note 1**).

5. Nonpreactivated amino acids with protection groups according to the Fmoc-protection strategy *(18,19)* (EMD Biosciences). Preactivated amino acid derivatives with protection groups according the Fmoc-protection strategy are usually pentafluorophenyl esters (OPfp ester; EMD Biosciences or Bachem) *(20)* (*see* **Note 1**).

6. Capping solution: 2% actetic anhydride (Sigma) in DMF. 2% ethyl-diisopropylamine (DIPEA, DIEA; Sigma) can be added to deprotonate the amino group and buffer the acetic acid being generated.

7. Staining solution: 0.002% bromophenol blue (BPB; Sigma) in methanol (20 mg in 1 L).

8. Cleavage solution I: Cleavage solution A: 90% trifluoroacetic acid (TFA; VWR) (v/v) + 5% dist. water (v/v) + 3% triisopropylsilane or triisobutylsilane (TIPS or TIBS; Fluka) (v/v) + 1% phenol (Sigma) (w/v) + 1% DCM (v/v); Cleavage solution II: 50% TFA (v/v) + 3% TIPS or TIBS (v/v) + 2% dist. water (v/v) + 1% phenol (w/v) + 44% DCM (v/v)

9. Ammonia gas (optional for releasing peptides from membrane).

2.2. Probing

2.2.1. General

1. 0.1 *M* Phosphate buffered saline (PBS), pH 7.2–7.4.
2. 50 m*M* Tris buffered saline (TBS), pH 8.0.
3. 0.1 *M* PBS with 0.2% Tween (PBS-T, T-PBS).
4. 50 m*M* TBS with 0.2% Tween (TBS-T, T-TBS).
5. Blocking buffer I: 5% Casein or skim milk (Sigma) and 4% sucrose (Sigma) in PBS-T or TBS-T.
6. Blocking buffer II: 5% Bovine serum albumin (BSA; Sigma) and 4% sucrose (Sigma) in PBS-T or TBS-T.
7. Protein solution (protein, antibody): Depending on the estimated strength of binding, 10 to 0.1 μg/mL corresponding to a dilution of 1:100 to 1:10,000 for a concentration of stock solution of 1.0 mg/mL (*see* **Note 2**). For a lower estimated affinity a higher concentration of the protein should be used. For protein mixtures (blood, plasma, cell extracts, etc.) estimation of the content of the target protein is necessary.
8. Regeneration buffer I: 8 *M* urea and 1% SDS in PBS or TBS.
9. Regeneration buffer II: 60% TFA, 30% ethanol, and 10% water. For commercially available acid-stable membranes the use of 90% aqueous TFA as regeneration buffer is recommended *(21)*.

2.2.2. HRP-Labeled Protein and Chemiluminescence

1. Horseradish peroxidase (HRP, POD)-labeled or unlabeled antibody against the probed protein: depending on the estimated strength of binding 1 to 0.1 μg/mL (in blocking buffer) corresponding to a dilution of 1:1,000 to 1:10,000 at a concentration of the stock solution of 1.0 mg/mL (*see* **Note 2**).
2. If the HRP-labeled antibody against the probed protein is not available, an additional incubation step with an HRP-labeled antibody against the first antibody is necessary. The concentration should be between 1 and 0.1 μg/mL (in blocking buffer) (*see* **Note 2**).
3. 10 mL Mixture for chemiluminescence detection: 1 mL 1 M Tris-HCl pH 8.5, 22 μL 80 mM p-coumaric acid in DMF, 50 μL 250 mM luminol in DMF, 3 μL 30% H_2O_2, 9 mL water.

2.2.3. HRP-Labeled Protein and Staining

1. For the protein solutions see **Subheading 2.2.1.**
2. Mixture for chemiluminescence detection (10 mL): 5 mg 4-chloro-1-naphthol dissolved in 1.7 mL methanol, 2.5 mL 200 mM Tris-HCl pH 7.4 (24.2 g/L), 100 mg NaCl; 5.8 mL H_2O, 5 μL 30% H_2O_2. First, dissolve the NaCl in the above amount of water and Tris-buffer (pH 7.4), followed by adding the methanolic chloronaphthol solution. Shortly before use mix this solution with the hydrogen peroxide.

2.2.4. AP-Labeled Protein and Staining

1. Alkaline phosphatase (AP)–labeled or unlabeled antibody against the probed protein: depending on the estimated strength of binding 0.1 to 1.0 μg/mL (in blocking buffer) corresponding to a dilution of 1:10,000 to 1:1,000 at a concentration of 1.0 mg/mL of the stock solution .(*see* **Note 2**).
2. If the AP-labeled antibody against the probed protein is not available, an additional incubation step with an AP labeled antibody against the first antibody is necessary. The concentration should be between 0.1 and 1.0 μg/mL (in blocking buffer) (*see* **Note 2**).
3. NBT stock solution: 0.5% (w/v) nitrotetrazolium chloride (NBT; Sigma) in 70% aqueous DMF; this solution can be stored cold (<8°C) for at least one year.
4. BCIP stock solution: 0.5% (w/v) 5-bromo-4-chloro-3-indolyl phosphate (BCIP; Fluka) in DMF; this solution can be stored cold (<8°C) for at least one year.
5. Tris buffer: 100 mM NaCl, 5 mM $MgCl_2$, 100 mM Tris; adjusted with HCl or NaOH to pH 9.5.
6. Detection solution: 660 μL NBT stock solution + 330 μL BCIP stock solution mixed in 10 mL Tris buffer. Always prepare this mixture fresh before use.

2.2.5. Biotin Labeling of Samples and Detection

1. Activated biotin derivatives: e.g., Biotin-ONp (EMD Biosciences) or sulfo-NHS-biotin (Pierce).

2. Biotinylation solution: 10 mM solution of the activated biotin in dimethyl sulfoxide (DMSO). Prepare this mixture always fresh before use.
3. Purification columns: PD-10 columns (Amersham).
4. HRP-labeled streptavidin.

3. Methods

If not noted elsewhere, all washing, incubation, and reaction steps could be performed using a rocking shaker.

3.1. Spot Synthesis of Macroarrays

Spot synthesis may be carried out manually, semi- or fully automated (*see* **Fig. 1**). Manual spot synthesis is convenient for the synthesis and screening of a relative small number of peptides (up to 100) and a large pipetting volume (>0.5 μL). For all other cases a semiautomated or fully automated synthesis is recommended. For synthesis of small spots with a pipetting volume of about 0.1 μL, spot synthesizers from Intavis AG (Köln, Germany) are very useful *(3)*.

If not noted elsewhere, all data correspond to the preparation of a 10 × 12 cm cellulose membrane (96-well plate size). For larger membranes preparation of related amounts of reagents is required. If we write about the use of amino acids, it should always include the use of other organic building blocks (e.g., PNA monomers, peptoidic elements, heterocycles) *(22,23)*, which can be used under spot synthesis conditions. Here we describe only the basic procedures for spot synthesis of linear peptides. For the synthesis of modified peptides, such as cyclization or side-chain modifications, *see* **ref.** *(24)*.

3.1.1. Preparation of Coupling Solutions

Preparation of the activated amino acid solutions can be performed according to two different methods.

One method involves the use of preactivated Fmoc-protected amino acids (e.g., pentafluorophenyl ester). This method has the advantage that only one reagent is necessary since preparation of the amino acid solutions is very simple and the likelihood of mistakes is low. One disadvantage is the higher price of the amino acid derivatives, but due to the small amount of activated amino acids used the absolute difference falls in the range of a few dollars for synthesis of an entire peptide membrane array. Another disadvantage lies in the fact that activated esters are only commercially available for the standard amino acids.

The other method requires the use of in situ activated amino acids. Activation of the amino acids is carried out by adding an activator and coupling reagent to the unactivated Fmoc-protected amino acid derivative. This method is more time

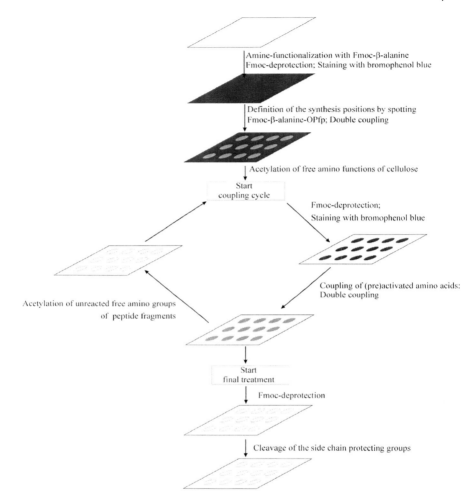

Fig. 1. Outline of spot synthesis (shown without washing steps).

consuming than the first method but can be applied for all building blocks available
for use in peptide synthesis according to the Fmoc-protecting group strategy.

3.1.1.1. METHOD 1: PREPARATION OF 0.3 *M* SOLUTIONS WITH PREACTIVATED AMINO ACID DERIVATIVES

Except for the arginine derivative, solutions of preactivated amino acids can
be prepared and used for an entire week. Dissolve most of the amino acid derivatives in amine-free NMP at a concentration of 0.3 *M*. Due to poor solubility,
derivatives of serine, threonine, tryptophan, and proline must be dissolved in
amine-free DMF. The solutions can be stored at −20°C for at least 1 wk. Use

the solutions for synthesis fresh every day, and discard those of the previous day. Due to the instability of dissolved preactivated arginine derivatives, this solution must be prepared fresh every day.

3.1.1.2. METHOD 2: PREPARATION OF SOLUTIONS WITH IN SITU ACTIVATED AMINO ACIDS OR OTHER BUILDING BLOCKS

Prepare a 0.9 *M* solution of HOBt in amine-free NMP. Dissolve the Fmoc-amino acids or desired protected building blocks with the HOBt-solution to a concentration of 0.45 *M*. Except for the arginine derivatives, these solutions can be stored at -20°C for at least a week. Each day, prepare a 20% mixture of DIC in amine-free NMP. Use the fresh amino acid/HOBt solution every day and discard the previous one. To these solutions add 20% DIC/NMP at a ratio of 3:1 (e.g., 75 µL amino acid solution and 25 µL DIC mixture prepared fresh every day) (*see* **Note 3**).

3.1.2. Preparation of the Membrane

1. Cut a piece of filter paper to the size required so that all peptide spots including controls can be accommodated. For 0.1 µL of coupling solution the distance should be at least 2.7 mm, and for 1 µL at least 7 mm.
2. Amine functionalization of filter paper: Dissolve 0.64 g Fmoc-β-alanine in 10 mL amine-free DMF. Add 374 µL DIC and 317 µL NMI and mix well before transferring this solution into a chemically resistant box with lid. Place the filter paper in the box while avoiding formation of air bubbles under the paper, and ensure that the surface of the membrane is slightly covered by the solution. Close the box. Allow the reaction to take place for at least 2 h or overnight (*see* **Note 4**).
3. Fmoc-deprotection: After the reaction is complete, wash the membrane three times with DMF for at least 30 s each. For storage, wash the modified membrane at least twice with methanol or ethanol and dry it in the air stream of a fume hood or using a hair dryer without heat. For resumption of synthesis after storage, treat the membrane once with DMF for 20 min (*see* **Note 5**). The Fmoc-deprotection is carried out by treatment of the membrane twice with 20% piperidine in DMF for at least 5 min each.
4. Staining (optional) *(25)*: Wash the membrane four times with DMF for at least 30 s each, followed by washing at least twice with methanol or ethanol for at least 30 s each. Treat the membrane with staining solution for at least 2 min until the filter paper shows a homogeneous blue color (*see* **Note 6**). After staining wash the membrane at least twice with methanol or ethanol, until the wash solution remains colorless and then dry the membrane.

3.1.3. Spot Synthesis of the Peptide Array

1. Definition of the synthesis pattern: Place the dried membrane on an inert surface (e.g., stainless steel or PP). Deliver the required volume of activated

Fmoc-β-alanine/DMSO solution to all spot positions (*see* **Note 4**). For manual synthesis we recommend marking the spot positions with a pencil and to use a multistep pipettor (*see* **Fig. 2**). Repeat the delivery after 20 min. After that allow the reaction to occur once more for at least 20 min.

2. Blocking nonspot areas: Place the membrane face-down in an appropriate amount of capping solution in the box. Ensure that there are no air bubbles under the surface. Do not shake! Allow the reaction to occur for at least 5 min. Remove the liquid. Add an appropriate amount of capping solution with 2% DIPEA. Place the membrane face-up in this solution and allow it to react for 20 min.

3. Fmoc-deprotection: Wash four times with DMF for at least 30 s each. Treat twice with 20% piperidine/DMF for 5 min each. Wash four times with DMF for at least 30 s each. Wash at least twice with methanol or ethanol.

4. Staining (optional) *(25)*: Stain the membrane by shaking with staining solution in a box. If the staining solution change its color to blue very rapidly, renew the solution. Let the reaction occur for at least 15 s until the spots are blue (*see* **Note 6**). Wash at least twice with methanol or ethanol until the wash solution remains colorless. If faster drying is necessary, an additional washing step with diethyl ether (twice) is possible.

5. Dry the membrane. The membrane is now ready for the first coupling cycle.

6. Coupling of activated amino acids: The stepwise buildup of peptides starts from the C-terminus. Deliver the prepared corresponding activated amino acid solutions to the corresponding positions on the membrane with the desired volume. Repeat the spotting after 20 min (*see* **Note 7**).

7. Blocking unreacted free amino groups on peptide fragments: Add an appropriate amount of capping solution to the box. Place the membrane face-down in this

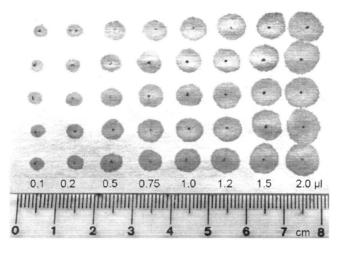

Fig. 2. Dependency of spot size on the manually delivered volume of Fmoc-β-Ala-OPfp/DMSO solution on Whatman 50 filter paper.

solution. Ensure that there are no air bubbles under the surface. Do not shake! Allow the reaction to occur for at least 5 min. Remove the liquid. Add an appropriate amount of capping solution with 2% DIPEA. Place the membrane face up in this solution. Allow the membrane to react with the solution for at least 5 min.

8. Building up the peptide chain: Except for the last coupling cycle, repeat **steps 3–11**. At the last coupling cycle repeat the above steps without capping and staining!

9. Final Fmoc-deprotection: Wash four times with DMF for at least 30 s each. Treat with 20% piperidine/DMF twice for 5 min. Wash four times with DMF followed by three times with DCM for at least 30 s each.

10. Final side-chain deprotection: Treat the membrane with at least 25 mL of cleavage solution I for 30 mins. The surface of the membrane must be well covered by the cleavage solution. Keep the box closed. Do not shake. Pour off the solution very carefully. Wash the membrane five times with DCM for at least 1 min each. Treat with at least 25 mL of cleavage solution B for 3 h. The surface of the membrane must be well covered by the cleavage solution. Keep the box tightly closed. Do not shake. Pour off the solution very carefully. Wash the membrane five times for at least 1 min each with DCM. After TFA treatment, the membrane may become very soft. Do not try to lift the membrane out until it becomes harder and less likely to break apart during the washing steps (around the last DCM washing step)! Wash the membrane five times with methanol or ethanol for at least 1 min each. Dry the membrane in the air stream of a fume hood or with a hair dryer without heat.

11. Sometimes it may be necessary to use the peptides solubilized and unbound to the cellulose membrane. If it is necessary to release the peptides from the membrane, one method is to expose the entire dry membrane overnight in a glass desiccator containing ammonia gas. The strong basic environment leads to a break of the ester bond to the cellulose by forming a C-terminal amide (*see* **Note 8**). The next day, punch out the spots and transfer the discs into wells of microtiter plates (MTPs) or into vials in which you can dissolve and test the released peptides (*see* **Note 9**). If a free carboxy terminal is desired on the peptides, then do not treat the membrane with ammonia gas, punch out the spots and expose them to a basic aqueous solution, for example, ammonium hydroxide or sodium hydroxide *(26)*.

Another method is to release the peptide by forming a C-terminal diketopiperazine. In this case, couple Boc-Lys(Fmoc)-OH instead of the β-alanine spacer (Fmoc-β-Ala-OH), followed by coupling of Fmoc-proline as the first coupled amino acid. After TFA treatment the spots can be punched out and transferred to a MTP or vials (*see* **Note 9**). The peptides can be released by overnight treatment with basic aqueous buffers (pH ≥ 7.5) *(1)*.

3.2. Screening Techniques

In this chapter we present some array techniques to screen for active peptides (*see* **Note 10**). Most of these screening techniques were developed before the first report on spot synthesis and can be used for other solid phases *(27)*. For instance, a peptide scan and substitution analysis was described following peptide synthesis on polypropylene rods *(28)*, and various combinatorial libaries have been used in resin-based peptide synthesis *(29)*. Unfortunately, the number and variety of these methods are too large to describe in detail here. Nevertheless, we will explain the most common as well as some other interesting special methodologies for screening for peptides with activity and provide corresponding references. Activity: this means, for example, binding (e.g., small organic molecules, metal ions, proteins like antibodies, or organisms like phages) *(16,30–32)*, reaction (e.g., with kinases *(33,34)*), inhibition (e.g., bacterial growth *(35)* or cell growth *(36)*) or enhancement (e.g., of cell growth or distinct cell activity) *(14,37)*. In addition to the common amino acids for screening, other amino acid derivatives or building blocks of interest can be used. Examples of modified amino acids or building blocks are phosphorylated *(38)*, glycosylated *(39)*, biotinylated, or fluorescence-labeled amino acids as well as peptoidic elements *(40)*, PNA monomers *(23)*, or other organic molecules. The peptides could have a linear, cyclic, or branched sequence *(41,42)*.

3.2.1. Peptide Scanning Library (Epitope Mapping, Peptide Scan)

Peptide scanning is a very useful tool for screening a known protein sequence for active regions (e.g., epitopes). It was first described by Geysen and coworkers for peptides bound to polypropylene rods *(28)*. The peptides are generated by shifting a frame of a distinct peptide length of a protein sequence of interest

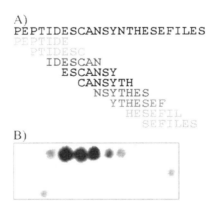

Fig. 3. Peptide scan. (**A**) Schema of sequences of a peptide scan. This peptide scan consists of peptides with a length of seven amino acids and a frameshift of two amino acids. Dark letters represent the region with activity. (**B**) Image of a peptide scan after incubation with a protein and detection of the bound protein.

(*see* **Fig. 3**) *(13,16,43)*. In general, a length between 10 and 15 amino acids is usual. Shifting of the frame should be between one and five amino acids, where the smaller the steps are, the more precise will be the identification of the minimal peptide length required for activity.

A special application of the peptide scan is the hybritope scan (hybrid-epitope) *(44,45)*. In order to increase the affinity by extention of the peptide sequence in a hybritope scan, several mixtures of amino acids at positions flanking the related sequence of the protein are introduced. Another strategy to optimize a peptide scan is the so-called duotope scan (duo-epitope) *(46)*. For a duotope scan one generates the peptide scan twice and combines the sequence of both arrays linked via a spacer, such as two β-alanine molecules. In this way it is possible to avoid possible steric hindrance and to increase activity by extention of the active sequence. The pattern of such a duotope scan resembles that of a combinatorial library with the exception that each field represents a combination of two sequences rather than two amino acids.

3.2.2. Substitution Analysis (Replacement Analysis, Mutational Analysis)

Substitution analyses, developed by Geysen and coworkers for peptides bound to polypropylene rods *(28)*, are used for investigation of the importance or of amino acids in a known peptide sequence. Therefore, one can generate the sequences for synthesis by successive systematic substitution of each amino acid by other amino acids or building blocks of interest (*see* **Fig. 4**) (for instance D-amino acids *(36,47)* and peptoidic monomers *(40)*). Usually, only one single amino acid should be exchanged at one spot position *(43,48,49)*. However, 2D-substitution analysis is also possible, where two amino acids in a sequence are simultaneously exchanged on a single peptide spot. The pattern of such a 2D-substitution analysis is similar to that of a combinatorial library *(40)*.

Special applications of substitution analysis include the so-called amino acid walks or scans (e.g., alanine scan, glycine walk) *(50,51)*. In this type of screening one replaces each single amino acid of the sequence with only one distinct amino acid (usually alanine or glycine). The progressive amino acid scan is achieved by a stepwise increase in the number of positions exchanged by the distinct amino acid *(52)*.

3.2.3. Combinatorial Peptide Library

A very powerful method for screening active peptides without knowing the actual sequence is the combinatorial peptide library. This method was first described by Houghten and coworkers for peptides synthesized on a polymer resin *(53)*. Using combinatorial libraries, screening begins in theory

A)

wt	A	C	D	E	...	W	Y
PEPTIDE	AEPTIDE	CEPTIDE	DEPTIDE	EEPTIDE	...	WEPTIDE	YEPTIDE
PEPTIDE	PAPTIDE	PCPTIDE	PDPTIDE	PEPTIDE	...	PWPTIDE	PYPTIDE
PEPTIDE	PEATIDE	PECTIDE	PEDTIDE	PEETIDE	...	PEWTIDE	PEYTIDE
PEPTIDE	PEPAIDE	PEPCIDE	PEPDIDE	PEPEIDE	...	PEPWIDE	PEPYIDE
PEPTIDE	PEPTADE	PEPTCDE	PEPTDDE	PEPTEDE	...	PEPTWDE	PEPTYDE
PEPTIDE	PEPTIAE	PEPTICE	PEPTIDE	PEPTIEE	...	PEPTIWE	PEPTIYE
PEPTIDE	PEPTIDA	PEPTIDC	PEPTIDD	PEPTIDE	...	PEPTIDW	PEPTIDY

B)

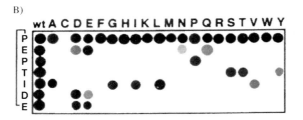

Fig. 4. Example of a substitution analysis. (**A**) Sequences resulting from substitution analysis of a peptide with the sequence PEPTIDE (wild type, wt). (**B**) Resulting image of the substitution analysis after incubation with a protein and detection of the bound protein (fictive).

with the entire pool of possible peptide sequences. Due to the impossibility of synthesizing all peptides as single sequences (e.g., all 6-mers of the common 20 L-amino acids would be 64,000,000 peptides), during the synthesis coupling at most positions introduces a mixture of all possible amino acids and building blocks of interest (amino acids *(54)*, PNA *(23)*, peptoids *(55)*). As a result one can synthesize a mixture of all possible peptides on each spot. In practice, for the first round of synthesis and screening one or two positions have defined amino acids or building blocks of interest (*see* **Notes 11** and **12**). Following that, each peptide consists of a sequence with several mixed positions and one or two defined amino acids. The advantage of substituting two mixed positions at once is the possibility of obtaining stronger signals during the screening. If one wants to substitute two mixture positions systematically, one has to combine the exchanges. For instance, if two mixture positions are exchanged with the common 20 L-amino acids one would obtain 20 × 20 = 400 possibilities. For a better overview the spots in an array can be arranged in a chessboard pattern with 20 rows and 20 columns. In each single row synthesize all substitutions at one mixture position and combine it with the column-wise substitution for the other mixture position (*see* **Fig. 5**). By screening one would obtain signals on spots with active sequence patterns (*see* **Note 13**). These sequence patterns represent the groups of all possible peptides of distinct length with one or two

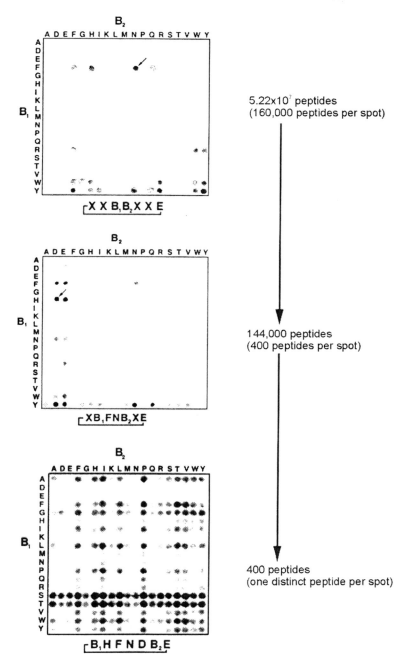

Fig. 5. Principle of a combinatorial library using as an example, the tAb2-binding combinatorial heptamer peptide library. The peptides were cyclized via the N-terminal

amino acids defined. After the first synthesis and screening round there are two possibilities: Either one can combine different sequence patterns to a complete sequence *(56,57)* or choose the sequence pattern with the strongest activity. The next combinatorial library is assembled with the sequence pattern conserved and the deconvolution of the other one or two mixture positions. This repeating of deconvolution, synthesis, and screening cycles proceeds until one obtains a complete sequence in which all amino acid positions are defined *(43,48,54)*.

A special type of combinatorial library is the positional scanning library *(43, 49,58)*. This array contains a set of combinatorial libraries with one or more (two positions = dual positional scan *(48,59)*)deconvoluted positions. In each of the synthesized libraries the deconvoluted positions are different. After screening one can combine the positions of the spot with the peptides of highest activity of each library to obtain the complete sequence of a peptide with most likely the greatest activity.

In order to reduce the number of possible combinations, combine several amino acids of similar properties in clusters (e.g., hydrophobicity—isoleucine, leucine, valine; or steric similarity—serine, cysteine, aminobutyric acid). These clustered amino acid peptide libraries provide easier and faster screening of large peptide libaries *(60)*. It is of course necessary to resolve these clusters at the end of the screening process.

3.2.4. Others

3.2.4.1. RANDOM PEPTIDE LIBRARY

Another possibility to screen for active peptides without prior knowledge of a starting sequence is use of a random peptide library. This array contains randomized peptide sequences *(61)*. Compared with the combinatorial peptide

◀―――

Fig. 5. (Continued) α-amino group and the carboxy side chain function of the C-terminal glutamic acid. In the first library only two positions of the peptide sequence were defined (B_1 in rows and B_2 in columns; cysteine was omitted). At all other positions, mixtures of all common L-amino acids were coupled (X). From the resulting pattern, the strongest spot was chosen for defining the sequence motif for the next screening cycle (B_1 = F and B_2 = N; see arrow). At the second screening cycle two of the positions with mixtures were deconvoluted, now (B_1, B_2). The sequence motif FN was synthesized in all peptide sequences. After probing, one of the strongest signals was chosen for the extended sequence motif (B_1 = H and B_2 = D, resulting in HFND; see arrow). With this sequence motif the third coupling cycle was performed to determine the remaining undefined positions. In this screening cycle each spot contains one single peptide with a defined sequence.

library, the advantage of the random peptide array is that the peptides on each spot are unique, providing the possibility of higher individual activity. The disadvantage is in the lower number of synthesized peptides available for screening.

3.2.4.2. Truncation Analysis (Length Scan)

To investigate the minimum possible length of an active peptide while maintaining activity, variations of the peptide sequence are synthesized by systematic shortening at the C-terminus, N-terminus, or both simultaneously *(48,62,63)*.

3.2.4.3. Loop Scan (Cysteine Scan)

In order to stabilize loop structures or to increase their resistance to proteolytic digestion, it is convenient to cyclize peptides. There are two main types of cyclization: cyclization via cysteines forming a disulfide bridge and cyclization via a pair of peptide amino and carboxy groups to form an amide bond. In both cases, a pair of amino acids is involved. If they are not present in the peptide sequence, they must be inserted or two existing amino acids should be replaced, which could lead to a loss of activity. Therefore it is necessary to investigate the effect of cyclization on the activity of the peptide *(64)*. For a cysteine loop scan, generate, synthesize, and test a set of all possible combinations of insertions or replacements using a pair of cysteines (*see* **Fig. 6**). For amide cyclization it is important that you have an amino group–containing amino acid such as lysine and a carboxy group–containing one like aspartic acid. That affords a higher number of possible combinations.

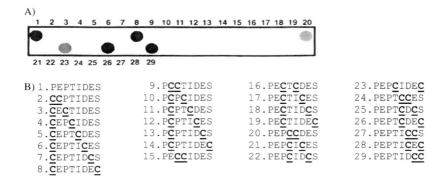

Fig. 6. Principle of a loop scan on the example of the peptide PEPTIDES. The amino acids were replaced by a pair of cysteines at all possible positions. (**A**) Image of the loop scan (fictive). (**B**) List of the sequences.

3.3. Probing

The number of probing possibilities is huge and even includes testing the activity of peptides on living cells *(14)* or bacteria *(35)*. Here we describe the use of labeled proteins such as antibodies. The labels described are horseradish peroxidase, alkaline phosphatase, and biotin. However, in general, most labeled proteins can be used according the method described in **Subheading 3.3.1.** Differences are in label-specific detection of the bound labeled protein.

3.3.1. HRP-Labeled Proteins and Chemiluminescence

1. If the membrane is dry, wash twice for 10 min each with methanol (or ethanol). Then wash three times with T-TBS or T-PBS for at least 5 min each.
2. Blocking: To decrease unspecific binding treatment with blocking buffer I or blocking buffer II is necessary (*see* **Note 2**). Blocking should be carried out for at least 2 h at room temperature. Treatment overnight at room temperature or over the weekend at 4°C is also possible.
3. Incubation with protein sample: Wash once with T-TBS or T-PBS for at least 5 min. After washing, incubate the membrane with the protein sample for at least 2 h at 37°C. Treatment can also be carried out overnight at room temperature.
4. Incubation with a target protein binding protein (e.g., primary antibody): Wash three times with T-TBS or T-PBS for at least 5 min each. After washing, incubate the membrane with the (HRP-labeled) primary antibody solution for 2 h at room temperature or 37°C.
5. Incubation with secondary antibody directed against the protein used in **step 4** (if no HRP-labeled primary antibody was used): Wash three times with T-TBS or T-PBS for at least 5 min each. After washing, incubate the membrane with the HRP-labeled secondary antibody solution for at least 1 h at room temperature or 37°C.
6. Detection: Wash at least three times for at least 5 min with TBS or PBS. Preparation of the chemiluminescence solution and detection (*see* **Note 14**): 22 μL of 80 mM p-coumaric acid in DMF and 50 μL of 250 mM luminol in DMF should be mixed with 1 mL 1 M Tris-HCl pH 8.5. Add to this mixture the corresponding amount of water (*see* **Materials**). The staining solution has to be activated by adding 3 μL of 30% H_2O_2 shortly before use. Excess washing buffer on the membrane has to be removed by placing the membrane on paper towels (*see* **Note 15**). Use forceps for all manipulations of the membrane. Never use fingers! The membrane then has to be placed on a plastic sheet (PP or PE). The detection buffer is poured over the whole membrane. To ensure homogeneous distribution the membrane has to be gently shaken on the plastic sheet. A reaction time of approximately 1 min is recommended. Detection of the chemiluminescence can be carried out using X-ray film or a chemiluminescence imager *(62)*.

7. Regeneration (stripping) of the membrane: The removal of bound proteins should be carried out in two steps. In the first step the membrane should be treated overnight at 37°C with regeneration buffer I (*see* **Note 16**). After washing twice with water the membrane should be treated for at least 1 h at room temperature using regeneration buffer II. After finishing the second regeneration step the membrane has to be washed at least three times with PBS, followed by washing twice with methanol and twice with water. The membrane could be stored at −80°C after washing with methanol and drying on air or could be used again for probing (*see* **Notes 5** and **17**).

3.3.2. HRP-Labeled Proteins and Staining

1. For treatment of the membrane with the protein sample and the detection antibody, follow the instructions in **steps 1–5** of **Subheading 3.3.1.**
2. Detection: Wash at least three times for at least 5 min with T-TBS or T-PBS. Preparation of the staining solution and detection of the bound protein: Activate the staining solution (*see* **Materials**) by adding the hydrogen peroxide shortly before use. Treat the membrane with the activated mixture with shaking. Spots containing HRP-labeled proteins should turn a violet or brown color. The staining reaction can be stopped by intensive washing with distilled water.

3.3.3. AP-Labeled Proteins and Staining

1. For treatment of the membrane with the protein sample and the detection antibody, follow the instructions in **steps 1–5** of **Subheading 3.3.1.** with the exception that AP-labeled proteins are used (antibodies) instead of HRP-labeled proteins (antibodies).
2. Detection: Wash at least three times for at least 5 min with T-TBS or T-PBS. Prepare the detection solution shortly before use. Treat the membrane with the detection solution (*see* **Materials**) until staining of the spots is strong enough. Stop the staining reaction by washing very thoroughly with distilled water.

3.3.4. Biotin Labeling of Protein Samples and Detection

1. Biotin labeling: Prepare the biotinylation solution shortly before use. Dissolve your protein in PBS (*see* **Note 18**). Deliver as much volume of the biotinylation solution to your protein solution as needed to obtain a molar ratio of 20:1 (biotinylation agent: protein), i.e., if you have 1 mL of a protein solution at a concentration of 10 μM, then you need at least 50 μL of 10 mM biotinylation solution. Let the mixture react for at least 30 min.
2. Purification of the biotinylated sample: To remove uncoupled biotin and side products, use desalting columns (e.g., PD-10 columns from Amersham) and follow the instructions of the manufacturer. Elute the sample with PBS, pH 7.4. After the purification, calculate the resulting protein concentration.

3. For incubation of the membrane with the biotinylated protein sample and streptavidin, follow the instructions of **steps 1–4** in **Subheading 3.3.1.** with the exception of using the HRP-labeled streptavidin at a concentration of 0.1 μg/mL instead of the HRP-labeled proteins (antibodies).

4. Detection: Follow **step 6** of **Subheading 3.3.1.** or **step 2** of **Subheading 3.3.2.**

4. Notes

1. Reagents must be protected from moisture. To avoid condensation, reagent bottles stored in the fridge or freezer should be unopened and warmed up to room temperature for approx 30 min before use.

2. If the binding is expected to be very weak the use of PBS, TBS, PBS-T, or TBS-T instead of blocking buffer is possible. In that case, due to possible unspecific binding, a higher background signal during detection may occur.

3. To avoid problems due to formation of poor solubility in the urea, mix the in situ activated solutions for about 30 min. Then centrifuge the mixtures and transfer the supernatant to vials for the synthesis.

4. Because of the flexibility and linear structure of the molecule, β-alanine is commonly used for modification of the cellulose as well as a spacer to achieve a higher distance of the amino functions from the cellulose matrix. However, other functionalizations are also possible *(65)*.

5. The membrane can be stored at −80°C for several months until needed. For longer storage a slight loss of loading is possible. Storage of the membranes at −80°C is recommended. The membranes can also be stored at −20°C or +4°C, but there is a higher risk of losing activity over a long storage time.

6. Do not stain the spots too strongly because if the the amount of absorbed bromophenol blue is too high, some of this dye could be incorporated in the peptide. In that case removal of the color is almost impossible and could affect the detection. Because bromophenol blue has a blue color in a basic environment, e.g., primary amino functions, a difference in the intensity of the color of the spots is normal due to different acidity of the coupled amino acids.

7. To cover the whole area of large spots, it is recommended to use at least 20% more amino acid solution volume than in **step 5** of **Subheading 3.1.**

8. Release of peptides with ammonia might work only on homemade membranes. Commercially available membranes with amino functional groups often have other than ester bonds between the spacer and cellulose. For such membranes the use of additional linkers might be necessary (e.g., thioester *(66)*, HMB linker *(67)*, Rink linker *(68)*).

9. To minimize possible contamination with side products from the edge of the spot, the diameter of the punched-out discs should be smaller than the spot diameter.

10. In general, the peptide sequences should be between 6 and 15 amino acids in length; spot syntheses of longer (e.g., 34 and 44 amino acids *(69,70)*) and shorter (e.g., 2 amino acids *(71)*) peptides are also described. With shorter peptides, the measurable activity might be too low. For the synthesis of long peptides, the

following rule must be kept in mind: the longer the peptide, the more difficulties arise during coupling—the number of possible side reactions increases and peptide purity decreases dramatically!

11. Since it is unlikely for one to be able to synthesize all possible sequences on a single spot, do not start synthesis of a combinatorial library with more than 6 mixture positions. If necessary, start to screen 8mer peptides (6 mixture + 2 deconvoluted positions) and extend the peptide length with each synthesis round by extending the deconvoluted positions and maintaining the number of mixed positions (e.g., 6 mixture + 4 deconvoluted positions).

12. Because of possible strong side reactions during the probing, avoid the use of cysteines in the first synthesis and screening rounds at defined amino acid positions.

13. After incubation, if a spot signal also shows a corresponding spot signal by reflection on a virtual diagonal (Pos.1,1 to pos. x,x), it is possible that these two signals are caused by unspecific binding.

14. Avoid exposure to daylight at all steps using the chemiluminescence mixture. In particular, when using X-ray film for detection a red light room should be used.

15. Use of towels made of nonrecycled paper is recommended. Using recycled towels could lead to disturbances in detection by reactions with trace materials in the paper.

16. After probing, the membrane must remain wet for regeneration. A previously dried membrane is more difficult to regenerate than a wet one.

17. From time to time incomplete regeneration could occur. Therefore, it is recommended to always use new membranes; if not, then the membrane should be probed first with the control and then with the protein sample. For further probing with the same membrane, completeness of the regeneration should be checked by repeating the detection without the protein sample. All detectable signals should be found in the previous control. Additional signals seen with the sample probe should have disappeared. Stained spots are generally very hard to regenerate.

18. Commonly used biotinylating reagents are usually amine reactive. For biotinylation of your protein, do not dissolve your protein sample in amine-containing buffers, such as TBS, TBS-T, or PBS-T.

References

1. Frank, R. (1992) Spot-synthesis: An easy technique for the positionally addressable, parallel chemical synthesis on a membrane support. *Tetrahedron* **48**, 9217–9232.
2. Kramer, A. and Schneider-Mergener, J. (1998) Synthesis and application of peptide libraries bound to continuous cellulose membranes. *Methods Mol. Biol.* **87**, 25–39.
3. Gausepohl, H. and Behn, C. (2002) Automated synthesis of solid-phase bound peptides. In: *Peptide Arrays on Membrane Support*, eds. J. Koch and M. Mahler, pp. 55–68. Berlin: Springer-Verlag.
4. Molina, F., Laune, D., Gougat, C., Pau, B., and Granier, C. (1996) Improved performances of spot multiple peptide synthesis. *Peptide Res.* **9**, 151–155.

5. Kramer, A., Reineke, U., Dong, L., et al. (1999) Spot-synthesis: observations and optimizations. *J. Peptide Res.* **54**, 319–327.
6. Frank, R. (2002) The SPOT-synthesis technique. Synthetic peptide arrays on membrane supports-principles and applications. *J. Immunol. Methods* **267**, 13–26.
7. Blackwell, H.E. (2006) Hitting the SPOT: small-molecule macroarrays advance combinatorial synthesis. *Curr. Opin. Chem. Biol.* **10**, 203–212.
8. Frank, R., Hoffmann, S., Overwin, H., Behn, C., and Gausepohl, H. (1996) Easy preparation of synthetic peptide repertoires for immunological studies utilizing SPOT synthesis. In: *Peptides in Immunology*, ed. C.H. Schneider, pp. 197–204. New York: John Wiley & Sons, Ltd.
9. Wenschuh, H., Volkmer-Engert, R., Schmidt, M., Schulz, M., Schneider-Mergener, J., and Reineke, U. (2000) Coherent membrane supports for parallel microsynthesis and screening of bioactive peptides. *Biopolymers (Peptide Science)* **55**, 188–206.
10. Reineke, U., Volkmer-Engert, R., and Schneider-Mergener, J. (2001) Applications of peptide arrays prepared by the SPOT-technology. *Curr. Opin. Biotechnol.* **12**, 59–64.
11. Frank, R. and Schneider-Mergener, J. (2002) SPOT synthesis—scope of applications. In: *Peptide Arrays on Membrane Support*, eds. J. Koch and M. Mahler, pp 1–22. Berlin Heidelberg: Springer-Verlag.
12. Hilpert, K., Winkler, D.F.H., and Hancock, R.E.W. (2007) Cellulose-bound peptide arrays: Preparation and applications. *Biotechnol. Genetic Eng. Rev.* **24**, 31–106.
13. Reineke, U., Sabat, R., Volk, H.-D., and Schneider-Mergener, J. (1998) Mapping of the interleukin-10/interleukin-10 receptor combining site. *Protein Sci.* **7**, 951–960.
14. Otvos Jr., L., Pease, A.M., Bokonyi, K., et al. (2000) In situ stimulation of a T helper cell hybrodoma with a cellulose-bound peptide antigen. *J. Immunol. Methods* **233**, 95–1051.
15. Bräuning, R., Mahler, M., Hügle-Dörr, B., Blüthner, M., Koch, J., and Petersen, G. (2002) Immobilized peptides to study protein-protein interactions—potential and pitfalls. In: *Peptide Arrays on Membrane Support*, eds. J. Koch and M. Mahler, pp.153–163. Berlin: Springer-Verlag.
16. Martens, W., Greiser-Wilke, I., Harder, T.C., et al. (1995) Spot synthesis of overlapping peptides on paper membrane supports enables the identification of linear monoclonal antibody binding determinants on morbillivirus phosphoproteins. *Vet. Microbiol.* **44**, 289–298.
17. Santona, A., Carta, F., Fraghi, P., and Turrini, F. (2002) Mapping antigenic sites of an immunodominant surface lipoprotein of Mycoplasma agalactiae, AvgC, with the use of synthetic peptides. *Infect. Immun.* **70**, 171–176.
18. Fields, G.B. and Noble, R.L. (1990) Solid phase synthesis utilizing 9-fluorenylmethoxycarbonyl amino acids. *Int. J. Peptide Protein Res.* **35**, 161–214.
19. Zander, N. and Gausepohl, H. (2002) Chemistry of Fmoc peptide synthesis on membranes. In: *Peptide Arrays on Membrane Support*, eds. J. Koch and M. Mahler, pp. 23–39.Berlin: Springer-Verlag.

20. Atherton, E. and Sheppard, R.C. (1989) 7.2. Activated esters of Fmoc-amino acids. In: *Solid Phase Peptide Synthesis—A Practical Approach*, pp.76–78. Oxford: IRL press at Oxford University Press.

21. Zander, N. (2004) New planar substrates for the in situ synthesis of peptide arrays. *Mol. Divers.* **8**, 189–195.

22. Ast, T., Heine, N., Germeroth, L., Schneider-Mergener, J., and Wenschuh, H. (1999) Efficient assembly of peptomers on continuous surfaces. *Tetrahedron Lett.* **40**, 4317–4318.

23. Weiler, J., Gausepohl, H., Hauser, N., Jensen, O.N., and Hoheisel, J.D. (1997) Hybridisation based DNA screening on peptide nucleic acid (PNA) oligomer arrays. *Nucleic Acids Res.* **25**, 2792–2799.

24. Hilpert, K., Winkler, D.F.H., and Hancock, R.E.W. (2007) Peptide arrays on cellulose support: SPOT synthesis - a time and cost efficient method for synthesis of large numbers of peptides in a parallel and addressable fashion. *Nature Protocols* **2**, 1333–1349.

25. Krchnak, V., Wehland, J., Plessmann, U., Dodemont, H., Gerke, V., and Weber, W. (1988) Noninvasive continuous monitoring of solid phase peptide synthesis by acid-base indicator. *Collect. Czechoslovak Chem. Commun.* **53**, 2542–2548.

26. Bhargava, S., Licha, K., Knaute, T., et al. (2002) A complete substitutional analysis of VIP for better tumor imaging properties. *J. Mol. Recognition* **15**, 145–153.

27. Jung, G. and Beck-Sickinger, A.G. (1992) Multiple peptide synthesis methods and their applications. *Angew. Chem. Int. Ed. (English)* **31**, 367–383.

28. Geysen, H.M., Meloen, R.H., and Barteling, S.J. (1984) Use of peptide synthesis to probe viral antigens for epitopes to a resolution of a single amino acid. *Proc. Natl. Acad. Sci. USA* **82**, 3998–4002.

29. Lebl, M. (1999) Parallel personal comments on "classical" papers in combinatorial chemistry. *J. Combin. Chem.* **1**, 3–24.

30. Maier, T., Yu, C., Külleritz, G., and Clemens, S. (2003) Localization and functional characterization of metal-binding sites in phytochelatin synthases. *Planta* **218**, 300–308.

31. Malin, R., Steinbrecher, R., Jannsen, J., et al. (1995) Identification of Technetium-99m binding peptides using combinatorial cellulose-bound peptide libraries. *J. Am Chem. Soc.* **117**, 11821–11822.

32. Bialek, K., Swistowski, A., and Frank, R. (2003) Epitope-targeted proteome analysis: towards a large-scale automated protein-protein-interaction mapping utilizing synthetic peptide arrays. *Anal. Bioanal. Chem.* **376**, 1006–1013.

33. Buss, H., Dörrie, A., Schmitz, M.L., et al. (2004) Phosphorylation of serine 468 by GSK-3β negatively regulates basal p65 NF-κB activity. *J. Biol. Chem.* **279**, 49571–49574.

34. Schutkowski, M., Reineke, U., and Reimer, U. (2005) Peptide arrays for kinase profiling. *ChemBioChem* **6**, 513–521.

35. Hilpert, K., Elliott, M.R., Volkmer-Engert, R., et al. (2006) Sequence requirements and an optimization strategy for short antimicrobial peptides. *Chem. Biol.* **13**, 1101–1107.

36. Piossek, C., Thierauch, K.-H., Schneider-Mergener, J., et al. (2003) Potent inhibition of angiogenesis by D, L-peptides derived from vascular endothelial growth factor receptor 2. *Thromb Haemost.* **90**, 501–510.

37. Grogan, J.L., Kramer, A., Nogai, A., et al. (1999) Cross-reactivity of myelin basic protein-specific T cells with multiple microbial peptides: Experimental autoimmune encephalomyelitis induction in TCR transgenic mice. *J. Immunol.* **163**, 3764–3770.

38. Frese, S., Schubert, W.-D., Findeis, A.C., et al. (2006) The phosphotyrosine peptide binding specificity of Nck1 and Nck2 SH2 domains. *J. Biol. Chem.* **281**, 18236–18245.

39. Jobron, L. and Hummel, G. (2000) Solid-phase synthesis of unprotected N-glycopeptide building blocks for SPOT synthesis of glycopeptides. *Angew. Chem. Intl Ed.* **39**, 1621–1624.

40. Zimmermann, J., Kühne, R., Volkmer-Engert, R., et al. (2003) Design of N-substituted peptomer ligands for EVH1 domains. *J. Biol. Chem.* **278**, 36810–36818.

41. Hahn, M., Winkler, D., Welfle, K., et al. (2001) Cross-reactive binding of cyclic peptides to an anti-TGFα antibody Fab fragment. An X-ray structural and thermo-dynamic analysis. *J. Mol. Biol.* **314**, 293–309.

42. Welschof, M., Reineke, U., Kleist, C., et al. (1999) The antigen binding domain of non-idiotypic human anti-F(ab´)2 autoantibodies: Study of their interaction with IgG hinge region epitopes. *Human Immunol.* **60**, 282–290.

43. Reineke, U., Kramer, A., and Schneider-Mergener, J. (1999) Antigen sequence- and library-based mapping of linear and discontinuous protein-protein interaction sites. *Curr. Topics Microbiol. Immunol.* **243**, 23–36.

44. Reineke, U., Ehrhard, B., Sabat, R., Volk, H.-D., and Schneider-Mergener, J. (1998) Novel strategies for the mapping of discontinuous epitopes using cellulose-bound peptide and hybritope scans. In: *Peptides 1996: Proceedings of the 24th European Peptide Symposium*, eds. R. Ramage and R. Epton, pp.751–752. Kingswinford: Mayflower Scientific Ltd.

45. Reineke, U., Sabat, R., Kramer, A., et al. (1996) Mapping protein-protein contact sites using cellulose-bound peptide scans. *Mol. Divers.* **1**, 141–148.

46. Reineke, U., Sabat, R., Hoffmüller, U., et al. (2002) Identification of minipro-teins using cellulose-bound duotope scans. In: *Peptides for the New Millenium. Proceedings of the 16th American Peptide Symposium*, eds. G.B. Fields, J.P. Tam, and G. Barany, pp. 167–169. Springer Netherlands.

47. Kramer, A., Stigler, R.-D., Knaute, T., Hoffmann, B., and Schneider-Mergener, J. (1998) Stepwise transformation of a cholera toxin an a p24 (HIV-1) epitope into D-peptide analogs. *Protein Eng.* **11**, 941–948.

48. Oggero, M., Frank, R., Etcheverrigaray, M., and Kratje, R. (2004) Defining the antigenic structure of human GM-CSF and its implications for the receptor inter-action and therapeutic treatments. *Mol. Divers.* **8**, 257–269.

49. Kramer, A., Keitel, T., Winkler, K., Stöcklein, W., Höhne, W., and Schneider-Mergener, J. (1997) Molecular basis for the binding promiscuity of an anti-p24 (HIV-1) monoclonal antibody. *Cell* **91**, 799–809.

50. Bolger, G.B., Baillie, G.S., Li, X., et al. (2006) Scanning peptide array analyses identify overlapping binding sites for the signaling scaffold proteins, β-arrestin and RACK1 in the cAMP-specific phosphodiesterase, PDE4D5. *Biochem. J.* **398**, 23–36.

51. Liang, M., Mahler, M., Koch, J., et al. (2003) Generation of an HFRS patient-derived neutralizing recombinant antibody to Hantaan virus G1 protein and definition of the neutralizing domain. *J. Med. Virol.* **69**, 99–107.

52. Espaniel, X. and Sudol, M. (2001) Yes-associated protein and p53-binding protein-2 interact through their WW and SH3 domains. *J. Biol. Chem.* **276**, 14514–14523.

53. Houghten, R.A., Pinilla, C., Blondelle, S.E., Appel, J.R., Dooley, C.T., and Cuervo, J.H. (1991) Generation and use of synthetic peptide combinatorial libraries for basic research and drug discovery. *Nature* **354**, 84–86.

54. Tegge, W., Frank, R., Hofmann, F., and Dostmann, R.G. (1995) Determination of cyclic nucleotide-dependent protein kinase substrate specificity by the use of peptide libraries on cellulose paper. *Biochemistry* **34**, 10569–10577.

55. Heine, N., Ast, T., Schneider-Mergener, J., Reineke, U., Germeroth, L., and Wenschuh, H. (2003) Synthesis and screening of peptoid arrays on cellulose membranes. *Tetrahedron* **59**, 9919–9930.

56. Frank, R., Kiess, M., Lahmann, H., Behn, C., and Gausepohl, H. (1995) Combinatorial synthesis on membrane supports by the SPOT technique. In: *Peptides 1994: Proceedings of the 23rd European Peptide Symposium*, ed. H.L.S. Maia, pp. 479–480. Leiden: ESCOM.

57. Frank, R. (1995) Simultaneous and combinatorial chemical synthesis techniques for the generation and screening of molecular diversity. *J. Biotechnol.* **41**, 259–272.

58. Schneider-Mergener, J., Kramer, A., and Reineke, U. (1996) Peptide libraries bound to continuous cellulose membranes: tools to study molecular recognition. In: *Combinatorial Libraries: Synthesis, Screening and Application Potential*, ed. R. Cortese. New York: Walter de Gruyter.

59. Frank, R. and Overwin, H. (1996) SPOT synthesis. Epitope analysis with arrays of synthetic peptides prepared on cellulose membranes. In: *Epitope Mapping Protocols*, ed. G.E. Morris, pp. 149–169. Totowa, NJ: Humana Press.

60. Kramer, A., Vakalopoulou, E., Schleuning, W.-D., and Schneider-Mergener, J. (1995) A general route to fingerprint analyses of peptide-antibody interactions using a clustered amino acid peptide library: comparison with a phage display library. *Mol. Immunol.* **32**, 459–465.

61. Reineke, U., Ivascu, C., Schlief, M., et al. (2002) Identification of distinct antibody epitopes and mimotopes from a peptide array of 5520 randomly generated sequences. *J. Immunol. Methods* **267**, 37–51.

62. Pulli, T., Lankinen, H., Roivainen, M., and Hyypiä, T. (1998) Antigenic sites of coxsackievirus A9. *Virology* **240**, 202–212.

63. Mahler, M., Kessenbrock, K., Raats, J., Williams, R., Fritzler, M.J., and Blüthner, M. (2003) Characterization of the human autoimmune response to the major C-terminal epitope of the ribosomal P proteins. *J. Mol. Med.* **81**, 194–204.

64. Winkler, D., Stigler, R.-D., Hellwig, J., Hoffmann, B., and Schneider-Mergener, J. (1996) Determination of the binding conformation of peptide epitopes using cyclic peptide libraries. In: *Peptides: Chemistry, Structure and Biology: Proceedings of the 14th American Peptide Symposium*, eds. P.T.P. Kaumaya and R.S. Hodges, pp. 315–316. Kingswinford: Mayflower Scientific Ltd.

65. Kamradt, T. and Volkmer-Engert, R. (2004) Cross-reactivity of T lymphocytes in infection and autoimmunity. *Mol. Divers.* **8**, 271–280.

66. Boisguerin, P., Leben, R., Ay, B., et al. (2004) An improved method for the synthesis of cellulose membrane-bound peptides with free C termini is useful for the PDZ domain binding studies. *Chem. Biol.* **11**, 449–459.

67. Volkmer-Engert, R., Hoffmann, B., and Schneider-Mergener, J. (1997) Stable attachment of the HMB-linker to continuous cellulose membranes for parallel solid phase spot synthesis. *Tetrahedron Lett.* **38**, 1029–1032.

68. Scharn, D., Wenschuh, H., Reineke, U., Schneider-Mergener, J., and Germeroth, L. (2000) Spatially addressed synthesis of amino- and amino-oxy-substituted 1,3,5-triazine arrays on polymeric membranes. *J. Combin. Chem.* **2**, 361–369.

69. Przezdziak, J., Tremmel, S., Kretzschmar, I., Beyermann, M., Bienert, M., and Volkmer-Engert, R. (2006) Probing the ligand-binding specificity and analyzing the folding state of SPOT-synthesized FBP28 WW domain variants. *ChemBioChem* **7**, 780–788.

70. Toepert, F., Pires, J.R., Landgraf, C., Oschkinat, H., and Schneider-Mergener, J. (2001) Synthesis of an array comprising 837 variants of the hYAP WW protein domain. *Angew. Chem. Int. Ed.* **40**, 897–900.

71. Münch, G., Schicktanz, D., Behme, A., et al. (1999) Amino acid specificity of glycation and protein-AGE crosslinking reactivities determined with a dipeptide SPOT library. *Nature Biotechnol.* **17**, 1006–1010.

5

Analysis of Aβ Interactions Using ProteinChip Technology

Eleni Giannakis, Lin-Wai Hung, Keyla Perez Camacaro, David P. Smith, Kevin J. Barnham, and John D. Wade

Summary

Aβ peptides are now acknowledged to play a central role in the pathogenesis of Alzheimer's disease. Their generation results from the sequential cleavage of amyloid precursor protein by β and γ secretases. The resulting peptide fragments impart toxicity via their ability to form soluble oligomers and bind to cell membranes. In this chapter we describe the use of ProteinChip® technology to study the physicochemical behaviour of Aβ and its mechanisms of toxicity. These include analyzing *(1)* Aβ processing and quantitation of peptide fragments, *(2)* Aβ aggregation and the quantitation of oligomers, and *(3)* Aβ–lipid interactions.

Key Words: Alzheimer's disease; Aβ; amyloid precursor protein; SELDI-TOF MS; ProteinChip technology.

1. Introduction

1.1. Alzheimer's Disease and Aβ Peptides

Alzheimer's disease (AD) is a progressive, neurodegenerative disorder, which is characterized by the presence of amyloid plaques in the cerebral cortex and hippocampus *(1)*. Although these plaques contain several different proteins, it is primarily composed of amyloid-β (Aβ) peptide fragments found principally in an aggregated filamentous form *(2)*. The generation of these fragments results from sequential cleavage of the transmembrane glycoprotein, amyloid precursor protein (APP, 590-680 αα), by α-, β-, and γ-secretases **(Fig. 1)** *(3,4)*.

From: *Methods in Molecular Biology, vol. 494: Peptide-Based Drug Design*
Edited by: L. Otvos, DOI: 10.1007/978-1-59745-419-3_5, © Humana Press, New York, NY

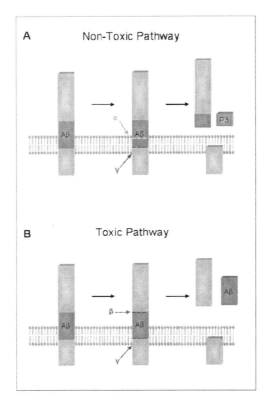

Fig. 1. Cleavage of APP via nontoxic and toxic pathways. Sequential cleavage of APP by α and γ secretases results in the production of the nonpathological fragment, P3 (**A**). Cleavage by β and γ secretases results in the production of toxic Aβ fragments (**B**). (From **ref. 3**.)

α-Secretase cleaves at position 17 of the Aβ domain *(5)*. β-Secretase cleaves at positions 1 and 11, which removes the large hydrophilic extracellular domain of APP. γ-Secretase, although not sequence specific, primarily cleaves at the C-terminal end at positions 40 and 42, located within the hydrophobic trans-membrane domain of APP *(6)*. Sequential cleavage by α/γ-secretases results in the generation of the nonpathogenic fragment called P3 *(6)*. Cleavage by β- and γ-secretases produces several variants of 39–43 residues, the most abundant products being the Aβ peptide fragments 1-40 and 1-42, the latter of which is particularly hydrophobic, resulting in a greater propensity to aggregate/fibrilize and form plaques than the Aβ 1-40 isoform *(7)*. The process of fibrilization initially results in the formation of soluble oligomers, believed to consist mainly of trimers and multiples of these oligomers, e.g., hexamers *(8)*. These oligomers are then converted into mature fibrils, which adopt cross-β-pleated sheets *(9)* and eventually fibrils *(10)*. Interestingly, it is the intermediate products, rather than the mature fibrils that have been shown to be the most neurotoxic *(8,10,11)*.

The mechanisms by which Aβ induce toxicity include *(1)* oxidative stress *(12)*, *(2)* direct disruption of membrane integrity *(13)*, and *(3)* alteration in Ca^{2+} homeostasis *(14)*. Oxidative stress is initiated by binding interactions between Aβ and metal ions (Cu^{2+}, Fe^{3+}) via histidine residues 6, 13, and 14 *(15)*. This interaction leads to the reduction in the oxidation state of metal ions and generation of H_2O_2 with the subsequent generation of free radicals that induce oxidative damage, including lipid peroxidation and the subsequent disruption of the cellular membrane. Methionine 35 (Met35) of Aβ is also thought to play a critical role since oxidized Met35 Aβ products have been found in postmortem AD plaques *(16)*. Furthermore, peptides lacking Met35 have a decreased capacity to reduce Cu^{2+} and generate H_2O_2 *(17)*. Aβ has also been demonstrated to directly interact with membrane lipids to form cation selective channels *(18)*. It is thought that these channels disrupt ion homeostasis *(18)*, leading to an accumulation of intracellular Ca^{2+} levels and the subsequent induction of apoptosis *(19)*.

The study of Aβ and the mechanisms described, including *(1)* production of toxic Aβ fragments, *(2)* generation of aggregates, and *(3)* interactions with membranes is essential to further understand the pathology of disease. ProteinChip® technology is a method that facilitates the analysis of all these processes. Information acquired from such analysis will not only help to further elucidate the precise role of Aβ in AD, but may have wider implications for the development of an anti-Aβ therapeutic that blocks the toxic effects of the molecule. Furthermore, this technology provides a valuable means to assess the effectiveness of therapeutic intervention. This chapter will describe ProteinChip technology, its advantages over traditional techniques, and how it can be used for the analysis of the mechanisms described.

1.2. ProteinChip Technology

ProteinChip technology employs surface enhanced laser desorption/ ionization time-of-flight mass spectrometry (SELDI-TOF MS), which combines two well-established methods of solid phase chromatography and TOF-MS into an integrated platform *(20)*. The proprietary ProteinChip arrays distinguish this technology from other MS-based systems. The arrays possess chromatographic surfaces including hydrophobic, hydrophilic, anion exchange, cation exchange, and immobilized-metal affinity and are utilized to enrich for subsets of the proteome with common biochemical properties. Furthermore, we have adapted this technology to create a synthetic solid phase membrane layer on the array surface to analyze membrane binding proteins. Proteins retained on the chip surface are resolved via TOF-MS, which displays the mass-to-charge value and signal intensities of the proteins *(20)*.

ProteinChip arrays with preactivated surfaces are also available for covalently coupling of a specific "bait" molecule (protein, DNA, RNA). This allows the investigation of specific protein interactions such as DNA–protein, receptor–ligand, and antibody-antigen (Ab-Ag), the latter of which permits the generation of a standard curve and hence quantitation studies *(21)*.

1.3. Quantitation of Aβ

Traditional methods employed to examine Aβ processing include enzyme linked immunosorbant assays (ELISA) and SDS-PAGE/Western blotting. ELISA generates an average signal of all peptide species bound to the antibody Ab and hence has poor assay specificity. Analysis of Aβ by SDS-PAGE/Western blotting exhibits poor discriminatory power between the variant forms of the peptide. In addition, molecular weights of peptides cannot be accurately determined. Hence, using these techniques for the analysis of multiple Aβ species can be difficult.

In contrast, ProteinChip technology allows the detailed analysis of Aβ fragments and their modifications *(22)*. Peptides are captured on arrays coated with anti-Aβ Ab, washed to remove nonspecifically bound fragments, and finally detected by SELDI-TOF MS. The resolution of the technique allows the discrimination of peptides of similar mass and modified products, e.g., oxidized peptides versus native peptides, which differ by only 16 Da. ProteinChip technology has been used extensively to detect Aβ fragments in various samples, including cell culture supernatants, serum, plasma, and cerebrospinal fluid (CSF) *(21–23)*.

1.4. Aβ Aggregation and Interactions with Lipids

Analysis of Aβ aggregation typically involves performing circular dichoism (CD) spectroscopy, thioflavin T assays, electron microscopy, SDS-PAGE, and chromatography. These techniques describe the aggregation of Aβ in a general sense without the ability to accurately quantitate the generation of specific oligomeric species. Photo-induced cross-linking of unmodified proteins (PICUP) assays have also been used for assessing oligomeric profiles. This method employs photolysis to induce covalent bonds between transiently formed oligomers. The PICUP method, however, is prone to artifacts by cross-linking transitory interactions. In addition, PICUP oligomers are assessed via SDS-PAGE/Western blotting, the limitations of which have been previously described. In contrast, SELDI-TOF MS can be used to generate oligomeric profiles and assess relative changes in stable oligomeric species.

Analysis of Aβ–lipid interactions involves performing ultracentrifugation or size filtration chromatography of vesicles exposed to Aβ, CD spectroscopy, or surface plasmon resonance (SPR) technology. Chromatography and ultracentrifugation require several hours, during which time the peptides can further aggregate, complicating the interpretation of the data. CD indicates structural changes to the peptide upon interactions with lipids, but quantitative interpretation is difficult. SPR allows lipid-binding affinities to be assessed, but determination of the specific oligomeric species interacting with the lipids is not possible. Consequently, we have developed novel aggregation lipid-binding assays using SELDI-TOF MS, in which vesicles are adsorbed to hydrophobic H50 ProteinChip arrays, forming a supported lipid monolayer with the hydrophilic heads facing the solution. This assay, unlike conventional techniques, allows the rapid detection and quantitation of individual oligomeric species which are bound specifically and directly to the lipid-coated array.

2. Materials

2.1. Capture and Quantitation of Aβ Using Antibody-Coated ProteinChip Arrays

1. ProteinChip arrays: PS10, PS20 (Bio-Rad Laboratories, Hercules, CA).
2. Ab: Anti-human Aβ (0.5 mg/mL) in Tris-free buffer (*see* **Table** 1 for list of commonly used Abs).
3. Aβ peptide standards.
4. Phosphate-buffered saline (PBS): 137 m*M* NaCl, 10 m*M* phosphate buffer, pH 7.4.
5. Deactivation buffer: 0.5 *M* ethanolamine in PBS, pH 8.0.
6. PBST: PBS, 0.5% Triton X-100.
7. 1 m*M* hydroxyethyl piperazine ethane sulfonate (HEPES), pH 7.2.

Table 1
Common Ab Used for Detection of Aβ, Including Information Pertaining to the Type (mAb/pAb) and Specificity of the Ab (epitope recognition site)

Antibody	Type (mAb/pAb)	Specificity (Ref.)
4G8	mAb	17-24 aa *(25)*
6E10	mAb	1-12 aa *(25)*
G210, G211	mAb	C-terminal residues of Aβ-40 and -42, respectively *(26)*
WO2	mAb	4-10 aa *(27)*
Anti-Aβ 40	pAb	704-711 aa of APP *(28)*
Anti-Aβ 42	pAb	706-713 aa of APP *(28)*

8. Acetonitrile (ACN), HPLC grade.
9. Trifluoroacetic acid (TFA), HPLC grade.
10. Matrix/Energy Absorbing Molecule (EAM): α-cyano-4-hydroxy cinnamic acid (CHCA), 5 mg/tube (Bio-Rad Laboratories).
11. ProteinChip reader (Bio-Rad Laboratories).
12. Bioprocessor (optional) (Bio-Rad Laboratories).

2.2. Capture of Aβ Aggregates on ProteinChip Arrays

1. ProteinChip arrays: H50 (Bio-Rad Laboratories).
2. Aβ peptides.
3. Hexafluoroisopropanol.
4. 20 mM NaOH.
5. 10 mM phosphate buffer, pH 7.4 (PB).
6. 1 mM HEPES, pH 7.4.
7. 20 μM minisart RC4 filters (Sartorius, Göttingen, Germany).
8. ACN.
9. TFA.
10. Matrix/EAM: CHCA.
11. UV spectrophotometer.
12. ProteinChip reader.

2.3. Capture of Aβ on Lipid-Coated ProteinChip Arrays

1. ProteinChip arrays: H50.
2. *1*-Palmitoyl-2-oeoyl-*sn*-glycero-3 [phospho-L-serine] (POPS) and *1*-palmitoyl-2-oleoyl-*sn*-glycero-3-phosphocholine (POPC) (Avanti Polar Lipids, Inc., Alabaster, AL).
3. Aβ peptides.
4. Ethanol-free chloroform.
5. Acid-washed glass beads, G1152 (Sigma-Aldrich, Poole, UK).
6. Liquid nitrogen.
7. 0.05 μM filters and spacers for extruder (Nuclepore Corp., Pleasanton, CA).
8. PB.
9. CHAPS (30 mg/mL).
10. 1 mM HEPES, pH 7.4.
11. ACN.
12. TFA.
13. Matrix/EAM: CHCA.
14. Sonicator.
15. Extruder (Avanti Polar Lipids).
16. ProteinChip reader.

3. Methods

3.1. Capture and Quantitation of Aβ Using Antibody-Coated ProteinChip Arrays

Prior to initiating Ab capture experiments, it is important to consider the following parameters to ensure that the assay yields the desired results: *(1)* chip selection, *(2)* Ab selection, and *(3)* sample preparation and storage.

3.1.1. Experimental Parameters

3.1.1.1. CHIP SELECTION

There are several preactivated ProteinChip arrays available for direct Ab coupling, including *(1)* PS20 arrays, which possess an epoxide active group and *(2)* PS10 and RS100 arrays, which have carbonyldiimidazole (CDI) functional groups. Abs covalently bind to these arrays via amine groups. PG20 arrays, which are PS20 arrays precoupled with Protein G, are also available for Ab capture. Protein G selectively binds to Abs via the Fc region (*see* **Note 1**).

3.1.1.2. ANTIBODY SELECTION

The preferred Abs for ProteinChip experiments are Ag affinity purified polyclonal Abs (pAb), they exhibit a high avidity and slow off rate. In addition, Ab must be pure and intact (~150 kDa), as contaminating proteins or free Ab light/heavy chains (~25 and ~50 kDa, respectively) will compete for active sites on the preactivated array and diminish the specific signal between the intact Ab and Ag (*see* **Note 2**). Further to this, it is important to assess the Ab buffer constituents; the buffer must be free of amines (Tris, ethanolamine, azide) as this will also compete for the active sites and diminish the specific signal of the Ab/Ag interaction.

If the Ab preparation contains contaminating proteins, it is recommended to purify the Ab using Protein G/A beads or spin columns, while free amines in the buffer can be removed by dialysis. Alternatively, PG20 arrays can be used to selectively capture the Ab "on chip." Ab bound to PG20 arrays will be orientated in the same direction, unlike the PS10, PS20, and RS100 arrays, in which Abs are randomly bound via amines; this can be advantageous and increase the specific signal (*see* **Note 3**).

3.1.1.3. PREPARATION AND STORAGE OF SAMPLES

Samples should be prepared keeping in mind that Aβ has a high propensity to aggregate. Furthermore, it can bind many serum proteins and other molecules, which may obscure the Ag recognition site on the Ab, thereby decreasing the specific signal. Consequently, it is recommended to add a nonionic detergent (e.g., CHAPS) and a denaturing agent (e.g., urea or guanidine isothiocyanate) to the samples, which will disrupt protein–protein interactions, exposing antigenic epitopes. Samples should be aliquoted, stored at −80°C and used as required, discarding remaining material. Aliquots are essential as repeated freeze–thaw cycles can be problematic and lead to the generation of oxidative modification, which can reduce the sensitivity of the assay.

3.1.2. Processing Arrays

3.1.2.1. DIRECTLY COUPLING ANTIBODIES TO PROTEINCHIP ARRAYS

1. Load 2 μL of anti-Aβ Ab or control Ab (0.1–0.5 mg/mL) to each spot of a preactivated ProteinChip array. As PS10 and PS20 arrays are commonly used for Aβ capture experiments, analysis of these chip types will be described.
2. Place arrays in a humidity chamber to prevent the spots from drying and incubate overnight at 4°C.
3. Carefully wick off the Ab using a Kimwipe®, being sure not to touch the spot. Load 5 μL of deactivation buffer to each spot to block remaining exposed active sites. Incubate for 30 min at room temperature.
4. Wash the arrays twice with 5 μL of PBST for 5 min on a shaking table, followed by two PBS washes.

3.1.2.2. ANTIGEN BINDING

1. Load 5 μL of sample per spot. Employing a bioprocessor allows larger volumes of up to 200 μL to be processed, which can be beneficial if the sample is dilute, e.g., cell culture supernatants. Once the sample has been loaded into the bioprocessor, it is important to centrifuge the unit (1 min, 1000 rpm) to ensure the sample is in contact with the array surface. Including replicates is also important to assess the reproducibility of the assay.
2. Incubate the sample on a shaking table for a minimum of 30–60 min. Generally samples are incubated for 1–4 h at room temperature or overnight at 4°C on a shaking table (*see* **Note 4**).

3.1.2.3. POST–ANTIGEN-BINDING WASHES

1. Wick off sample using a Kimwipe and wash spots twice with 5 μL of PBST for 5 min on a shaking table, followed by two PBS washes (*see* **Note 5**).
2. Wash array twice for 1 min with 1 mM HEPES, pH 7.2 (*see* **Note 6**).
3. Dry array for approx 10 min in preparation for matrix addition (*see* **Note 7**).

3.1.2.4. EAM ADDITION

1. Combine 200 μL of 1% TFA with 200 μL of ACN to make 50% ACN, 0.5% TFA.
2. Add 200 μL of 50% ACN, 0.5% TFA to a tube containing 5 mg of CHCA to make 100% saturated solution.
3. Vortex and incubate for 5 min at room temperature.

4. Centrifuge for 1 min at 13,000 rpm and combine 100 μL of 100% saturated CHCA with 100 μL of 50% ACN, 0.5% TFA to make 200 μL of 50% saturated CHCA. Prepare CHCA in an opaque tube as it is light sensitive.
5. Load 1 μL of 50% CHCA to each spot and allow to air-dry. Repeat this step, ensuring spots are dry before addition of the second layer of CHCA (*see* **Note 8**).
6. See **Fig. 2** for a summary of the Aβ capture assay.

3.1.2.5. DATA ACQUISITION SETTINGS

Laser and sensitivity settings need to be optimized for each new set of experimental conditions. Excess laser energy will generate flat-topped broad peaks

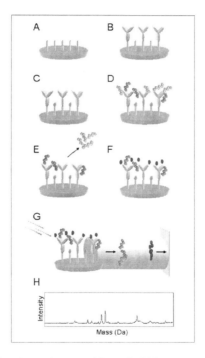

Fig. 2. Capture of Aβ using Ab-coated ProteinChip arrays. Preactivated arrays with exposed active sites are shown in (**A**). Ab are covalently bound to the array via amine groups (**B**). Exposed active sites are blocked using Tris-containing buffer (**C**). Samples are added to the array, incubated for 1–4 h (**D**), and washed to remove nonspecifically bound proteins (**E**). EAM is applied to each spot (**F**), and the array is introduced into the ProteinChip reader where the laser is fired onto the chip surface. Proteins retained on the array surface are subsequently resolved via TOF-MS, which displays the mass-to-charge value and signal intensities of each protein (**G, H**). (Adapted from ProteinChip technology training course, Bio-Rad Laboratories.)

and off-scale readings, under which conditions the detector will be saturated and quantitation is not possible. This will also negatively affect the reproducibility of the assay. In such cases, consider decreasing the laser intensity value until peaks are sharp and well resolved (*see* **Note 9**).

Calibrate the spectra to ensure reliable mass accuracy. Generate a calibration equation using calibrants that cover the mass range of interest, and then match the calibrant and sample matrix. External mass calibration can subsequently be applied to the relevant spectra.

3.1.2.6. QUANTITATIVE ASSESSMENT OF THE DATA

Relative changes in the concentration of Aβ can be determined simply by assessing the differences in peak intensity (signal-to-noise/area under the curve) between the sample groups. In order to quantitate the amount of Aβ present in the sample, a standard curve must be generated. Aβ standards of known concentration should be spiked into a control sample, which has been treated in the exact same manner as the remaining control and test samples, including final buffer constituents, total protein concentration, storage, etc.

The sensitivity of the assay is peptide specific, consequently it is important to generate a standard curve for each peptide fragment. Threefold dilutions starting from 100 to 0.13 nM are a recommended starting point, although depending on the sensitivity of the assay, it may be necessary to either decrease or increase the initial starting concentrations. Once the data have been collected, export the peak information into an Excel spread sheet, including signal-to-noise values/area under the curve, and use this information to plot the standard curve. From this curve the R^2 value and slope-intercept equation ($y = mx + b$) can be generated. The R^2 value is an indicator of the models "goodness of fit"; it is expressed as a fraction between 0.0 and 1.0, with high values reflecting a good quality model (*see* **Note 10**). The slope-intercept equation is used to determine the value of the test sample, where y is the unknown value, x is the signal-to-noise of the sample, m is the slope, and b is the y-intercept value.

3.2. Capture of Aβ Aggregates on ProteinChip Arrays

3.2.1. Preparation of Aβ Aggregates

1. Dissolve lyophilized Aβ in hexafluoroisopropanol to a final concentration of 1 mg/mL to induce a monomeric and helical conformation of the peptides.
2. Aliquot 100 μL of peptides into Eppendorf tubes and dry using a speed vacuum system.
3. As required, resuspend 100-μg aliquots in 50 μL of 20 mM NaOH and sonicate for 15 min.
4. Dissolve Aβ solution into 50 μL of PB and 400 μL of H_2O.

5. Filter the solution to ensure preformed aggregates are removed (20 μm Minisart RC4, Sartorius).
6. To account for the loss of peptide resulting from the filtering process, it is necessary to reassess the protein concentration by UV spectrophotometer. The extinction coefficient for Aβ fragments is determined by measuring light absorbance of Aβ dissolved in a high-pH solution (e.g., NaOH at pH 11 to allow for complete solubility).
7. Adjust samples to 10 μ*M* and incubate at 37°C with shaking (200 rpm) to induce aggregation.

3.2.2. ProteinChip Analysis

1. Place H50 ProteinChip arrays in a humidity chamber.
2. Pre-equilibrate the arrays in the sample binding buffer; load each spot with 5 μL of PB and incubate for 5 min on a shaking table. Repeat for a total of two PB washes.
3. Adjust samples to 10 μ*M* in PB and load 5 μL per spot.
4. Incubate for 2 h at room temperature on a shaking platform.
5. To remove nonspecifically bound peptide, wash spots three times with 5 μL of PB for 5 min on a shaking platform.
6. Finally wash arrays twice with 1 m*M* HEPES, pH 7.4 for 1 min.
7. Dry arrays and load 1 μL of 50% CHCA. Allow the array to air-dry and repeat the CHCA application.
8. Collect data at the low and high mass range.
9. Export data into an Excel spread sheet and determine the signal-to-noise/area under the curve for each oligomeric species (*see* **Note 11**).

3.3. Capture of Aβ on Lipid-Coated ProteinChip Arrays

3.3.1. Vesicle Preparation

1. Combine equal amounts of *1*-Palmitoyl-2-oeoyl-*sn*-glycero-3 [phospho-L-serine] (POPS) and *1*-palmitoyl-2-oleoyl-*sn*-glycero-3-phosphocholine (POPC) (16 mg) in a 10-mL glass conical flask and dissolve in 2 mL of ethanol-free chloroform.
2. Allow chloroform to evaporate for 1 h under a stream of nitrogen; this will create a thin layer of lipid in the bottom of the flask.
3. Add acid-washed glass beads and 1 mL of PB to the flask and incubate for 1 h at 37°C with agitation (220 rpm) to resuspend the lipid layer.
4. Transfer lipid to an Eppendorf tube, sonicate for 15 min, and subject sample to several freeze–thaw cycles (5×) using liquid nitrogen and a 37°C water bath.
5. Prepare the extruder apparatus according to the manufacturer's specifications.
6. Place apparatus on a 37°C heating block. Pass the lipid through a 0.05 μ*M* presoaked filter 11 times, after which time the solution will appear more translucent.
7. Dispense the solution into a glass jar and store under nitrogen at 4°C; use within 48 h.

3.3.2. Formation of a Lipid Monolayer on H50 ProteinChip Arrays

1. Place H50 ProteinChip arrays in a humidity chamber.
2. Load each spot with 5 μL of CHAPS (30 mg/mL), and immediately wick off using a Kimwipe.
3. Remove any residual detergent by washing with PB; load 5 μL of PB per spot and incubate for 2 min on a shaking platform; repeat for a total of three washes.
4. Load 5 μL of the 41 μM POPC/POPS mixture or PB to each spot and incubate for 2 h at 37°C.
5. To remove unbound lipid, wash spots with 5 μL of PB twice for 2 min on a shaking platform.

3.3.3. Protein Binding to the Lipid-Coated H50 ProteinChip Arrays

1. Prepare samples in PB to a final concentration of 50 μM and load 5 μL onto spots coated with either the POPC/POPS lipid mixture or PB alone.

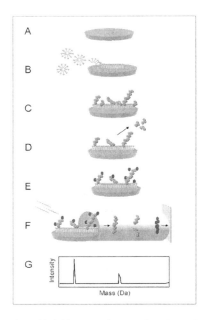

Fig. 3. Analysis of peptide–lipid interactions using ProteinChip arrays. Vesicles are absorbed to H50 ProteinChip arrays via interactions with C8 functional groups, creating a supported lipid monolayer on the chip surface (**A, B**). Samples are applied to the chip, incubated for 5 min (**C**), and washed to remove nonspecifically bound proteins (**D**). Matrix is added (**E**), and the array is introduced into the ProteinChip reader, where the laser is fired onto the chip surface. Peptides retained on the surface are finally resolved by TOF-MS, displaying mass-to-charge versus signal intensity (**F, G**). (Adapted from ProteinChip technology training course, Bio-Rad Laboratories.)

2. Incubate the samples for 5 min with agitation to allow the peptide to bind to the lipid layer (*see* **Note 12**).
3. To remove nonspecifically bound protein, wash arrays twice with PB for 2 min, followed by two 1-min washes with 1 mM HEPES, pH 7.2.
4. Allow the arrays to air-dry for approx 10 min and load 1 μL of 50% CHCA. Once the arrays are dry, load another application of 1 μL of 50% CHCA and dry again.
5. Collect data at low and high molecular weight range.
6. Export data into an Excel spread sheet and determine the lipid binding affinity of the oligomers by assessing the signal-to-noise/area under the curve (*see* **Note 11**).
7. *See* **Fig. 3** for a summary of the Aβ lipid-binding assay.

4. Notes

1. It is advantageous to initially screen all the available chip types, including PS10, PS20, RS100, and PG20, to determine which array provides the highest degree of sensitivity.
2. To assess the purity and quality of the Ab intended for use on preactivated arrays, screen the Ab on NP20 arrays (normal phase arrays); also refer to the product data sheet for the presence of carrier proteins such as BSA and gelatin.
3. The binding affinity of Protein G to Ab is species and isotype dependent; rabbit, goat, and bovine pAb all exhibit high affinity and hence are the preferred Ab for PG20 arrays. Mouse monoclonal Ab (mAb) IgG1, a common mAb isotype, exhibits low binding affinity to Protein G and is not recommended.
4. Allow the binding interaction to reach equilibrium (minimum of 30–60 min). This will minimize differences in the intensity and peak number among the replicates, improving reproducibility.
5. If nonspecific binding is obscuring Aβ detection, consider increasing the *(1)* wash time, *(2)* number of washes, *(3)* wash volume by using a bioprocessor, or *(4)* concentration of detergent/salts. Alternatively, detergent/salt can be added to the sample during the Ag-binding step.
6. A final 1mM HEPES, pH 7.2 wash is performed to remove contaminants such as salts and detergent, which may interfere with interpretation of spectra. Formation of salt adducts creates additional peaks with a mass shift of 22 and 38 Da for sodium and potassium, respectively, which can affect the sensitivity of the assay. Detergents generate a series of low molecular weight peaks, which can complicate analysis and decrease the specific signal.
7. Do not allow extended drying time (>20 min) as this can significantly reduce the reproducibility of the assay.
8. As EAM addition requires working with low volumes, it is critical to use a well-calibrated 2-μL pipet. Further to this, it is essential to be consistent as the application of EAM and drying time affect crystallization and hence the signal intensity.
9. It is advantageous to have two data acquisition settings: one for the low molecular weight range (0–20 kDa) and the other for the high molecular weight range

(>20 kDa). Laser settings for the higher mass range need to be increased as the larger molecules require more energy to "fly," and hence these setting may not be optimal for detection of smaller peptide fragments and yield off-scale readings.

10. Following the recommendations provided, a R^2 value of >0.95 is easily obtainable, demonstrating the highly quantitative nature of the technology.

11. The aggregation and lipid-binding assays do not exhibit the same degree of selectivity/specificity as the Ab capture experiments; hence these assays are useful for the examination of less complex systems, including synthetic peptides and cell culture supernatants.

12. During the lipid assay, Aβ binding time is kept to a minimum to prevent the peptide further aggregating in the presence of the lipid.

5. Acknowledgments

EG and JDW gratefully acknowledge the Ian Potter Foundation (Australia) for a grant for the establishment of a biomarker facility at the Howard Florey Institute. DPS is a Wellcome Trust Travelling Fellow.

References

1. Gorevic, P.D., Goni, F., Pons-Estel, B., Alvarez, F., Peress, N.S. and Frangione, B. (1986) Isolation and partial characterization of neurofibrillary tangles and amyloid plaque core in Alzheimer's disease: immunohistological studies. *J. Neuropathol. Exp. Neurol.* **45**, 647–664.

2. Masters, C.L., Simms, G., Weinman, N.A., Multhaup, G., McDonald, B.L. and Beyreuther, K. (1985) Amyloid plaque core protein in Alzheimer disease and Down syndrome. *Proc. Natl. Acad. Sci. USA* **82**, 4245–4249.

3. Selkoe, D.J. (2001) Alzheimer's disease: genes, proteins, and therapy. *Physiol. Rev.* **81**, 741–766.

4. Sisodia, S.S. and St George-Hyslop, P.H. (2002) Gamma-Secretase, Notch, Abeta and Alzheimer's disease: where do the presenilins fit in? *Nat. Rev. Neurosci.* **3**, 281–290.

5. Sisodia, S.S., Koo, E.H., Beyreuther, K., Unterbeck, A. and Price, D.L. (1990) Evidence that beta-amyloid protein in Alzheimer's disease is not derived by normal processing. *Science* **248**, 492–495.

6. Haass, C., Hung, A.Y., Schlossmacher, M.G., Teplow, D.B. and Selkoe, D.J. (1993) beta-Amyloid peptide and a 3-kDa fragment are derived by distinct cellular mechanisms. *J. Biol. Chem.* **268**, 3021–3024.

7. Jarrett, J.T., Berger, E.P. and Lansbury, Jr. P.T. (1993) The carboxy terminus of the beta amyloid protein is critical for the seeding of amyloid formation: implications for the pathogenesis of Alzheimer's disease. *Biochemistry* **32**, 4693–4697.

8. Lesne, S., Koh, M.T., Kotilinek, L., et al. (2006) A specific amyloid-beta protein assembly in the brain impairs memory. *Nature* **440**, 352–357.

9. Harper, J.D., Wong, S.S., Lieber, C.M. and Lansbury, P.T. (1997) Observation of metastable Abeta amyloid protofibrils by atomic force microscopy. *Chem. Biol.* **4**, 119–125.
10. Relini, A., Torrassa, S., Rolandi, R., et al. (2004). Monitoring the process of HypF fibrillization and liposome permeabilization by protofibrils. *J. Mol. Biol.* **338**, 943–957.
11. Hartley, D.M., Walsh, D.M., Ye, C.P., et al. (1999) Protofibrillar intermediates of amyloid beta-protein induce acute electrophysiological changes and progressive neurotoxicity in cortical neurons. *J. Neurosci.* **19**, 8876–8884.
12. Behl, C., Davis, J.B., Lesley, R. and Schubert, D. (1994) Hydrogen peroxide mediates amyloid beta protein toxicity. *Cell* **77**, 817–827.
13. Ambroggio, E.E., Kim, D.H., Separovic, F., et al. (2005) Surface behavior and lipid interaction of Alzheimer beta-amyloid peptide 1–42: a membrane-disrupting peptide. *Biophys. J.* **88**, 2706–2713.
14. Mattson, M.P., Cheng, B., Davis, D., Bryant, K., Lieberburg, I. and Rydel, R.E. (1992) beta-Amyloid peptides destabilize calcium homeostasis and render human cortical neurons vulnerable to excitotoxicity. *J. Neurosci.* **12**, 376–389.
15. Curtain, C.C., Ali, F., Volitakis, I., et al. (2001) Alzheimer's disease amyloid-beta binds copper and zinc to generate an allosterically ordered membrane-penetrating structure containing superoxide dismutase-like subunits. *J. Biol. Chem.* **276**, 20466–20473.
16. Dong, J., Atwood, C.S., Anderson, V.E., et al. (2003) Metal binding and oxidation of amyloid-beta within isolated senile plaque cores: Raman microscopic evidence. *Biochemistry* **42**, 2768–2773.
17. Pike, C.J., Walencewicz-Wasserman, A.J., Kosmoski, J., Cribbs, D.H., Glabe, C.G. and Cotman, C.W. (1995) Structure-activity analyses of beta-amyloid peptides: contributions of the beta 25–35 region to aggregation and neurotoxicity. *J. Neurochem.* **64**, 253–265.
18. Arispe, N., Rojas, E. and Pollard, H.B. (1993) Alzheimer disease amyloid beta protein forms calcium channels in bilayer membranes: blockade by tromethamine and aluminum. *Proc. Natl. Acad. Sci. USA* **90**, 567–571.
19. Guo, Q., Fu, W., Xie, J., et al. (1998) Par-4 is a mediator of neuronal degeneration associated with the pathogenesis of Alzheimer disease. *Nat. Med.* **4**, 957–962.
20. Wiesner, A. (2004) Detection of tumor markers with ProteinChip technology. *Curr. Pharm. Biotechnol.* **5**, 45–67.
21. Austen, B.M., Frears, E.R. and Davies, H. (2000) The use of seldi proteinchip arrays to monitor production of Alzheimer's betaamyloid in transfected cells. *J. Pept. Sci.* **6**, 459–469.
22. Bradbury, L.E., LeBlanc, J.F. and McCarthy, D.B. (2004) ProteinChip array-based amyloid beta assays. *Methods Mol. Biol.* **264**, 245–257.
23. Maddalena, A.S., Papassotiropoulos, A., Gonzalez-Agosti, C., et al. (2004) Cerebrospinal fluid profile of amyloid beta peptides in patients with Alzheimer's disease determined by protein biochip technology. *Neurodegener. Dis.* **1**, 231–235.

24. Bitan, G., Lomakin, A. and Teplow, D. B. (2001) Amyloid beta-protein oligomer-ization: prenucleation interactions revealed by photo-induced cross-linking of unmodified proteins. *J. Biol. Chem.* **276**, 35176–35184.
25. Pirttilä, T., Kim, K.S., Mehta, P.D., Frey, H. and Wisniewski, H.M. (1994) Soluble amyloid β-protein in the cerebrospinal fluid from patients with Alzheimer's disease, vascular dementia and controls. *J. Neurol. Sci.* **127**, 90–95.
26. Ida, N., Hartmann, T., Pantel, J., et al. (1996) Analysis of heterogeneous A4 peptides in human cerebrospinal fluid and blood by a newly developed sensitive Western blot assay. *J. Biol. Chem.* **271**, 22908–22914.
27. Jensen, M., Hartmann, T., Engvall, B., et al. (2000) Quantification of Alzheimer amyloid beta peptides ending at residues 40 and 42 by novel ELISA systems. *Mol. Med.* **6**, 291–302.
28. Barelli, H., Lebeau, A., Vizzavona, J., et al. (1997) Characterization of new polyclonal antibodies specific for 40 and 42 amino acid-long amyloid beta peptides: their use to examine the cell biology of presenilins and the immunohistochemistry of sporadic Alzheimer's disease and cerebral amyloid angiopathy cases. *Mol. Med.* **3**, 695–707.

6

NMR in Peptide Drug Development

Jan-Christoph Westermann and David J. Craik

Key Words: NMR; binding studies; solution structure; drug design; binding mode; screening; binding site.

1. Introduction

As is clear from other chapters in this book, peptides have a range of properties that make them excellent leads in drug design, including exquisite selectivity for their molecular targets and high potency. Peptides offer a particularly versatile platform in the drug-design process because many biological interactions are mediated by protein–protein interactions and peptides may be derived from knowledge of the protein sequence and binding motif to yield a starting point for drug design. The versatility of peptides derives from the chemical diversity of naturally occurring amino acids and the ready availability of chemically modified building blocks for peptide synthesis, featuring modifications on the peptide backbone and/or the side chain. With the development of high-yield solid phase synthesis procedures, these building blocks and amino acids offer a vast resource for exploring chemical space in terms of functionality and chirality. Additionally, there is also a huge repository of peptides isolated from natural sources offering a diverse range of biological activities. These easily accessible sources of diverse molecules offer the chance of designing peptide libraries suitable for several kinds of screening approaches, including high-throughput screening (HTS) of large compound libraries or a more focused screening, the latter particularly examining changes in the interaction between a given peptide and receptor after introducing point mutations.

From: *Methods in Molecular Biology, vol. 494: Peptide-Based Drug Design*
Edited by: L. Otvos, DOI: 10.1007/978-1-59745-419-3_6, © Humana Press, New York, NY

To facilitate applications of biologically active peptides in drug design, it is crucial to have a thorough understanding of their conformations and their modes of interactions with biological receptors *(1)*. Nuclear magnetic resonance (NMR) spectroscopy is the preeminent technique for determining the structures of molecules in the solution state and has been widely used for studies of peptides. A wide range of NMR techniques can be used to give valuable information about peptides and their intermolecular interactions. This chapter provides an overview of the use of NMR in peptide-based drug design. It is aimed at providing both a basic level of understanding of how peptides are studied by NMR and how various NMR techniques applied to peptides are used in drug design. For additional information the reader is referred to a number of excellent articles on NMR to study the structure elucidation of peptides and proteins *(2–6)*, the interactions of ligands and their receptors *(7–16)*, screening approaches to lead discovery *(17–24)*, lead optimization *(25–27)*, and pharmacokinetic profiling *(28,29)*. Several books also address the topic of NMR in drug design *(30–32)*.

2. Drug Design and the Role of NMR

Strategies for the discovery and development of new chemical entities serving as lead compounds or drugs can roughly be divided in two major categories: the rational (design-based) and the screening-based approach. Rational drug design can further be subdivided into two strategies: receptor-based design and ligand-based design. Generally both major drug-discovery approaches can be facilitated by a range of NMR methods.

The screening approach to drug discovery involves finding a lead structure from either natural sources or large compound repositories by screening for binding or biological activity against the target protein of interest. Most recently, this approach has been augmented by the use of fragment-like compounds in the screening library *(19)* exemplified by the so-called structure-activity relationships (SAR)-by-NMR method *(33)*. In the field of rational or structure-based design, the ligand-based approach utilizes information on the ligand conformation or its binding interaction mode with the target protein. Typical information retrieved on the ligand interaction ranges from determining the binding specificity, the thermodynamics of the interaction, the bound conformation, to identifying the binding face of the ligand. The receptor-based design strategy employs knowledge of the receptor structure and the localization of the binding site to design a ligand to specifically interact with these features. Both the three-dimensional (3D) structure of the protein and the residues involved in the binding process can be identified by NMR methods, but in contrast to NMR experiments based on the observation of ligand resonances, these techniques usually require isotopically labeled protein *(19)*. Furthermore, NMR is limited to the extent that it is not usually able to provide high-resolution structures for macromolecules or complexes larger than approx 100 kDa, the size of many typical receptors. Nevertheless, lesser degrees of information can in many cases be extremely helpful—for example, the conformation of a bound ligand, or even just knowledge of which residues on a ligand are involved in binding; i.e., the binding epitope are extremely useful.

3. NMR Parameters

NMR is perhaps unique among the spectroscopic methods in that it has an extremely large range of parameters that can be measured, and many of these give valuable information about peptides and their molecular interactions. We provide a brief summary of these parameters in this section, focusing on those most useful in the drug-development process. Readers familiar with NMR can skip this section.

3.1. Chemical Shift

The chemical shift is a measure of the resonance frequency of a given nucleus and reflects the local electronic environment. It is generally given the symbol δ and expressed in ppm relative to an internal or external standard. Typical reference standards used in peptide ^{1}H NMR in aqueous solution are TSP (3-trimethylsilylpropionate) or DSS (2,2-dimethyl-2-silapentane-5-sulfonate) *(34)*. The chemical shift of the αH signals of constituent amino acids is of particular value in studies of peptides. Specifically, the difference in chemical shift of αH for a particular amino acid in a given peptide and the average chemical shift for αH of the same amino acid in a random coil peptide is referred to as secondary shift. Secondary shifts are frequently used as quick yet robust ways to determine the secondary structure of peptides in solution *(35)*. Downfield secondary shifts correlate with β-structure and upfield shifts with helical structure. The chemical shift index (CSI) *(35,36)* is effectively a "digital" version of the secondary shift, given a value of -1, 1, or 0, depending on whether the shift is respectively upfield, downfield, or within 0.1 ppm of the random coil value.

The change in the chemical environment of a nucleus at the binding interface of a ligand–protein complex upon binding is likely to induce a change in chemical shift. This effect is used for example as a readout in the SAR by NMR approach to detect binding and thus identify lead molecules *(33)*. Heteronuclear chemical shifts (particularly ^{13}C and ^{15}N) are widely used in such experiments and are detected via HSQC-type (heteronuclear single quantum correlation) experiments.

3.2. Coupling Constants

Nuclei "connected" to each other via a limited number (≤ 4) of bonds are subject to scalar coupling interactions that cause a splitting in their resonance signals. Depending on the number of coupling partners, complex splitting patterns may occur. Coupling constants are reported in Hertz (Hz) and are independent of the magnetic field applied. They are valuable in peptide NMR for two main reasons. First, the existence of a coupling implies a proximal through bond connectivity—a condition that leads to cross peak in COSY or TOCSY spectra and is used in the assignment and structure elucidation processes. Second, the size of the observed coupling constant for vicinal nuclei depends on the dihedral angle between the nuclei involved and therefore yields angular information that is valuable in the structure-determination process *(37)*.

3.3. The Nuclear Overhauser Effect (NOE)

The nuclear Overhauser effect (NOE) is manifest as an intensity change in a one-dimensional (1D) spectrum or a cross peak in two-dimensional (2D) NOESY spectrum that reflects a through space dipolar coupling interaction between nuclei. The size of the NOE is proportional to the reciprocal of the distance of the two nuclei to the power of six and is a function of the correlation time τ_c of the molecule, as indicated in Eq. 1:

$$\text{NOE} \propto \frac{1}{r^6} f(\tau_c) \tag{1}$$

Based on this equation, the NOE gives a measure of the distance between two nuclei, typically up to a maximum observable distance of about 5 Å. The distances determined for proton pairs in a peptide can be used as restraints in simulated annealing approaches for the calculation of the 3D structure of the peptide *(38–40)*.

3.4. Relaxation Processes

NMR relies on the excitation of nuclear spins by radiofrequency (RF) pulses, after which they return to the ground state via one of several relaxation mechanisms. NMR relaxation times are a measure of the efficiency of these mechanisms. The spin-lattice relaxation time T_1 reflects return to the ground state by exchange of energy with the surroundings (longitudinal relaxation), while T_2 is a measure of the efficiency of exchange energy with other nuclei (transverse relaxation). For large proteins transverse relaxation is predominant, and since T_2 is proportional to the reciprocal to the line width of a resonance signal, with increasing protein size the peaks in NMR spectra become broader, ultimately limiting the size of molecules that can be studied. Nevertheless, T_2 effects can be used to detect binding events, since the relaxation properties of ligands are different from macromolecular targets. Upon binding, ligand resonances are broadened compared to the free state *(41)*.

^{15}N and ^{13}C labeling of peptides facilitates the study of their molecular dynamics in solution by measurements of relaxation parameters *(42,43)*. Heteronuclear relaxation times and heteronuclear NOEs are predominantly affected by the dipole–dipole interaction of the heteronucleus with the directly attached proton. Since the internuclear (i.e., chemical bonding) distances are known from the molecular geometry, correlation times for overall and internal motions can be determined.

3.5. Signal Intensities

The intensity of a signal in an NMR spectrum is proportional to the number of nuclei contributing to that signal. Intensity information is thus used implicitly during the spectral assignment process, but is also valuable when titration experiments are carried out— i.e., it is useful to determine the mole ratios of species present when studying protein–ligand interactions. In yet another application the intensity of signals from amide protons can be used to monitor exchange rates when the peptide is transferred from H_2O to D_2O and thereby deduce information about internal hydrogen bonding. Attenuation or loss of the amide proton resonance is an indicator of a solvent accessible amide, while protons that are not accessible to the deuterated solvent are not exchanged *(44–48)*. Signal intensities are also employed in the analysis of saturation transfer difference NMR spectra (**Subheading 4.4.2.**).

4. NMR Techniques

4.1. Overview

Having introduced the key NMR parameters, we now describe in more detail how NMR can contribute to the drug-design/-development process. **Figure 1** summarizes the ways in which various types of NMR experiments can contribute. We have arranged the diagram to reflect three elements: namely the peptide, the receptor, and their binding interaction.

The simplest experiments are those that focus just on the ligand. These are typically used to determine solution conformations or 3D structures of ligands. Homonuclear ^{1}H 1D or 2D NMR experiments are used mainly here. At the other end of the scale, experiments to study the macromolecular binding partner often require labeled protein and multidimensional NMR methods, as indicated on the right-hand side of **Fig. 1**. Finally, many NMR experiments provide information

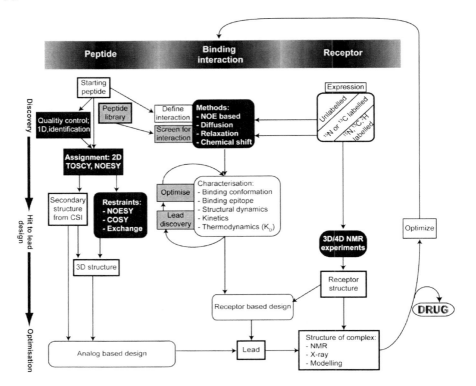

Fig. 1. Overview of the role of NMR in drug design and development. NMR methods are indicated in dark boxes. Screening-based approaches are shaded grey.

on binding events. These include the screening of compound libraries, determination of binding thermodynamics, determination of the binding mode of the ligand, and determination of the binding site on the receptor *(11)*. Importantly, some NMR experiments are able to detect binding of molecules to membrane-bound receptors *(49–51)*, reconstituted in vesicles or embedded on the surface of living cells. This capability gives NMR-based studies a significant advantage over x-ray studies in these cases, since membrane bound receptors are not easily accessible by x-ray crystallography.

In **Fig. 1** we have highlighted with a dark background the different types of NMR methods that are used in drug-discovery projects. These include basic 1D and 2D methods that are used to confirm the identity of peptides, determine their conformation, or derive restraint information used in 3D structure calculations (left side of **Fig. 1**). Methods to study binding interactions (middle section of **Fig. 1**) can be broadly categorized as being based on NOEs, diffusion, relaxation, or chemical shift changes. NOE-based methods include the transferred NOE and STD NMR experiments, described in more detail in **Subheadings 4.4.1.** and **4.4.2.** The relaxation behavior of nuclei changes when a ligand binds to a receptor, allowing insight to the dynamics of the complex, described in **Subheading 4.5.** Similarly, diffusion NMR experiments rely on a change in the diffusion coefficient upon binding of a smaller molecule to a significantly larger molecule, described in **Subheading 4.6.** Finally, the chemical environment of nuclei involved in the binding interface changes on binding, resulting in changes in chemical shifts. As noted earlier, $^1H/^{15}N$ or $^1H/^{13}C$ HSQC experiments are useful in the determination of binding sites based on chemical shift changes and are widely used in screening studies *(33,52)*. They will be described in **Subheading 4.3.** We now examine these various categories of experiments in more detail.

4.2. 1D/2D NMR of Peptides

1D NMR experiments are often used for quality-control purposes and can readily be used to confirm the purity of peptides. Simple 1D spectra are also often used to determine whether a peptide is structured or unstructured or whether aggregation is present. Dispersion of the amide chemical shifts is an indicator of the former, whereas a narrow distribution, in the range of 7.5–8.5 ppm, is characteristic of unstructured peptides. Aggregation leads to broad peaks and spectra of poor quality. Adjustment of conditions by varying pH, buffer, cosolvent (e.g., acetonitrile for hydrophobic peptides), or peptide concentration and monitoring the effects on 1D spectra is often used to find optimum conditions.

Although 1D experiments are useful as a first step in the analysis of peptides, 2D NMR is really the workhorse technique of peptide NMR, with DQF-COSY, TOCSY, and NOESY being the main techniques used for homonuclear studies and HSQC for heteronuclear studies.

4.2.1. Assignment Methods

Since the assignment of spectral peaks to specific nuclei in a peptide is an essential prerequisite to utilizing the information available in NMR spectra, we discuss it here before examining techniques relevant in drug development. Strategies for assigning homonuclear ^1H NMR spectra of peptides were developed in the mid-1980s and have been widely applied since *(6)*. For small peptides they typically involve the use of 2D NMR spectroscopy. As the number of residues in a peptide increases beyond around 50, it becomes increasingly difficult to use 2D homonuclear techniques and techniques that involve isotopic enrichment are generally used.

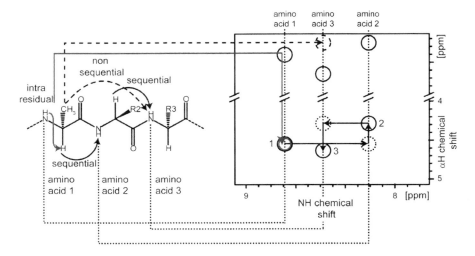

Fig. 2. Schematic representation of NOESY cross peaks used in peptide assignments.

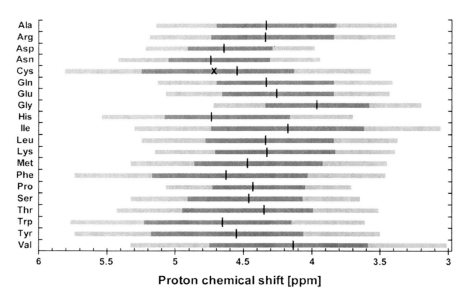

Fig. 3. Range of chemical shifts for αH in amino acids. The bars are centered at the average shift value for the respective amino acid (BioMagResBank [http://www. bmrb.wisc.edu/] *(54)*). The dark grey bars represent the 68% and the light grey bars the 95% confidence interval. The values for diamagnetic proteins were taken from the BioMagResBank in mid-February 2007. The vertical black lines highlight the random coil shift. The x marks the random coil shift for oxidized cysteine.

The basis of the sequential assignment approach is to use a combination of TOCSY/COSY spectra to assign peaks to residue type and then link those residue types sequentially together using NOESY spectra *(53)*. A schematic representation of some of the key NOESY connectivities used in assigning the spectra of peptides is given in **Fig. 2**.

It is worth noting here that despite an extensive database of chemical shifts being available for peptides and proteins (BioMagResBank (http://www.bmrb.wisc.edu/) *(54)*), chemical shifts by themselves cannot be used to make spectral assignments. This is illustrated in **Fig. 3**, which shows that the overlap in the range of chemical shifts seen for the αH protons in the 20 natural amino acids among known proteins is too large to be of diagnostic value.

Once the NMR spectra of a peptide are assigned the range of techniques mentioned earlier can be used to retrieve information that is valuable in the drug-design process. Clearly the 3D structure of the peptide is of vital importance, and we examine it first. Indeed, as is illustrated in **Fig. 4**, the assignment and structure determination processes are often completed together.

4.2.2. Peptide Structure Determination

Figure 4 shows the general workflow required for determining the structure of a small peptide by NMR, including the assignment step. Typically, structure determination requires the recording of 6–12 2D spectra under various conditions of temperature, pH, or spectral parameters such as mixing times. As is implied by the figure, temperature variation is often useful to remove ambiguities due to overlap, as the amide chemical shifts are differentially sensitive to temperature. Exchange of the sample into D_2O provides information about hydrogen-bonding networks.

It is convenient to illustrate the principles involved in the assignment and structure determination process by reference to a structure of a small bioactive peptide solved recently in our laboratory. **Figure 5** shows the sequential connectivity diagram (an overlay of 2D NOESY and TOCSY spectra) for Vc1.1, a 16-residue conotoxin-based peptide that is currently in clinical trials as a treatment for neuropathic pain *(55–57)*. In general, completion of this sequential walk, together with assignment of the side-chain protons, provides a basis for assessing whether a full 3D structure determination is warranted. At this stage it is also important to check the spectra for evidence of multiple conformations, which can occur due to the presence of *cis-trans* isomerization near proline residues. In the case of Vc1.1 the data are of good quality, and a full 3D structure was determined *(58)*.

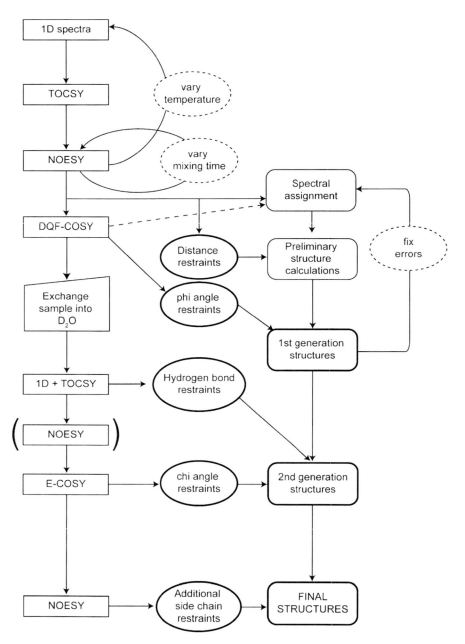

Fig. 4. Flowchart of assignment/structure determination process in peptide NMR.

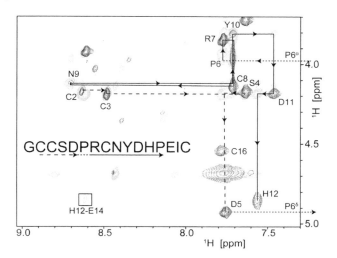

Fig. 5. Overlay of TOCSY (dark) and NOESY (gray) spectra of Vc1.1. The peaks for the αH chemical shift of the residues are labeled. The dashed lines and arrows show the sequential walk from Cys2 to Asp5. Here the sequential walk stops since the next contact is not in this section of the spectrum (dotted lines). The sequence is continuing with the peak for αH of Pro6 in the NH trace of Arg7. The solid lines and arrows show the sequential walk through the spectrum from Arg7 to His12. The sequence of Vc1.1 is given in the figure with the residues underlined in the respective style.

After the assignment process is complete, various types of NMR data can be used to build up a picture of the secondary structure of the peptide (based on coupling constants, chemical shifts, and NOEs) and then, if required, a full 3D structure can be determined. **Figure 6** shows some of this information for Vc1.1. As noted earlier, deviations from random coil chemical shifts provide an indication of secondary structure, and in the case of Vc1.1. a region of negative αH secondary shifts provides a first indication that the helix typically present in α-conotoxins is indeed present in Vc1.1 (**Fig. 6A**). This conclusion is supported by the CSI, coupling, and NOE data in **Fig. 6B**. Finally, the 3D structure is shown in various representations in the lower panels of the figure *(58)*.

Note that NMR structures are generally represented in three ways: an ensemble of the 20 lowest energy structures, usually drawn in stereo-view mode (**Fig. 6 C**), a cartoon representation of the lowest energy structure to highlight any secondary structure elements or disulfide bonds (**Fig. 6 D**), and a space-filling model to highlight any features that might have a bearing on biological activity (**Fig. 6 E**).

We now turn to an examination of a range of NMR methods for studying intermolecular interactions relevant in the drug-design process.

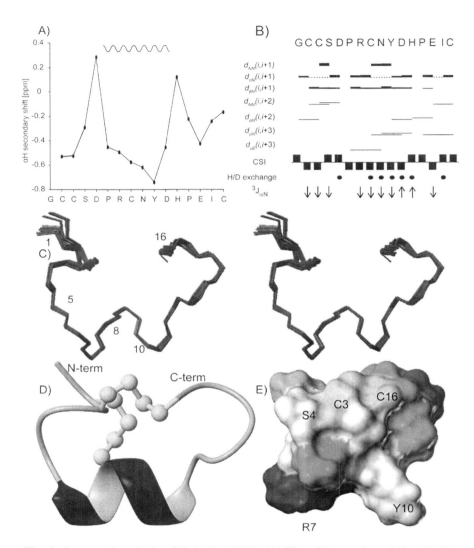

Fig. 6. Structural analysis of Vc1.1 by NMR. (**A**) The αH secondary shifts. (**B**) Structural information from NOEs, chemical shift index (CSI), amide exchange, and scalar coupling, are summarized (adapted from **ref. 58**). (**C**) A stereoview of the 20 low-energy structures. Some residues are numbered for orientation. (**D**) "Ribbon" representation of the peptide: the cysteine residues are in yellow. The N-terminus and C-terminus are marked N and C, respectively. (**E**) The surface of the peptide with selected residues labeled.

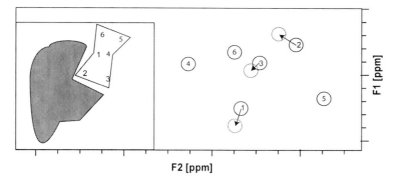

Fig. 7. Schematic representation of chemical shift changes. The scheme in the box represents the binding event observed. The unfilled shape represents a labeled substance, and three of its residues are in contact with a binding partner (grey). The solid line circles represent cross peaks in a HSQC spectrum of the unbound labeled compound. The F2 domain displays the ¹H chemical shift, the F1 domain either ¹³C or ¹⁵N chemical shift. The chemical shifts of the three positions in contact with the binding partner change upon binding, as indicated by the arrows and the new cross peaks shown in dotted lines.

4.3. Chemical Shift–Based Methods

A change from the free state to the bound state results in changes in the chemical environment of nuclei involved in binding interfaces, causing changes of their respective chemical shifts. This effect is commonly used to detect binding of molecules to their receptors, and ¹H/¹⁵N or ¹H/¹³C HSQC experiments are particularly useful for these studies. A schematic representation of the changes in an HSQC spectrum on binding is given in **Fig. 7**. Residues proximal to the binding interface are affected, while those more distant are not.

For the ¹H/¹⁵N experiment, uniformly labeled substance is obtainable at a reasonable cost, but uniform ¹³C labeled proteins are more expensive. However, in some cases selective labeling of the methyl groups in valine, leucine, or isoleucine residues of a protein proves sufficient for screening purposes *(59)*. The employment of HSQC experiments to detect binding is especially well exemplified in the technique termed SAR by NMR *(33)*.

4.4. NOE-Based Experiments

NOEs are a particularly powerful NMR tool, and several NOE-based methods are applied in drug design. It should be noted NOEs are dependent on correlation times and can change from positive, to zero, to negative values depending on the size of the peptide being studied. The correlation time at which a molecule displays negative NOEs depends on the temperature and the field strength of the spectrometer used. As a guideline, peptides around 1 kDa display only weak or no NOEs on a 500 MHz spectrometer, and ROESY experiments *(60)* may be necessary for peptides of this size. Notwithstanding this complication, NOEs are extremely useful.

4.4.1. Transferred NOE

In this method the different correlation times and NOE build-up rates of free and bound states of a ligand are used to detect NOEs on the free ligand that reflect its bound conformation. This class of experiment usually is conducted by performing transient NOE experiments by means of 2D NOESY spectra. The basis of the technique is that small molecules show positive NOEs while macromolecules show negative NOEs and that NOEs build up/decay quickly in macromolecules but slowly in small molecules. Upon binding of a small molecule to a macromolecule, the motional characteristics of the small molecule change and it shows negative NOEs. After dissociation from the complex, information on the negative NOE is transferred back to the free ligand, which is usually in large excess and readily detectable *(12)*.

Besides intramolecular transferred NOEs, intermolecular transferred NOEs are of interest when studying molecular interactions *(61,62)*. In this case NOEs build up between protons of the ligand during its residence time in the binding pocket. This gives information on the orientation and conformation of the ligand in the binding site. The transferred NOE approach for structural characterisation of the bound conformation of a ligand is schematically illustrated in **Fig. 8**

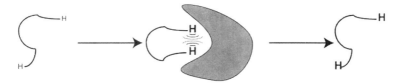

Fig. 8. Schematic representation of the trNOE effect. On the left the conformation of the unbound ligand does not allow the buildup of a NOE because the two protons are too far away. In the middle representation the bound conformation orients the two protons at a distance that allows the buildup of a NOE. After dissociation of the complex the information of spatial proximity of the two protons in the bound conformation is transferred back into the solution conformation (right), where the NOE still is observable.

4.4.2. Saturation Transfer Difference (STD) NMR

The STD NMR experiment employs a train of low-power pulses to selectively saturate proton resonances of a receptor *(63)*. If the receptor exceeds ~10 kDa, it is usually subject to spin diffusion, which causes the saturation to spread uniformly over all resonances of the receptor. The saturation can then be transferred to the parts of a small molecule directly bound to the receptor, as illustrated in **Fig. 9**. The saturation of the bound molecule's proton resonances near the binding site is detected after dissociation of the ligand into solution as attenuation of the respective signals.

A typical STD NMR spectrum is achieved by subtracting the so-called on-resonance spectrum from the off-resonance spectrum, where no saturation is applied to the protein. The latter spectrum is usually acquired by applying saturation pulses far outside the spectral window of the receptor to maintain relaxation and thermal equality for the experiments. To average out effects occurring over time, the two spectra are usually recorded in an interleaved fashion. The generation of the difference spectrum is highlighted in **Fig. 10**. STD NMR experiments can be used to determine both the orientation of a ligand in a binding site and the dissociation constant of a ligand-receptor complex via titration experiments. STD NMR can be performed as a 1D experiment or as 2D correlation spectra if more dispersion is required *(63,64)*.

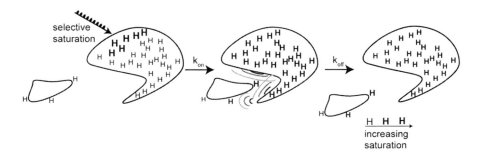

Fig. 9. Schematic representation of the saturation transfer from receptor to a bound ligand. The protons of the receptor are saturated by a train of selective pulses. Upon binding (middle) the saturation is transferred to protons of the ligand. The different emphasis on the ligand protons represents the effectiveness of the transfer depending on proximity of ligand protons to protons on the receptor surface. This information is transferred back into solution upon dissociation of the complex.

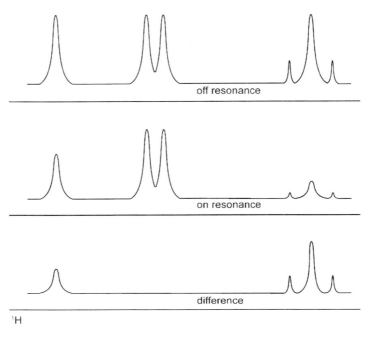

¹H

Fig. 10. Schematic representation of the creation of the STD or difference spectrum. In the off resonance or reference spectrum, all protons have normal intensity. Depending on the vicinity of the respective proton to the receptor surface, the resonance signal intensity is attenuated in the on resonance spectrum. The difference spectrum is created by subtraction of the on resonance from the off resonance spectrum and displays only resonances of signals of protons in vicinity to the receptor. This gives indications on the binding mode and can be employed for screening purposes, as nonbinding molecules do not display and intensities in the difference spectrum.

4.5. Determination of Spin-Lattice and Spin-Spin Relaxation Times

Relaxation parameters provide valuable information about molecular motions. The spin-lattice relaxation time T_1 is usually determined by the so-called inversion recovery pulse sequence (65). The experiment comprises a set of spectra with different interpulse delays, and T_1 is determined by fitting the signal intensities for a given nucleus to Eq. 2, where A and B are constants, τ is the respective interpulse delay, and I_t is the intensity measured at that delay:

$$I_t = A + Be^{-\tau/T1} \tag{2}$$

Spin-spin relaxation times T_2 are determined by the Carr-Purcell-Meiboom-Gill spin echo pulse sequence and provide information about slower molecular motions (66,67).

4.6. Diffusion NMR

The development of gradient NMR spectroscopy led to the development of methods for determining translational diffusion coefficients in solution *(68)*. In turn, these provide valuable information about various intermolecular interactions, including self-association phenomena and ligand–protein binding. The diffusion NMR experiment consists of a series of spectra acquired with variable gradient strengths, the diffusion constant being determined from the slope of a plot of $\ln(I_G/I_0)$ versus G^2, where I_G and I_0 are the signal intensities at a given gradient strength or no gradient, respectively, and G is the gradient strength.

4.7. 3D and 4D Methods

For peptides larger than approx 80–100 amino acids, 2D methods become increasingly cumbersome, and larger proteins are assigned using 3D or four-dimensional (4D) methods that require the protein to be uniformly enriched with ^{13}C, ^{15}N, or both. These techniques will not be described here because we rather focus on peptide NMR, but the reader is referred to excellent articles on this topic *(69–72)*.

5. Examples of NMR Applications in Peptide-Based Drug Design

In this section we describe some applications of the various NMR techniques to peptides and their binding partners.

5.1. TrNOESY Experiments

Several groups have demonstrated the use of trNOESY experiments to determine the structure of peptides bound to their receptor. For example, Kisselev et al. determined the 3D solution structure of the transductin α-subunit bound to light-activated rhodopsin *(51)*. They found that the peptide IKENLKDCGLF formed an α-helix terminated by an open reverse turn **(Fig. 11)**, while in contrast the conformation of the peptide remained disordered when in contact with nonactivated rhodopsin. Their findings led to the development of derivatives of the peptide that maintained the binding conformation and had improved affinity *(73)*. Importantly, in their studies they used rhodopsin in a membrane environment (extracted from bovine rods), emphasizing the capabilities of the method for the study of integral membrane receptor directed inhibitors in their natural environment.

In the course of understanding the molecular and structural basis of HIV infection, the structural features of the HIV envelope glycoprotein gp120 have

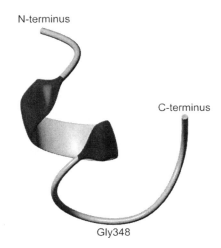

Fig. 11. Structure of IKENLKDCGLF bound to rhodopsin (PDB code: 1aqg). The backbone structure is shown in ribbon representation. The peptide forms an α-helical N-terminus that is terminated by a turn motif. The turn motif is centered on Gly348.

attracted substantial interest. Tugarinov et al. determined the 3D structure of a peptide, taken from the V3 loop the HIV-1$_{IIIB}$ virus strain, bound to a strain-specific antibody *(74)*, as indicated in **Fig. 12**. The bound 24-residue peptide forms two antiparallel β-strands consisting of residues Lys5-Ile9 and Arg16-Thr20. Five residues linking the two strands are forming a β-turn motif centered

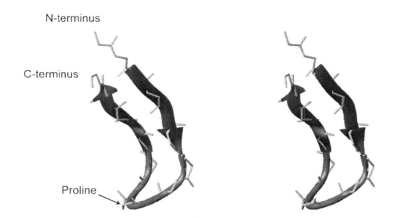

Fig. 12. Structure of an HIV V3 peptide (NNTRKSIRIQRGPGRAFVTIGKIG) in contact with a specific antibody (PDB code 1b03). The peptide forms two antiparallel β-strands with the turn centered at the GPG motif.

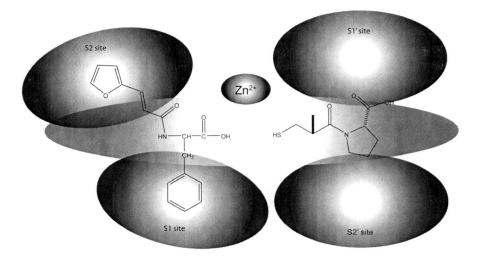

Fig. 13. Proposed distribution of furylacryloyl phenylalanine (left) and captopril (right) between the binding sites of ACE, as determined by trNOESY experiments. (Adapted from **ref. 76**).

at *cis* proline at position 13. The turn is oriented at ~90° in respect to the plane of the β-strands.

The use of trNOESY experiments for screening purposes was first demonstrated on carbohydrates binding to a lectin, an interaction that usually is of medium to low affinity *(75)*. Meyer et al. were able to identify a binding carbohydrate from two carbohydrate libraries. In another experiment Mayer et al. employed trNOESY experiments to determine the affinity and subsite selectivity of derivatives of dipeptides and amino acids binding to the angiotensin-converting enzyme (ACE). In their study they identified the highest-affinity ligand from a small library of compounds. Using mixtures of amino acid derivatives with and without additional dipeptides or one derivatized amino acid and captopril, they were also able to determine the distribution of the peptides between the different binding sites in the active site of ACE *(76)*. Their proposal for the distribution of furylacryloyl-phenylalanine and captopril between the binding sites of ACE is shown in **Fig. 13**.

5.2. Cyclic RGD Peptides Binding to Integrin Studied by STD NMR

The RGD motif has long been known to be the principal binding motif of several integrins *(77)*. Aumailly et al. showed that the cyclic peptide c(RGDfV), where f denotes D-phenylalanine (D-Phe), is a potent inhibitor of the integrin $\alpha_{IIb}\beta_3$ and determined the three-dimensional structure of the cyclic peptide from

NOESY experiments *(78)*. Meinecke et al. determined the thermodynamics and the binding epitope of c(RGDfV) bound to $\alpha_{IIb}\beta_3$, reconstituted in a lipid bilayer by STD NMR *(50)*. They showed that the protons of the aromatic ring of D-Phe received the most saturation, indicating close contact with the receptor. Interactions of medium strength were determined for the β-position of D-Phe, arginine (a, β, and γ-positions), glycine, the β-position of Asp, and the γ-position of Val.

Using the saturation transfer double difference technique, Claasen et al. showed that c(RGDfV) binds to the integrin $\alpha_{IIb}\beta_3$ with even higher affinity, when the protein is embedded in the surface of living human trombocytes *(49)*. The applicability of STD NMR to the investigation of integral membrane proteins in their natural cellular environment makes it a very useful tool for studying these receptors.

5.3. Diffusion NMR for the Study of Peptide-Membrane Interactions

Diffusion NMR was recently used to study the interaction of bradykinin with micelles formed from ganglioside monosialyated-1 (GM1), a glycosphingolipid *(79)*. Gangliosides make up a significant fraction of lipids in the brain. The change in the peptide fold when partitioning from the aqueous to the membrane phase and the structure of bradykinin in the presence of GM1 was solved by NMR. From NOEs the formation of a β-turn–like structure for residues Ser6 to Arg9 in the presence of the micelle was deduced. STD NMR experiments were performed to directly detect the association of bradykinin with the micelle. Additionally, the proportion of peptide bound to the micelle was investigated by diffusion NMR. The free peptide showed a diffusion coefficient of 3.5×10^{-10} m^2/s at 300 K. At the same temperature and pH in the presence of GM1 micelles, the diffusion coefficient decreased to 2.8×10^{-10} m^2/s, a sure sign of incorporation of the peptide into the micelle. From their findings Chatterjee and Mukhopadyay deduced that approximately 30% of the peptide bound to the micelles with an association constant $K_{eq} = 0.19 \times 10^3$ M^{-1}. It was concluded that bradykinin needs to be incorporated into the GM1 micelle to adopt an ordered structure.

5.4. ^{15}N Relaxation Methods to Determine the Dynamics of a Peptide Bound to an RNA Hairpin

Stuart et al. employed relaxation rate measurements of ^{15}N in the backbone of a uniformly ^{15}N-labeled peptide representing the N-terminal 49 residues of the Nun protein from lambdoid phage HK002 to determine the dynamics of the peptide when bound to a 17-residue RNA hairpin (BoxB17) *(80)*. Nun protein prevents λ-phage superinfection by blocking gene expression. They found residues Ser24 to K42 formed an α-helix when in complex with the RNA construct, as seen from secondary shifts, while the free Nun protein was unstructured. The N-terminal residues did not show secondary shifts or homonuclear NOEs to indicate the presence of a stable secondary structure on the μs-ms time scale. However, small heteronuclear ^{1}H-^{15}N NOEs and ^{15}N relaxation rates ranging from 3.5 to 5.3 s^{-1} for these residues indicated the presence of a limited set of backbone conformations in the ps-ns time scale. From increased relaxation rates of $R_2 > 11$ s^{-1} for residues Ser24, Arg25, Asp26, Ala31, Trp33, Ile37, and Ala38, all part of the α-helix, the presence of chemical exchange in the μs-ms time scale was deduced. The dissociation of the complex is slow on the chemical shift time scale, and therefore these relaxation rates must be a result of intramolecular conformational changes in the complex. Stuart et al. assumed that restricted flexibility of the N-terminus may play a role in the formation of larger molecular assemblies involved in transcription termination. These results show that NMR can reveal the presence of preferred conformations, even if secondary structural motifs are not present in the peptide.

5.5. Chemical Shifts

5.5.1. Description of a Peptide–Protein Complex

Using isotopically labeled peptides, Stamos et al. determined the structure of the high-affinity IgE receptor FCεRI in complex with a β-hairpin IGE32 peptide and helical zeta peptide e117. The study is a nice example of the use of NMR techniques for analyzing peptide–receptor complexes *(81)*. For the bound state Stamos et al. found dramatic changes in the ^{1}H chemical shifts of Pro16 in the zeta peptide, ranging from 1.7 to 3.1 ppm. Shift changes of this magnitude are a result of close vicinity to aromatic side chains in the receptor, in a sandwich-like packing. This finding was in agreement with an x-ray structure of FCεRIα in complex with an Fc fragment of IgE *(82)* where Pro426 of IgE is placed between Trp87 and Trp110 of FCεRIα. Combining mutagenesis data and intermolecular NOEs, it was confirmed that the zeta peptide bound in site 2 of FCεRI. These data were used to model the complex of FCεRI and e117 zeta peptide on the basis of the crystal structure of FCεRIα and zeta peptide e131 *(83)*, revealing high similarity between the structures. Using the same NMR experimental approach, the structure of the hairpin peptide–receptor complex was determined and shown to retain the β-hairpin structure upon binding. Intermolecular NOEs were used to orient the ligand in binding site 2 of the receptor with Pro9 of the peptide packed between Trp87 and Trp110, indicated as before by dramatic shift changes of the proline residue.

5.5.2. Discovery of Small Molecule Ligands Inhibiting Protein–Peptide Interactions Through SAR-by-NMR

The SAR-by-NMR *(33)* strategy has not yet been widely applied in screening peptides despite that fact that numerous publications demonstrate the use of this technique in the discovery of nonpeptide compounds as inhibitors of protein–protein interactions. Petros et al. demonstrated the application of SAR-by-NMR to discover small molecule inhibitors for interaction of the anti-apoptotic protein Bcl-x_L with a peptide derived from its endogenous binding partner Bad *(52)*. The fluorescein-labeled peptide NLWAAQRYGRELRRMSDKFVD showed a dissociation constant (K_d) of 20 nM. SAR-by-NMR screening of a 10,000 compound library identified a fluorobiaryl acid compound. Using an excess of this compound in the screening mixture, a second molecule was identified, binding in a site in close vicinity to the first ligand. A linking strategy was devised and a molecule featuring a biphenyl system linked to the biaryl acid via a trans-olefin linker was developed, exhibiting an IC_{50} of 1.4 μM. Optimization of the lead compound yielded a small molecule inhibitor of the protein–protein interaction with an IC_{50} of 36 nM.

6. Conclusions

NMR has proven to be a versatile and robust method for the study of biomolecular interactions involving peptides. The information gained from NMR experiments ranges from the identification of binding partners to determination of conformations in the bound state and characterization of the binding interface between ligand and receptor. This information is extremely valuable for the design of new drugs from bioactive peptides.

References

1. Fry, D. and Sun, H. (2006) Utilizing peptide structures as keys for unlocking challenging targets. *Mini. Rev. Med. Chem.* **6**, 979–987.
2. Valente, A. P., Miyamoto, C. A. and Almeida, F. C. (2006) Implications of protein conformational diversity for binding and development of new biological active compounds. *Curr. Med. Chem.* **13**, 3697–3703.
3. Craik, D. J. and Clark, R. J. (2005) Structure-based drug design and NMR-based screening. In: Encyclopedia of Molecular Cell Biology and Molecular Medicine (Meyers R. A., ed.). 2nd ed. Wiley-VCH, Weinheim, pp. 517–605.
4. Fu, R. and Cross, T. A. (1999) Solid-state nuclear magnetic resonance investigation of protein and polypeptide structure. *Annu. Rev. Biophys. Biomol. Struct.* **28**, 235–268.
5. Guthrie, D. J. (1997) 1H nuclear magnetic resonance (NMR) in the elucidation of peptide structure. *Methods Mol. Biol.* **73**, 163–184.

6. Wuthrich, K. (1986) *NMR of Proteins and Nucleic Acids.* Wiley Interscience, New York.
7. Sillerud, L. O. and Larson, R. S. (2006) Nuclear magnetic resonance-based screening methods for drug discovery. *Methods Mol. Biol.* **316**, 227–289.
8. Pellecchia, M. (2005) Solution nuclear magnetic resonance spectroscopy techniques for probing intermolecular interactions. *Chem. Biol.* **12**, 961–971.
9. Homans, S. W. (2005) Probing the binding entropy of ligand-protein interactions by NMR. *Chembiochem* **6**, 1585–1591.
10. Homans, S. W. (2004) NMR spectroscopy tools for structure-aided drug design. *Angew. Chem. Int. Ed. Engl.* **43**, 290–300.
11. Salvatella, X. and Giralt, E. (2003) NMR-based methods and strategies for drug discovery. *Chem. Soc. Rev.* **32**, 365–372.
12. Meyer, B. and Peters, T. (2003) NMR spectroscopy techniques for screening and identifying ligand binding to protein receptors. *Angew. Chem. Int. Ed.* **42**, 864–890.
13. Fejzo, J., Lepre, C. and Xie, X. (2003) Application of NMR screening in drug discovery. *Curr. Top. Med. Chem.* **3**, 81–97.
14. Coles, M., Heller, M. and Kessler, H. (2003) NMR-based screening technologies. *Drug Discov. Today* **8**, 803–810.
15. Pellecchia, M., Sem, D. S. and Wuthrich, K. (2002) NMR in drug discovery. *Nat. Rev. Drug. Discov.* **1**, 211–219.
16. Hajduk, P. J., Meadows, R. P. and Fesik, S. W. (1999) NMR-based screening in drug discovery. *Q. Rev. Biophys.* **32**, 211–240.
17. Zartler, E. R. and Shapiro, M. J. (2006) Protein NMR-based screening in drug discovery. *Curr. Pharm. Des.* **12**, 3963–3972.
18. Leone, M., Freeze, H. H., Chan, C. S. and Pellecchia, M. (2006) The Nuclear Overhauser Effect in the lead identification process. *Curr. Drug Discov. Technol.* **3**, 91–100.
19. Zartler, E. R. and Shapiro, M. J. (2005) Fragonomics: fragment-based drug discovery. *Curr. Opin. Chem. Biol.* **9**, 366–370.
20. Schade, M. and Oschkinat, H. (2005) NMR fragment screening: tackling protein-protein interaction targets. *Curr. Opin. Drug Discov. Devel.* **8**, 365–373.
21. Lepre, C. A., Moore, J. M. and Peng, J. W. (2004) Theory and applications of NMR-based screening in pharmaceutical research. *Chem. Rev.* **104**, 3641–3676.
22. Vogtherr, M. and Fiebig, K. (2003) NMR-based screening methods for lead discovery. In: *Modern Methods of Drug Discovery* (Hillisch, A. and Hilgenfeld, R., ed.). Birhäuser Verlag, Switzerland, pp. 183–202.
23. Jahnke, W., Florsheimer, A., Blommers, M. J., et al. (2003) Second-site NMR screening and linker design. *Curr. Top. Med. Chem.* **3**, 69–80.
24. Wyss, D. F., McCoy, M. A. and Senior, M. M. (2002) NMR-based approaches for lead discovery. *Curr. Opin. Drug. Discov. Devel.* **5**, 630–647.
25. Villar, H. O., Yan, J. and Hansen, M. R. (2004) Using NMR for ligand discovery and optimization. *Curr. Opin. Chem. Biol.* **8**, 387–391.
26. Pellecchia, M., Becattini, B., Crowell, K. J., et al. (2004) NMR-based techniques in the hit identification and optimisation processes. *Expert Opin. Ther. Targets* **8**, 597–611.

27. Lepre, C. A., Peng, J., Fejzo, J., et al. (2002) Applications of SHAPES screening in drug discovery. *Comb. Chem. High Throughput Screen* **5**, 583–590.

28. Sun, C. and Hajduk, P. J. (2006) Nuclear magnetic resonance in target profiling and compound file enhancement. *Curr. Opin. Drug Discov. Devel.* **9**, 463–470.

29. Sun, C., Huth, J. R. and Hajduk, P. J. (2005) NMR in pharmacokinetic and pharmacodynamic profiling. *Chembiochem* **6**, 1592–1600.

30. Holzgrabe, U., Wawer, I. and Diehl, B. (1999) *NMR Spectroscopy in Drug Development and Analysis.* Wiley-VCH, Weinheim.

31. Craik, D. J. (1996) *NMR in Drug Design.* CRC, New York.

32. Hansdschumacher, R. E. and Armitage, I. M. (1989) *NMR Methods for Elucidating Macromolecule-Ligand Interactions: An Approach to Drug Design.* Pergamon Press, Oxford.

33. Shuker, S. B., Hajduk, P. J., Meadows, R. P. and Fesik, S. W. (1996) Discovering high-affinity ligands for proteins: SAR by NMR. *Science* **274**, 1531–1534.

34. Reid, D. G., MacLachlan, L. K., Edwards, A. J., Hubbard, J. A. and Sweeney, P. J. (1997) Introduction to the NMR of proteins. In: *Protein NMR Techniques* (Reid D. G., ed.). Humana Press, Totowa, NJ, pp. 1–28.

35. Wishart, D. S., Sykes, B. D. and Richards, F. M. (1992) The chemical shift index: a fast and simple method for the assignment of protein secondary structure through NMR spectroscopy. *Biochemistry* **31**, 1647–1651.

36. Wishart, D. S., Bigam, C. G., Holm, A., Hodges, R. S. and Sykes, B. D. (1995) 1H, 13C and 15N random coil NMR chemical shifts of the common amino acids. I. Investigations of nearest-neighbor effects. *J. Biomol. NMR* **5**, 67–81.

37. Karplus, M. (1963) Vicinal proton coupling in nuclear magnetic resonance. *J. Am. Chem. Soc.* **85**, 2870–2871.

38. Lautz, J., Kessler, H., Blaney, J. M., Scheek, R. M. and Van Gunsteren, W. F. (1989) Calculating three-dimensional molecular structure from atom-atom distance information: cyclosporin A. *Int. J. Pept. Protein. Res.* **33**, 281–288.

39. Ottiger, M., Zerbe, O., Guntert, P. and Wuthrich, K. (1997) The NMR solution conformation of unligated human cyclophilin A. *J. Mol. Biol.* **272**, 64–81.

40. Weber, C., Wider, G., von Freyberg, B., et al. (1991) The NMR structure of cyclosporin A bound to cyclophilin in aqueous solution. *Biochemistry* **30**, 6563–6574.

41. Palmer, A. G. (2004) NMR Characterization of the dynamics of biomacromolecules. *Chem. Rev.* **104**, 3623–3640.

42. O'Sullivan, D. B., Jones, C. E., Abdelraheim, S. R., et al. (2007) NMR characterization of the pH 4 beta-intermediate of the prion protein: the N-terminal half of the protein remains unstructured and retains a high degree of flexibility. *Biochem. J.* **401**, 533–540.

43. Renisio, J. G., Perez, J., Czisch, M., et al. (2002) Solution structure and backbone dynamics of an antigen-free heavy chain variable domain (VHH) from Llama. *Proteins* **47**, 546–555.

44. Feng, L., Orlando, R. and Prestegard, J. H. (2006) Amide proton back-exchange in deuterated peptides: applications to MS and NMR analyses. *Anal. Chem.* **78**, 6885–6892.

45. Morris, K. F., Gao, X. and Wong, T. C. (2004) The interactions of the HIV gp41 fusion peptides with zwitterionic membrane mimics determined by NMR spectroscopy. *Biochim. Biophys. Acta* **1667**, 67–81.

46. D'Amelio, N., Bonvin, A. M., Czisch, M., Barker, P. and Kaptein, R. (2002) The C terminus of apocytochrome b562 undergoes fast motions and slow exchange among ordered conformations resembling the folded state. *Biochemistry* **41**, 5505–5514.

47. Liepinsh, E., Otting, G. and Wuthrich, K. (1992) NMR spectroscopy of hydroxyl protons in aqueous solutions of peptides and proteins. *J. Biomol. NMR* **2**, 447–465.

48. Englander, S. W., Downer, N. W. and Teitelbaum, H. (1972) Hydrogen exchange. *Annu. Rev. Biochem.* **41**, 903–924.

49. Claasen, B., Axmann, M., Meinecke, R. and Meyer, B. (2005) Direct observation of ligand binding to membrane proteins in living cells by a saturation transfer double difference (STDD) NMR spectroscopy method shows a significantly higher affinity of integrin α(IIb)β3 in native platelets than in liposomes. *J. Am. Chem. Soc.* **127**, 916–919.

50. Meinecke, R. and Meyer, B. (2001) Determination of the binding specificity of an integral membrane protein by saturation transfer difference NMR: RGD peptide ligands binding to integrin alphaIIbbeta3. *J. Med. Chem.* **44**, 3059–3065.

51. Kisselev, O. G., Kao, J., Ponder, J. W., Fann, Y. C., Gautam, N. and Marshall, G. R. (1998) Light-activated rhodopsin induces structural binding motif in G protein α subunit. *Proc. Natl. Acad. Sci.* **95**, 4270–4275.

52. Petros, A. M., Dinges, J., Augeri, D. J., et al. (2006) Discovery of a potent inhibitor of the antiapoptotic protein Bcl-x¡sub¿L¡/sub¿ from NMR and parallel synthesis. *J. Med. Chem.* **49**, 656–663.

53. Basus, V. J. (1989) Proton nuclear magnetic resonance assignments. *Methods Enzymol.* **177**, 132–149.

54. Seavey, B. R., Farr, E. A., Westler, W. M. and Markley, J. L. (1991) A relational database for sequence-specific protein NMR data. *J. Biomol. NMR* **1**, 217–236.

55. Gayler, K., Sandall, D., Greening, D., et al. (2005) Molecular prospecting for drugs from the sea. Isolating therapeutic peptides and proteins from cone snail venom. *IEEE Eng. Med. Biol. Mag.* **24**, 79–84.

56. Livett, B. G., Gayler, K. R. and Khalil, Z. (2004) Drugs from the sea: conopeptides as potential therapeutics. *Curr. Med. Chem.* **11**, 1715–1723.

57. Sandall, D. W., Satkunanathan, N., Keays, D. A., et al. (2003) A novel α-conotoxin identified by gene sequencing is active in suppressing the vascular response to selective stimulation of sensory nerves in vivo. *Biochemistry* **42**, 6904–6911.

58. Clark, R. J., Fischer, H., Nevin, S. T., Adams, D. J. and Craik, D. J. (2006) The synthesis, structural characterization, and receptor specificity of the α-conotoxin Vc1.1. *J. Biol. Chem.* **281**, 23254–23263.

59. Hajduk, P. J., Augeri, D. J., Mack, J., et al. (2000) NMR-based screening of proteins containing 13C-labeled methyl groups. *J. Am. Chem. Soc.* **122**, 7898–7904.

60. Bothner-By, A. A., Stephens, R. L., Lee, J., Warren, C. D. and Jeanloz, R. W. (1984) Structure determination of a tetrasaccharide: transient nuclear Overhauser effects in the rotating frame. *J. Am. Chem. Soc.* **106**, 811–813.

61. Johnson, M. A. and Pinto, B. M. (2004) NMR spectroscopic and molecular modeling studies of protein-carbohydrate and protein-peptide interactions. *Carbohydr. Res.* **339**, 907–928.

62. Johnson, M. A., Rotondo, A. and Pinto, B. M. (2002) NMR studies of the antibody-bound conformation of a carbohydrate-mimetic peptide. *Biochemistry* **41**, 2149–2157.

63. Mayer, M. and Meyer, B. (1999) Characterization of ligand binding by saturation transfer difference NMR spectroscopy. *Angew. Chemie Int. Ed.* **38**, 1784–1788.

64. Mayer, M. and Meyer, B. (2001) Group epitope mapping by saturation transfer difference NMR to identify segments of a ligand in direct contact with a protein receptor. *J. Am. Chem. Soc.* **123**, 6108–6117.

65. Vold, R. L., Waugh, J. S., Klein, M. P. and Phelps, D. E. (1968) Measurement of spin relaxation in complex systems. *J. Chem. Phys.* **48**, 3831–3832.

66. Carr, H. Y. and Purcell, E. M. (1954) Effects of diffusion on free precession in nuclear magnetic resonance experiments. *Physical Rev.* **94**, 630–638.

67. Meiboom, S. and Gill, D. (1958) Modified spin-echo method for measuring nuclear relaxation times. *Rev. Sci. Instrum.* **29**, 688–691.

68. Stejskal, E. O. and Tanner, J. E. (1965) Spin diffusion measurements: spin echoes in the presence of a time-dependent field gradient. *J. Chem. Phys.* **42**, 288–292.

69. Foster, M. P., McElroy, C. A. and Amero, C. D. (2007) Solution NMR of large molecules and assemblies. *Biochemistry* **46**, 331–340.

70. Keeler, C., Dannies, P. S. and Hodsdon, M. E. (2003) The tertiary structure and backbone dynamics of human prolactin. *J. Mol. Biol.* **328**, 1105–1121.

71. Grace, C. R. R. and Riek, R. (2003) Pseudomultidimensional NMR by spin-state selective off-resonance decoupling. *J. Am. Chem. Soc.* **125**, 16104–16113.

72. Pervushin, K., Riek, R., Wider, G. and Wuthrich, K. (1997) Attenuated T2 relaxation by mutual cancellation of dipole-dipole coupling and chemical shift anisotropy indicates an avenue to NMR structures of very large biological macromolecules in solution. *Proc. Natl. Acad. Sci.* **94**, 12366–12371.

73. Arimoto, R., Kisselev, O. G., Makara, G. M. and Marshall, G. R. (2001) Rhodopsin-transducin interface: studies with conformationally constrained peptides. *Biophys. J.* **81**, 3285–3293.

74. Tugarinov, V., Zvi, A., Levy, R. and Anglister, J. (1999) A cis proline turn linking two beta-hairpin strands in the solution structure of an antibody-bound HIV-1IIIB V3 peptide. *Nat. Struct. Biol.* **6**, 331–335.

75. Meyer, B., Weimar, T. and Peters, T. (1997) Screening mixtures for biological activity by NMR. *Eur. J. Biochem.* **246**, 705–709.

76. Mayer, M. and Meyer, B. (2000) Mapping the active site of angiotensin-converting enzyme by transferred NOE spectroscopy. *J. Med. Chem.* **43**, 2093–2099.

77. D'Souza, S. E., Ginsberg, M. H. and Plow, E. F. (1991) Arginyl-glycyl-aspartic acid (RGD): a cell adhesion motif. *Trends Biochem. Sci.* **16**, 246–250.

78. Aumailley, M., Gurrath, M., Muller, G., Calvete, J., Timpl, R. and Kessler, H. (1991) Arg-Gly-Asp constrained within cyclic pentapeptides. Strong and selective inhibitors of cell adhesion to vitronectin and laminin fragment P1. *FEBS Lett.* **291**, 50–54.
79. Chatterjee, C. and Mukhopadhyay, C. (2005) Interaction and structural study of kinin peptide bradykinin and ganglioside monosialylated 1 micelle. *Biopolymers* **78**, 197–205.
80. Stuart, A. C., Gottesman, M. E. and Palmer, A. G., 3rd (2003) The N-terminus is unstructured, but not dynamically disordered, in the complex between HK022 Nun protein and λ-phage BoxB RNA hairpin. *FEBS Lett.* **553**, 95–98.
81. Stamos, J., Eigenbrot, C., Nakamura, G. R., et al. (2004) Convergent recognition of the IgE binding site on the high-affinity IgE receptor. *Structure* **12**, 1289–1301.
82. Garman, S. C., Wurzburg, B. A., Tarchevskaya, S. S., Kinet, J. P. and Jardetzky, T. S. (2000) Structure of the Fc fragment of human IgE bound to its high-affinity receptor Fc epsilonRI α. *Nature* **406**, 259–266.
83. Nakamura, G. R., Reynolds, M. E., Chen, Y. M., Starovasnik, M. A. and Lowman, H. B. (2002) Stable "zeta" peptides that act as potent antagonists of the high-affinity IgE receptor. *Proc. Natl. Acad. Sci. USA* **99**, 1303–1308.

7

Molecular Dynamics Simulations of Peptides

Jeffrey Copps, Richard F. Murphy, and Sandor Lovas

Key Words: Force field; Molecular dynamics; simulations; replica exchange; GROMACS; GROMOS96; folding; secondary structure

1. Introduction

Molecular dynamics (MD) simulations fill a significant niche in the study of chemical structure. While nuclear magnetic resonance (NMR) yields the structure of a molecule in atomic detail, this structure is the time-averaged composite of several conformations. Electronic and vibrational circular dichroism spectroscopy and more general ultraviolet/visible and infrared (IR) spectroscopy yield the secondary structure of the molecule, but at low resolution. MD simulations, on the other hand, yield a large set of individual structures in high detail and can describe the dynamic properties of these structures in solution. Movement and energy details of individual atoms can then be easily obtained from these studies.

In MD simulations, trajectories (configurations as a function of time) of individual atoms are generated by simultaneous integration of Newton's equation of motion. The forces acting on each atom are the negative derivative of the potential energy and are termed the "force field." Force fields are parametrized using physical data from x-ray crystallography, IR and Raman spectroscopy, as well as high-level quantum mechanical calculations with model compounds, to reproduce the vibrational and conformational characteristics of a wide variety of molecules. The potential energy is the sum of bond and angle energies, the energy of bond rotations, and the energy of nonbonded van der Waals and electrostatic interactions, as in the general class I force field equation

From: *Methods in Molecular Biology, vol. 494: Peptide-Based Drug Design*
Edited by: L. Otvos, DOI: 10.1007/978-1-59745-419-3_7, © Humana Press, New York, NY

$$E_{total} = \sum_{bonds} (\frac{1}{2} k_{ij}^b (r_{ij} - r_b)^2) + \sum_{angles} (\frac{1}{2} k_{ijk}^\theta (\theta_{ijk} - \theta_{ijk}^0)^2)$$

$$+ \sum_{dihedrals} (k_\phi [1 + \cos(n\phi - \phi_0)]) + \sum_{improper} (k_{ijkl}^\xi (\xi_{ijkl} - \xi_0)^2)$$

$$+ \sum_{i<j} [\frac{A_{ij}}{r_{ij}^{12}} - \frac{B_{ij}}{r_{ij}^6} + \frac{1}{4\pi \epsilon_0} \frac{q_i q_j}{\epsilon r_{ij}}]$$

Fig. 1. Potential energy function for general class I force field.

shown in **Fig. 1(1)**. In class I empirical force fields, bond stretching and angle bending relative to equilibrium radius and angle values are described by the classical, not quantum harmonic oscillator function, necessitating the use of constraints to approximate the quantum effects of vibrating bonds and bond angles. Bond rotation is described by a sinusoidal function that approximates the peak energy from repulsion when a torsional angle is in the *cis* configuration, and the minimum when it is in *trans* configuration. Finally, nonbonded interactions are described by Lennard-Jones as well as Coulombic potentials, which approximate the long-range forces between uncharged and charged atoms, respectively. However, they overestimate the effect of molecular dipoles and cannot simulate molecules with significantly different polar attributes simultaneously, and this affects real equilibrium distance values because of the movement of atoms based on polarization. Class I force fields also cannot compute properties which are far from equilibrium or accurately predict vibrational spectra and they are temperature dependent. Class II empirical force fields approximate the forces upon molecules using more descriptive, complicated functions, such as using a Morse potential in place of the harmonic oscillator in describing bond stretching and angle bending, and higher-order terms in describing nonbonded interactions. This increases the accuracy of prediction, but also increases the computation time, often prohibitively.

The choice of a particular force field depends on the type of system for which it has been designed. Several class I force fields have been designed for description of polypeptides, including AMBER *(2)*, CHARMM *(3)*, OPLS and OPLS-AA *(4,5)*, and GROMOS96 *(1)*. The computational study of peptides and proteins can yield information regarding the importance of residues and functional groups in determining the structure, folding and solubility in various environments. This information can then be applied to the study of the structure-activity relationships of those molecules in ligand–receptor complexes and aid in the design of new therapeutics.

GROMOS96 is a united atom force field, modeling only polar hydrogen atoms explicitly, while aliphatic hydrogens (such as methyl hydrogens) are grouped with attached carbons to form united atoms treated as single atoms. This small concession in modeling precision greatly accelerates computational time. GROMOS96 is implemented by the GROMACS MD simulation package (along with several other force fields) and was developed specifically for proteins *(5)*. Due to its greatly optimized code, GROMACS is currently the fastest MD simulation program available. As a class I force field, the accuracy of GROMOS96 is more limited than class II force fields. The speed of calculation, relatively accurate prediction of peptide conformation, ease of setup, and use of the GROMACS program and GROMOS96 force field have proven to be quite useful in conducting MD simulations, so both are used here to demonstrate a procedure for conducting the MD simulation of the structure of a peptide in an explicit solvent.

2. Methods for Standard MD Simulation (Fig. 2)

1. The first step in running an MD simulation is the generation of an input structure. Preferably, experimentally determined structures from NMR spectroscopy or x-ray crystallography studies should be available for the peptide/protein of interest. The RCSB Protein Databank (http://www.rcsb.org/) *(6)*, for example, has many such structures available for download. Alternatively, energy-minimized structures can be generated from the original sequence using theoretical methods such as homology modeling and simulated annealing *(7)*.

2. Once an input structure has been selected, it must be converted to the GROMACS file format. The program pdb2gmx converts the initial structure file to the GROMACS structure file (.gro) format and generates a system topology file based on predefined standard residue topologies describing the atoms, bonds, and torsional angles of the residues, as well as the force field (in this case, GROMOS96 with the 53a6 parameter set *(8)*) and protonation state of polar sidechains and of the N-terminus and C-terminus. If a standard topology is not available for a given residue, *ab initio* calculations may be required to generate one. A file that describes positional restraints on heavy atoms in the original .pdb file is also generated (*see* **Note 1**).

3. Create a solvent box in which to solvate the protein and conduct the simulation. This can be accomplished using the editconf program and specifying a box type and box dimensions (*see* **Note 2**). For peptides/proteins, select the standard rectangular box and set box dimensions. This command generates a modified .gro file, which now includes the box dimensions, centering the solute in the box unless otherwise specified.

4. The molecule can then be solvated with explicit solvent molecules by inputing the modified structure file into the genbox program. Specify a solvent model in .gro file format (e.g., spc216.gro for water) consisting of a small box containing a

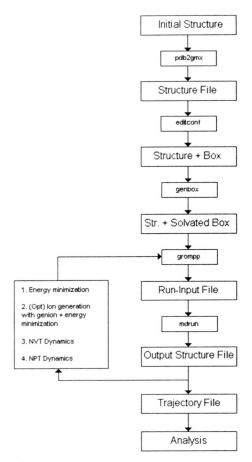

Fig. 2. Flowchart for MD simulation method.

selected concentration of solvent molecules, and modify the system topology file
to include a topology file for the solvent molecules (*see* **Note 3**). With this information, GROMACS will solvate the protein by completely filling the simulation
box with duplicates of the solvent box, leaving a small area around the solute free
of solvent molecules, and generating modified system topology and .gro structure
files to reflect these changes. If any atom of a solvent molecule is placed closer
to any atom of the solute molecule(s) than the sum of the van der Waals radii of
the two atoms, then that solvent molecule is removed (*see* **Notes 4–6**).

5. Create an index file using make_ndx after the solvation of the protein. Default
index files are generated by and also required by most GROMACS commands,
but it is generally desirable to create files with special index groups in order to
analyze the behavior of a particular subset of solute and/or solvent atoms.

6. After the addition of solvent molecules, the energy of the system should be minimized to decrease high potential energy, as well as to bring the starting configuration close to an equilibrium state (*see* **Note 7**). Use the steepest descent method as the algorithm to minimize the system energy, because it is a good approximation of the local minimum and fairly expedient compared to other methods (*see* **Note 8**). This method makes a user-defined number of steps towards the negative gradient, always moving towards it, but without factoring in the history of previous steps, and constantly adjusting the size of the steps to minimize the convergence time. An MD parameter file is needed, for the energy minimization run and this should include the type of minimization method desired. It should also list the number of steps for the method to take, the initial step size, and the tolerance (the user-defined value of the maximum force on any particular atom in the system under which the system is sufficiently minimized). The default initial step size of 0.01 nm and a tolerance of 0.05 kJ/mol/nm are typical values, while a run of 2000 steps is usually sufficient to minimize the energy of the system using the steepest descent method. Simple cut-off values for the long range nonbonded potential functions (van der Waals, Coulombic) should be included (*see* **Note 6**). This file, along with the .gro structure file and system topology file, are input into the grompp program, which combines the information and generates a binary run-input file. The run-input file can then be input to the mdrun program, which will run the minimization and output trajectory, structure, and simulation log files.

7. If the system contains charged particles, an equal number of ions of opposite charge must be introduced to neutralize the overall charge of the system (*see* **Note 9**). The genion program can be used to replace individual solvent molecules with favorable electrostatic potential (or at random) with monoatomic ions, usually Na^+ and Cl^-. Once this is completed, the ion topology file, the number of ions, and the ion charge must be included in the system topology file, while the replaced solvent molecules should be subtracted.

8. The system should then be minimized again, and a new index file should be generated so that there is no discrepancy between the numbers of molecules in the various system files.

9. After minimization, initial velocities must be generated for the solvent molecules for the final simulation, while the solute is held in place at the center of the system. This procedure is known as a positionally restrained MD simulation. In this case, the number of molecules, the volume of the system, and the temperature of the system will be held constant, hence the description NVT dynamics. A second parameter file needs to be created for this preliminary simulation. In addition to including the position restraint file (generated during the conversion of the initial structure file to the .gro structure file) to fix the solute in place, as well as the long-range cut-off values, the reaction field method should be specified as the cut-off algorithm type for nonbonded potentials. The file should include all bonds as constraints, and the LINCS constraint algorithm (*9*) should be used (*see* **Note 10**). The MD algorithm should be specified as the integrator, and the user should make sure to enable periodic boundary conditions.

10. In accordance with NVT conditions, the temperature of the system must be coupled to a virtual heat bath, with reference temperature (typically 300 K) set by the user. Solvent, solute, and ions should be coupled separately to a bath of the same reference temperature using Berendsen coupling *(10)*, with the same time constant (frequency of coupling, typically 0.1 ps) for each. Pressure is not coupled and held constant during the NVT simulation, in order to keep the box rigid and the volume constant. Dielectric constant and isothermal compressibility of the solvent are also important parameters to set here. Finally, enable generation of velocities for solvent atoms and ions. GROMACS will generate velocities using a Maxwell distribution at a user-defined temperature, which should be the same as the coupled temperature. Generate a run-input file using the grompp program as in the energy minimization step, input to the mdrun program, and start the simulation. A positionally restrained simulation of 100 ps with a time step of 2 fs is typical. Choose the number of processors on which to run the simulation, and if using more than one, use the MPI program described in the GROMACS manual *(11)* (*see* **Note 11**).

11. For the full NPT (constant pressure, temperature, and number of molecules) simulation run, the parameter file is unchanged from the NVT simulation, except that generation of velocities as well as position restraints should be turned off. Along with temperature coupling, pressure coupling should be enabled (and the system volume allowed to scale), with a reference pressure of typically 1 bar and a 1 ps time constant for coupling *(10)*. Dispersion corrections for the cut-off of the long-range Lennard-Jones potential should also be enabled for both energy and pressure. Once again, generate the final run-input file using grompp, input to mdrun and start the simulation. The user should specify a full simulation time and a time step (again, typically 2 fs). The number of processors used for the full simulation should be the same as used for the NVT simulation and optimal for speed and efficient use of computational resources (*see* **Note 11**).

12. When the simulation is complete, check the fidelity of the final trajectory file using the gmxcheck program and, if desired, convert to the less memory-consuming .xtc format. Subset group trajectories based on groups listed in the index file can also be written.

3. Replica Exchange Molecular Dynamics (REMD)

With standard MD simulations at low temperatures, an explicitly solvated protein or peptide generally becomes trapped in any of many local energy minima, prohibiting a representative sampling of the entire range of conformations. Of a few suggested solutions, REMD is least time-consuming, easiest to implement, and theoretically sound *(12–14)*. In REMD, multiple independent simulations ("replicas") are conducted, each at a different temperature in a limited range. At user-defined time steps, the trajectory coordinates of simulations of "neighboring" temperatures are either randomly exchanged or not exchanged depending on a probability equation, which is a function of the temperatures and instantaneous potential energies of the two systems. The algorithm attempts only the exchange of "odd pairs" and "even pairs" every other step to prevent the possibility of exchange being dependent on multiple systems. The exchange of replicas allows the possibility for a system to gain enough energy to escape from a local minimum into which it may have fallen and ensures better sampling of possible conformations.

In an REMD study of Chingnolin using the GROMACS package, van der Spoel and Seibert *(15)* successfully folded, predicted the folding and unfolding time constants, derived folding energies, and calculated a melting curve for the decapeptide.

4. Methods for REMD Simulation

1. In practice, the user must first decide on the number of replicas to use, as well as the temperature range. The smallest temperature replica should be low enough to sample states of lowest energy, while the highest temperature should be high enough to overcome the various energy barriers of the system. In addition, since the probability of exchange is based partly on the difference in temperature between the two systems being considered, and exchange probability falls off rapidly with difference (illustrating the importance of exchanging only neighboring replicas), temperatures should be chosen so as to facilitate the user's desired probability of exchange (*see* **Note 12**).
2. Methods for the REMD simulation are the same as in the standard case, until the actual full NPT simulation. A separate run-input file must be generated for each replica using the grompp program. Then the simulation run (using mdrun) with replica exchange enabled, the time step for exchange specified, and the number of processors to be used is specified (*see* **Note 13**).
3. When the simulation is complete, the relevant frames of each trajectory file of each replica must be "demultiplexed" according to a replica index file, which describes how the frames need to be ordered in the final composite trajectory file. The included GROMACS script demux.pl *(16)*, which takes as input one of the standard simulation log files, derives exchange data and generates the replica index file. This file then must be input to the trjcat program along with the trajectory files to generate the final trajectory file.
4. As with the standard simulation, check the fidelity of the final trajectory file using gmxcheck and convert to .xtc format if desired.

5. Analysis

Several analyses can be performed once the full simulation method has been completed. One of the most useful is secondary structure analysis using do_dssp. This program, in combination with the xpm2ps program, uses the Dictionary

Copps et al.

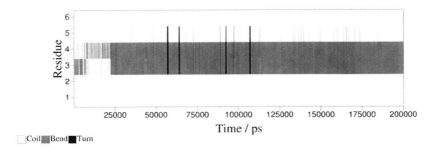

Fig. 3. DSSP for gastrin(1-6) in dimethyl sulfoxide.

of Protein Secondary Structure (DSSP) criteria for defining secondary structural forms to graphically represent the structure of the molecule throughout the simulation. An example the gastrin fragment G17(1-6) in dimethyl sulfoxide, showing a predominance of bend and turn structure is seen in **Fig. 3**.

A root-mean-square deviation (RMSD) analysis of the amide backbone atoms is often a strong indicator of conformational changes of a protein in solution. An RMSD example is shown in **Fig. 4**, once again of G17(1-6) in DMSO. Periods of conformational stability are indicated by stretches in which the RMSD does

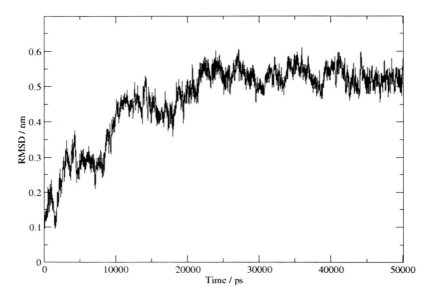

Fig. 4. RMSD for G17(1-6) in dimethyl sulfoxide.

not vary significantly, while variances of 1 Å or more indicate a large change in the fold of the protein.

Many other analyses can be made. The radius of gyration of a group of atoms as a function of time can be calculated. The dihedral angles of residues can be determined and Ramachandran plots made. The hydrogen bond or salt bridge propensity of atoms can be measured. As stated previously, by generating index files for the simulation, groups of atoms for analysis can be defined. Cluster analysis can be performed with a user-defined RMSD criterion, grouping similar structures along the simulation trajectory into clusters. This feature is particularly useful in determining which conformations of the protein are most predominant during the simulation.

6. Notes

1. Because GROMOS is a united atom force field, either all nonpolar hydrogen atoms must be edited out of the input structure or the user must specify that they be ignored by GROMACS. Aromatic hydrogens are an exception, as GROMACS cannot model aromatic–aromatic interactions with the united atom simplification. Hydrogens are then added to the structure topology by GROMACS as united atoms.

2. GROMACS offers a variety of box types: the standard cubical/rectangular box, the truncated octahedron, the hexagonal prism, and the rhombic dodecahedron. As the shape of the box can affect the simulation, a box type that best reflects the geometry of the solute should be used. For example, the truncated octahedron approximates a sphere, so it is the optimal choice for globular proteins and other roughly spherical molecules.

3. These solvent models are generally developed with a specific force field in mind, so using a solvent model with a force field for which it was not intended may require the inclusion or exclusion of certain parameters.

4. The user should (1) check that the number of solvent molecules added is suitable for the density of the solvent, (2) edit the output system topology file to include the number of solvent molecules added, and (3) use a visualization program to check that the new structure file does not have any solvent molecules placed too close to the solute.

5. The effect of the surface of the box on the solute is of major importance in the simulation of systems such as the one described here. The sudden cut-off of long-range nonbonded potentials at the box surface (beyond which is vacuum) would have an unnatural effect on the dynamics of the simulation. Only an extremely large system size could ensure a small influence of this surface effect on the solute. The computational cost of such a large system would be prohibitive. For this reason, periodic boundary conditions are used. The image of the simulation box is translated repeatedly to form an "infinite" lattice. When a particle in the simulation box moves, the image in all other translated boxes moves correspond-

ingly with the same orientation. Thus, when a particle leaves the simulation box, its translated image enters through the opposite face of the box. While this solution is not strictly accurate for nonperiodic systems (i.e., liquids and solutions), the errors generated are less likely to be as severe as those caused by the sudden cut-off of potentials at the simulation box surface.

6. The long range of nonbonded interactions can also be a problem in periodic systems. Interactions which do not decay faster than r-n, where *n* is the box dimensionality, can often have a range greater than half the box length, which may allow a particle to interact with itself or allow it to interact with another particle twice. A parameter must therefore be set to prevent these forces from extending past half the box (cubic) length. However, a simple cut-off is also problematic, especially for Coulombic interactions, both because of the discontinuity in the potential energy and force intro- duced (and the subsequent movement of ions), and the tendency of the potential energy to diverge if it does not decay faster than r^{-3}. GROMACS has implementable options to remedy this situation, including shift and switch functions *(17)*, the particle mesh Ewald method *(18,19)*, and reaction- field methods *(20)*, each of which modifies the simple cut-off scheme with varying success. These options are implemented in the MD parameter file.

7. Forces on systems far from equilibrium may be extremely large and cause the simulation to fail. Thus, it is important to bring the system as close to the nearest local minimum of the potential energy of the system as possible before commencing MD simulations. GROMACS cannot, in all likelihood, find the global minimum of the potential energy function, but it can find the nearest local minimum.

8. The conjugate gradient method and the L-BFGS minimizer, both of which incor- porate gradient information, can usually bring the potential energy closer to the local minimum than does steepest descent. However, convergence can take far longer.

9. If this is not done, a large dipole moment will be induced on the system, causing incorrect energy calculations and other erroneous behavior.

10. As an alternative, the SHAKE algorithm *(21)* is used if angle constraints are implemented or the LINCS algorithm otherwise fails.

11. Depending on available computational resources, it may be useful to do a short full simulation run to determine the optimal number of processors to use, since beyond a certain number of processors, speed of the simulation is not improved and resources are wasted.

12. The GROMACS manual *(11)* suggests a relationship of T2 = (1 + 1/sqrt(Natoms)T1, where Natoms is the number of atoms of the system, and T1 and T2 are neighboring temperatures for a protein/water system, giving an exchange probability of 13.5%. More detailed discussion is given in Nguyen and associates *(14)*.

13. REMD demands parallel computing. As GROMACS is currently implemented, each replica must be run on a separate processor.

References

1. van Gunsteren, W.F. and Berendsen, H.J.C. (1987) *Groningen Molecular Simulation (GROMOS) Library Manual*. Biomos, Groningen.
2. Weiner, P.K. and Kollman, P.A. (1981) AMBER: Assisted model building with energy refinement. A general program for modeling molecules and their interactions. *J. Comput. Chem.* **2**, 287–303.
3. Brooks, B.R., Bruccoleri, R.E., Olafson, B.D., States, D.J., Swaminathan, S., and Karplus, M. (1983) A program for macromolecular energy minimization and dynamics calculations. *J. Comput. Chem.* **4**, 187–217.
4. Jorgensen, W.L. and Tirado-Rives, J. (1988) The OPLS potential functions for proteins. energy minimizations for crystals of cyclic peptides and Crambin. *J. Am. Chem. Soc.* **110**, 1657–1666.
5. Jorgensen, W.L., Maxwell, D.S., and Tirado-Rives, J. (1996) Development and testing of the OPLS All-Atom Force Field on conformational energetics and properties of organic liquids. *J. Am. Chem. Soc.* **118**, 11225–11236.
6. Berman, H.M., Westbrook, J., Feng, Z., et al. (2000) The Protein Data Bank. *Nucleic Acids Res.* **28**, 235–242.
7. Lovas, S. and Murphy, R.F. (1997) Molecular modeling of neuropeptides. *Methods Mol. Biol.* **73**, 209–217.
8. Oostenbrink, C., Villa, A., Mark, A.E., and van Gunsteren, W.F. (2004) A biomolecular force field based on the free enthalpy of hydration and solvation: the GROMOS forcefield parameter sets 53A5 and 53A6. *J. Comput. Chem.* **13**, 1656–1676.
9. Hess, B., Bekker, H., Berendsen, H.J.C., and Fraaije, J.G.E.M. (1997) LINCS: A linear constraint solver for molecular simulations. *J. Comput. Chem.* **18**, 1463–1472.
10. Berendsen, H.J.C., Postma, J.P.M., DiNola, A., and Haak, J.R. (1984) Molecular dynamics with coupling to an external bath. *J. Chem. Phys.* **81**, 3684–3690.
11. van der Spoel, D., Lindahl, E., Hess, B., et al. (2005) Gromacs User Manual version 3.3, http://www.gromacs.org.
12. Sugita, Y. and Okamoto, Y. (1999) Replica-exchange molecular dynamics method for protein folding. *Chem. Phys. Lett.* **314**, 141–151.
13. Baumketner, A. and Shea, J.-E. (2005) Free energy landscapes for amyloidogenic tetrapeptides dimerization. *Biophys. J.* **89**, 1493–1503.
14. Nguyen, P.H., Mu, Y., and Stock, G. (2005) Structure and energy landscape of a photoswitchable peptide: A replica exchange molecular dynamics study. Proteins **60**, 485–494.
15. Seibert, M.M., Patriksson, A., Hess, B., and van der Spoel, D. (2005) Reproducible polypeptide folding and structure prediction using molecular dynamics simulations. *J. Mol. Biol.* **354**, 173.
16. http://www.gromacs.org/pipermail/gmx-revision/2006-August/000152.html
17. van der Spoel, D. and van Maaren, P.J. (2006) The origin of layer structure artifacts in simulations of liquid water. *J. Chem. Theor. Comp.* **2**, 1–11.
18. Darden, T., York, D., and Pedersen, L. (1993) Particle mesh Ewald: An N-log(N) method for Ewald sums in large systems. *J. Chem. Phys.* **98**, 10089–10092.

19. Essman, U., Perera, L., Berkowitz, M.L., Darden, T., Lee, H., and Pedersen, L.G. (1995) A smooth particle mesh ewald potential. *J. Chem. Phys.* **103**, 8577–8592.
20. Tironi, I.G., Sperb, R., Smith, P.E., and van Gunsteren, W.F. (1995) A generalized reaction field method for molecular dynamics simulations. *J. Chem. Phys.* **102**, 5451–5459.
21. Ryckaert, J.P., Ciccotti, G., and Berendsen, H.J.C. (1977) Numerical integration of the cartesian equations of motion of a system with constraints; molecular dynamics of n-alkanes. *J. Comput. Phys.* **23**, 327–341.

8

Short Linear Cationic Antimicrobial Peptides: Screening, Optimizing, and Prediction

Kai Hilpert, Christopher D. Fjell, and Artem Cherkasov

Summary

The problem of pathogenic antibiotic-resistant bacteria such as *Staphylococcus aureus* and *Pseudomonas aeruginosa* is worsening, demonstrating the urgent need for new therapeutics that are effective against multidrug-resistant bacteria. One potential class of substances is cationic antimicrobial peptides. More than 1000 natural occurring peptides have been described so far. These peptides are short (less than 50 amino acids long), cationic, amphiphilic, demonstrate different three-dimensional structures, and appear to have different modes of action. A new screening assay was developed to characterize and optimize short antimicrobial peptides. This assay is based on peptides synthesized on cellulose, combined with a bacterium, where a luminescence gene cassette was introduced. With help of this method tens of thousands of peptides can be screened per year. Information gained by this high-throughput screening can be used in quantitative structure-activity relationships (QSAR) analysis.

QSAR analysis attempts to correlate chemical structure to measurement of biological activity using statistical methods. QSAR modeling of antimicrobial peptides to date has been based on predicting differences between peptides that are highly similar. The studies have largely addressed differences in lactoferricin and protegrin derivatives or similar de novo peptides. The mathematical models used to relate the QSAR descriptors to biological activity have been linear models such as principle component analysis or multivariate linear regression. However, with the development of high-throughput peptide synthesis and an antibacterial activity assay, the numbers of peptides and sequence diversity able to be studied have increased dramatically. Also, "inductive" QSAR descriptors have been recently developed to accurately distinguish active from inactive drug-like activity in small compounds. "Inductive" QSAR in combination with more complex mathematical modeling algorithms such as artificial neural networks (ANNs) may yield powerful new methods for in silico identification of novel antimicrobial peptides.

From: *Methods in Molecular Biology, vol. 494: Peptide-Based Drug Design*
Edited by: L. Otvos, DOI: 10.1007/978-1-59745-419-3_8, © Humana Press, New York, NY

Key Words: Screening; high-throughput; SPOT synthesis; cellulose; luminescence; *Pseudomonas aeruginosa*; QSAR; antimicrobial; antibacterial; peptide; machine learning; artificial neural networks.

1. Cationic Antimicrobial Peptides

Cationic antimicrobial peptides, also called cationic host defense peptides, are present in virtually every form of life, from bacteria and fungi to plants, invertebrates, and vertebrates *(1)*. They support bacteria and fungi in the defense of their ecological niche. In addition to ribosomally produced peptides, many microbes are able to synthesize powerful antimicrobial peptides by using multienzyme complexes *(2)*. These complexes are able to build peptides with nonproteogenic amino acids or catalyze unusual modifications. Antimicrobial peptides like polymyxin B and gramicidin S are examples of such modified peptides. Antimicrobial peptides have been found in plants that were active against bacteria and fungi. Until now only peptides with a β-sheet globular structure have been isolated from plants *(3)*. More structurally diverse peptides have been found in invertebrates. Invertebrates are reliant on the innate immunity, of which antimicrobial peptides are a critical part. In invertebrates, antimicrobial peptides are found in the hemolymph, in phagocytic cells, and in different epithelial cells and show activities against bacteria, yeast, and viruses *(4)*. Intensive studies have been conducted on antimicrobial peptides and their role in innate immune system on the invertebrates: fruit fly, as well as horseshoe crabs *(5–7)*. It has only recently become apparent that the adaptive immune system arose in jawed vertebrates, but little is known about the deeper origins of this system or the relationship with the innate immune system in invertebrates *(8,9)*. In vertebrates, antimicrobial peptides are found mainly at the skin and mucosal surfaces and within granules of immune cells *(10,11)*. For example, amphibian skin is a rich source of antimicrobial peptides, e.g., the skin of the frog *Odorrana grahami* contains 107 different antimicrobial peptides *(12)*. Studies on antimicrobial peptides of mammals, including human, have revealed that besides the direct antimicrobial activity, some peptides also influence the immune system. These particular peptides are able to counter sepsis/entotoxemia, enhance phagocytosis, recruit various immune cells, and increase or decrease the production of cytokines and chemokines in different cell types *(13)*.

Despite the variety of different sources, sequences and structures, all cationic antimicrobial peptides share some common features: small size (12–50 amino acids), positive net charge (+2 to +9), amphiphilic (≥30% hydrophobic amino acids) and antimicrobial and/or immunomodulatory activity. Some cationic antimicrobial peptides also show antiviral *(14)* and/or anticancer activity *(15)* as well as wound healing properties *(16)*. More than 1000

Table 1
Selected Natural Cationic Antimicrobial Peptides

Name	Source	Sequence	Remark	MIC (µg/mL)		Ref.
				E. coli	*S. aureus*	
Bac7	Bovine neutrophils	RRIRPRPPRLPRPRPLPF PRPGPRPIPRPLPFPRPGP TPIPRPLPFPRPGPRPI PRPL-NH$_2$	Unusual amino acid composition	12	200	*(37,38)*
Bactenecin	Bovine neutrophils	RLCRIVVIRVCR	β–Turn	8	32–64	*(39)*
Cecropin B	Hemolymph of *Hyalophora cecropia*	KWKVFKKIEKMGRNIRN GIVKAGPAIAVLGEAKAL-NH$_2$	α-Helical	1.7 (µM)	>100 (µM)	*(40)*
Cecropin P$_1$	Porcine small intestine	SWLSKTAKKLENSAKKR ISEGIAIAIQGGPR	α-Helical	3.3 (µM)	>100 (µM)	*(40,41)*
α-Defensin-1 (HNP-1)	Human neutrophils	ACYCRIPACIAGERRYG TCIYQGRLWAFCC	Cystine-rich	18.4	25	*(42)*
β-Defensin-12	Bovine neutrophils	GPLSCGRNGGVCIPIRC PVPMRQIGTCFGR-PVKCCRSW	Cystine-rich	18 (µM)	22 (µM)	*(43,44)*

(Continued)

Table 1
(Continued)

Name	Source	Sequence	Remark	MIC (μg/mL) E. coli	MIC (μg/mL) S. aureus	Ref.
Gramicidin S	Bacillus brevis	cyclo-((D)Phe-Pro-Val-Orn-Leu)$_2$ (Orn is Ornithine)	Unusual amino acid composition	200	3.1	(45,46)
Indolicidin	Bovine neutrophil	ILPWKWPWWPWRR-NH$_2$	Unusual amino acid composition	7.7	3.8	(47,48)
Lactoferricin H-20	Human, derived from Lactoferrin	KCFQWQRNMRKVRGPPVSCI	β–Hairpin peptide (amino acids 19–38 of Lactoferrin)	256	128	(49,50)
Lactoferricin P-20	Porcine, derived from Lactoferrin	KCRQWQSKIRRTNP IFCIRR	β-Sheet (amino acids 18–37 of Lactoferrin)	64	32	(50)
LL37	Human neutrophils and various epithelial cells	LLGDFFRKSKEKIGKEF KRIVQRIKDFLRNLVPRTES	α-Helical	>32 (μM)	>32 (μM)	(51,52)
Magainin2	Xenopus laevisskin	GIGKFLHSAKKFGKAFVGEIMNS	α-Helical	11.6	64	(53–55)

Nisin	*Lactococcus lactis*	IOA*IULA*A+PGA+KA+GALMGA+NMKA+AA+A+HA+HVUK	Unusual amino acid composition	Inactive	2	(56–58)
Polymyxin B	*Bacillus polymyxa*	R-Dab-Thr-Dab-Dab-Dab-(D)Phe-Leu-Dab-Dab-Thr (Dab is diaminobutyric acid)	Cyclic lipopeptide	1.8	26	(59–61)
PR-39	Pig intestine	RRRPRPPYLPRPRPPPFF PPRLPPRIPPGFPPRF PPRFP-NH$_2$	Unusual amino acid composition	4	250	(62,63)
Protegrin-1	Porcine leukocyte	RGGRLCYCRRRFCVCVGR	β-Sheet	0.75–11	1.7–8	(64–66)
Tachyplesin I	Horseshoe crab	KWCFRVCYRGICYRRCR	β-Sheet	6.3–12.5	3.1–6.3	(67,68)

A*, part of lanthionine; A+, part of 3-methyllanthionine; U, dehydroalanine; O, dehydrobutyrine.

different naturally occurring peptides have been discovered so far; about 900 of these are described in a database for eukaryotic host defense peptides (http://www.bbcm.units.it/~tossi/pag1.htm). Selected examples of natural cationic antimicrobial peptides are provided in **Table 1**. For all these peptides it was proposed that permeabilization of the cytoplasmic membrane of the microbe was the cause of the antimicrobial activity. There are many different models that try to explain the detailed steps of the interaction of cationic antimicrobial peptides with microbial membranes. The most prominent ones are barrel-stave, carpet, toroidal pore, and aggregate models *(14)*. Charge and hydrophobicity of the peptides support the interaction with the microbial cytoplasmic membrane: the positive charge amino acids interact with the anionic lipid head and the hydrophobic amino acids with the lipid core *(17)*. Solid-state NMR and attenuated total reflectance–Fourier transform infrared spectroscopy (ATR-FTIR) studies showed that peptides bind in a membrane parallel orientation, interacting only with the outer membrane layer. Only at higher concentrations (more than needed for killing the microbes) can membrane disruptions or pore formations be detected *(18–20)*. However, in the last decade it has become evident that some antimicrobial peptides are not disrupting the cytoplasmic membrane, but seem to interact with different internal targets, e.g., protein or RNA synthesis *(21)*.

2. Screening and Optimizing Peptides for Antimicrobial Activity

Biological or chemical libraries can be used to synthesize and screen large numbers of peptides. Using biological techniques, such as phage *(22)*, bacterial *(23)*, or ribosome display *(24)*, screening peptide libraries for antimicrobial activity is applicable and led, for example, to moderately active peptides with minimal inhibitory concentrations (MIC) against *Escherichia coli* of 500 μg/mL for linear 10mer peptides obtained from a phage display *(25)* or 25 μg/mL from ribosomal display *(26)*. One main advantage to these approaches is that the peptides are synthesized by a biological process and, therefore, the cost of the peptides is low. In addition, repetitive rounds of enrichment may increase the chance of discovering highly active peptides. On the other hand, using biological approaches, only the gene-encoded amino acids can be used, limited numbers of sequences permit only partial information, the biological peptide libraries are tricky to handle, and fusion peptides rather than isolated molecules are created. To synthesize large amounts of peptide chemically, several different modified peptide synthesis procedures have been developed, e.g., tea bag synthesis *(27)*, digital photolithography *(28)*, pin synthesis *(29)*, and SPOT™ synthesis on cellulose *(30)*. All these methods can incorporate more than 600 commercially available building blocks, and it is possible to systematically investigate the interaction of interest. Up to now, only the SPOT synthesis on cellulose (*see also* Chapter 4) has been utilized to screen large amounts of short peptides for antimicrobial activity against microbes of choice *(31)*. We will outline in the following how the cellulose peptides were used to screen for antimicrobial activity and what results were gained from this approach.

The peptides were synthesized as large spots (diameter about 0.6 cm) at high density, which can reach up to 1.9 μmol/cm^2 *(32)*. About 1000 peptides were synthesized per 20 × 29 cm cellulose sheet. After the final side-chain deprotection step, the membrane was washed and the peptides were cleaved from the membrane *(33)*. At this point, the peptides still remain adsorbed onto the membrane but will dissolve once a liquid is added. The peptides were then punched out with a normal single-hole puncher, and the spots were transferred into a 96-well microtiter plate. Distilled water was added in each well to dissolve the peptides.

The next stage involved introducing bioluminescence into *P. aeruginosa* to increase the sensitivity and high throughput readout of the assay. Alternatively, unmodified microbes can be used as well *(31)*. Bioluminescence was gained by cloning the *luxCDABE* in *P. aeruginosa*. The bioluminescence reaction involves oxidation of reduced riboflavin mononucleotide (FMNH$_2$) and a long-chain fatty aldehyde. The genes *luxCDE* provide the synthesis of a fatty aldehyde and *luxAB* encode the luciferase subunits. In order to generate light, the bacterium must provide reduced flavin mononucleotide and molecular oxygen, and both are abundant as long as the bacteria are energized. Any event that changes the energy level of the bacteria, and therefore FMNH$_2$ concentration, can be monitored by detection changes in light intensity. Luminescence was measured by a luminometer in a multiplex format (e.g., microtitre tray). *P. aeruginosa* isolate, strain H1001, with a *Tn5-luxCDABE* transposon inserted into the *fliC* *(34)* grows normally and expresses luminescence in an abundant and constitutive fashion. This strain, together with buffer and a carbon source, was transferred in each well of a 96-well microtiter plate, suitable for luminescence measurements. In each well of the first row of the microtiter plate, dissolved peptides (see previous paragraph) were added and a serial dilution was performed. After 4 h of incubation time, the luminescence was measured, and the effect of each peptide at different concentration was evaluated. It was shown that the loss in luminescence parallels with killing of bacteria *(35)*.

Using the Spot synthesis strategy, a complete substitution analysis of a 12mer peptide was performed *(35)*. Each amino acid of the original 12mer peptide Bac2A was substituted by all other 19 gene-encoded amino acids (**Fig. 1** shows the principle of this approach). Substitution analysis is a powerful tool for investigating the importance of each amino acid for the interaction of interest, and it can help identify key amino acids or important patterns for the interaction.

wt	A	C	D	E	F	...	...	W	Y
MIDI	AIDI	CIDI	DIDI	EIDI	FIDI			WIDI	YIDI
MIDI	MADI	MCDI	MDDI	MEDI	MFDI			MWDI	MYDI
MIDI	MIAI	MICI	**MIDI**	MIEI	MIFI			MIWI	MIYI
MIDI	MIDA	MIDC	MIDD	MIDE	MIDF			MIDW	MIDY

Fig. 1. Principle of a substitution analysis of the 4 mer peptide MIDI. The first column represents the parent or wild-type (wt) peptide, which is used to compare the effects of the substitution variants. All other columns describe the substituted amino acids at each position of the peptide. All underlined positions show the actual substituted amino acid in the sequence. The wild-type sequence is bolded.

Based on the substitution analysis of Bac2A (RLARIVVIRVAR-NH$_2$), different 12mer peptides with superior activity against different human pathogen were developed *(35)*. For example, peptide Sub3 showed MIC values against *P. aeruginosa* and *S. aureus* of 2 μg/mL, *E. coli* of 0.5 μg/mL, and *Staphylococcus epidermidis* of 0.2 μg/mL. The data of the complete substitution analysis were also used to develop shorter peptides with strong antimicrobial activity. One 8mer peptide was described showing MIC values of 2 μg/mL against *E. coli* and *S. aureus*. Results also showed that changes in any single position of the peptide may affect other residues at all other positions in the peptide; therefore, each single substituted peptide variant may lead to different preferred substitutions at any other position *(36)*.

To investigate the flexibility of amino acid arrangements for creating active peptides, the Bac2A sequence was scrambled *(36)*. Consequently 49 peptide variants of Bac2A were created, all of which were composed of the same amino acids and had the same length, net charge, and amount of hydrophilic and hydrophobic amino acids, and thus were ideally suited for testing the importance of the primary amino acid sequence for activity. It has been demonstrated that scrambling of a sequence destroys sequence-specific interactions, as observed, e.g., with antibodies recognizing linear epitopes. In this case, scrambled peptides are normally used as a negative control to proof the sequence specificity. Conversely, if amino acid composition was the sole determinant of killing activity, all scrambled peptides should be active. Experiments using 49 scrambled Bac2A peptides revealed peptides ranging in activity from superior antimicrobial activity up to inactive, indicating that activity was not solely dependent on the composition of amino acids or the overall charge or hydrophobicity, but rather required particular linear sequence patterns *(36)*. Therefore, suitable amino acid composition including charge and ratio of hydrophobic to hydrophilic amino acids is necessary but not sufficient for antimicrobial activity. In the case of the scrambled peptide set, it was shown that the inactive peptides were not able to form similar structures in lipids compared to the active peptides. In addition, depolarization experiments of the cytosolic membrane of tested bacteria showed only minor depolarization effects of the inactive peptides compared to active peptide sets. Both results—structural features and depolarization—suggest that the inactive peptides are hampered in their ability to interact with lipids/membranes *(36)*. In addition, the screening of 1400 biased-random peptides once again confirmed the importance of the balance between charge and hydrophobicity, as well as the fact that the "right" composition alone is not sufficient to create an active peptide (K. Hilpert et al., manuscript in preparation).

3. Predicting Antimicrobial Peptide Activity Using Quantitative Structure-Activity Relationships

3.1. Introduction

A QSAR seeks to relate quantitative properties (descriptors) of a compound with other properties such as drug-like activity or toxicity. The essential assumption of QSAR is that quantities that can be conveniently measured or calculated for a compound can be used to accurately predict another property of interest (e.g., antibacterial activity) in a nontrivial way. QSAR has become an integral part of screening programs in pharmaceutical drug-discovery pipelines of small compounds and more recently in toxicological studies *(69)*. However, the use of QSAR modeling applied to the search for antimicrobial peptides is relatively recent. Advances in this area are reviewed in brief here.

There are two separate but interrelated aspects to QSAR modeling of antibacterial peptides: the choice of QSAR descriptors and the choice of numerical analysis techniques used to relate these values to antibacterial activity. A simple example of a QSAR descriptor is the total charge of a peptide. A large number of QSAR descriptors is available for small compounds in the literature and from commercial software products that may be considered. A smaller subset is used in QSAR studies of antibacterial peptides and may be separated into two categories: descriptors based on empirical values and calculated descriptors. An example of an empirical value is HPLC retention time, which is a surrogate measure of solubility or hydrophilicity/hydrophobicity. An example of a calculated descriptor is total peptide charge at pH 7.

In addition to the choice of QSAR descriptors, many statistical learning methods are available to relate the descriptors to the predicted value. There are two main categories of prediction to answer two different questions: regression models (for predicting the activity of a peptide as a continuous variable such as MIC or a surrogate such as in the luminescence assay) or classification where the model is trained to classify as simply active or inactive. Historically, linear methods primarily have been used in determination of QSAR for antimicrobial peptides: principal component analysis (PCA) and a similar method, projections to latent structures (PLS). The advantage of these techniques is they can be used where multiple regression fails: when there are fewer peptides than variables in the model. Choices of a prediction technique also involve trade-offs between model accuracy and meaningfulness. For the purpose of classification, the nonlinear techniques of support vector machines and ANNs are considered to give superior results, but at the cost of introducing rather opaque models that cannot easily be used to shed light on the underlying mechanisms involved. Linear models such as multiple linear regression and principle component analysis, on the other hand, result in models that explicitly relate the input descriptors to the output prediction of activity, with fairly easy-to-interpret parameters. However, while more intuitively satisfying, this clarity comes at the cost of poorer performance *(70)*.

3.2. Previous QSAR Analysis

3.2.1. QSAR of Lactoferricin Derivatives

Several studies have examined the activities of lactoferricin derivatives against bacteria targets *(71–73)* and herpes simplex virus *(74)*. These studies examined the impact of changing specific amino acids in derivatives of lactoferricin, a peptide fragment of lactoferrin that has antimicrobial activity. Strom et al. *(73)* modeled a set of 20 peptides with QSAR descriptors. (See **Table 2** for explanation of descriptors used.) The 20 synthetic peptides were based on derivatives of lactoferricin and included bovine lactorferricin, murine lactoferricin, and 18 variants of murine lactoferricin created by substituting one of three amino acids at four different positions. (See **Table 3** for detailed descriptions of all models and their accuracy.) A total of 20 QSAR descriptors were used. These included α-helicity determined from circular dichroism spectroscopy in SDS and HFIP or calculated by three different means. Other empirical descriptors included HPLC retention time (an indication of hydrophilicity/hydrophobicity), calculated net charge, micelle affinity, three calculated values for α-helicity, calculated hydrophobicity, molecular surface, and two moments due to nonsymmetrical charge and hydrophobicity distribution. In a principal component analysis, the charge-related descriptors and hydrophobicity descriptors had the highest "loadings," indicating that they were most important in describing variation among the peptides.

A later study using the same data set of 20 peptides showed that similar results could be obtained using a set of three descriptors, z values, based on an earlier analysis of changes in peptide properties due to amino acid substitutions. These were found using principal component analysis *(75)* of the effects of

Table 2
QSAR Descriptors Used for Antimicrobial Peptide Studies

Ref.	Descriptor symbols[a] (number)	Explanation
Strom et al., 2001	(12)	
	α-SDS	Measured α-helicity in 30 mM SDS
	α-HFIP	Measured α -helicity in 50% HFIP in water
	Rt	Measured retention time on RP-HPLC C18 column
	Ch	Net charge at pH 7
	MA	Micelle affinity: 1-[(α-HFIP - α-SDS) / αHFIP]
	Ea	Eisenberg α-helix propensity
	Ga	Garnier α-helix propensity
	C-Fa	Chou-Fasman α-helix propensity
	K-D H	Kyte-Doolittle hydrophobicity
	ESI	Emini surface index
	M	Mean hydrophobic moment
	C	Mean charge moment
Hellberg et al., 1987	(29)	
	MW	Molecular weight
	pKCOOH	COOH on Cα
	pKNH2	
	pI	NH2 on n Cα
	—	pH at the isoelectric point
	—	substituent Van der Waals volume
	—	1H NMR for Cα-H (cation)
	—	1H NMR for Cα-H (dipolar)
	—	1H NMR for Cα-H (anion)
	—	13C NMR for C==O
	—	13C NMR for Cα-H
	—	13C NMR for C==O in tetrapeptide
	—	13C NMR for Cα-H in tetrapeptide
	Rf	Rf for l-N-(4-nitrobenzofurazono)amino acids in ethyl acetate/pyridine/water
	dG	Slope of plot 1/(Rf - 1) vs. mol % H2O in paper chromatography
	Rf	dG of transfer of amino acids from organic solvent to water

(Continued)

Table 2
(Continued)

Ref.	Descriptor symbols[a] (number)	Explanation
	log P	Hydration potential or free energy of transfer from vapor phase to water
	logD	Salt chromotography
	dG	Partition coefficient for amino acids in octanol/water
	—	Partition coefficient for acetylamide derivatives ofamino acids in octanol/water
		dG = RTln f; f = fraction buried/accessible amino acids
		HPLC retention times for nine combinations of three different pH and three eluent mixtures (9 descriptors)
Lejon et al., 2001	(3)[a]	
Lejon et al., 2004		
Jenssen et al., 2005	z1	First principal component of Hellberg et al., 1987, related to hydrophilicity
	z2	Second principal component of Hellberg et al., 1987, related to bulk
	z3	Second principal component of Hellberg et al., 1987, related to electronic properties
Frecer et al., 2004	(3)	
	Q_m	Total charge
	AI	Amphipathicity
	$\prod_{o/w}$	Lipophilicity index
Frecer, 2006	(25)	
	Q	Total charge
	L	Overall lipophilicity
	P, N	Lipophilicity of polar and nonpolar faces
	S_p, S_n	Surface area of polar and nonpolar faces
	M_w, M_{wP}, M_{wN}	Molecular mass of polar and nonpolar faces
	C_{SL}, C_{HL}, C_{AR}	Count of small lipophilic, highly lipophilic and aromatic residues forming the nonpolar face total number of hydrogen bond donor and acceptor centers

	HB_{don}, HB_{acc}	Number of hydrogen bond donors and acceptors
	RotBon	Total number of rotatable bonds
	P/L, P/N, L/N, Q/L, Q/N, S_P/S_N, M_{wP}/M_{wN},,Q/C_{SL}, Q/C_{HL} and Q/C_{AR}	A-phipathicity descriptors based on other descriptors
Ostberg and Kaznessis, 2005	(18)	
	MWEI	Molecular weight
	CHRG	Formal charge
	SASA	Solvent accessible surface area
	FOSA	Hydrophobic component of SASA
	NESA	Negative component of SASA
	POSA	Positive component of SASA
	EELE	Electrostatic portion of potential energy
	ESOL	Energy change of solvation in water
	EVDW	Van der Waals portion of potential energy
	FLEX	KierFlex; flexibility of the peptide
	DIPO	Dipole moment of peptide
	PMOI	Principle moment of inertia
	HBAC	Number of hydrogen bond acceptors
	HBDN	Number of hydrogen bond donors
	MVOL	Molecular volume
	DENS	Density of the peptide
	GLOB	Globularity of peptide
	LOGP	Octanol/water partition coefficient
Hilpert et al.,2006	(51)	
	D1, D2, D3	Distance between the first and second arginines, the second and third, and the third and fourth, measured as number of residues
	D_{max}, D_{avg}	Maximum and average distance between any two consecutive arginines
	ASA_H	Water accessible surface are of all hydrophobic atmons
	S3x, S4x, S5x	Hydrophobicity using Eisenberg scale over 3, 4, or 5 residues, for each residue position (x varies from 1 to 12, the length of the peptides)
	S3max, S3min, S4max,	Hydrophobicity maximum and minimum values using Eisenberg scale
	S4min, S5max, S5min	Maximal stretch of hydrophobic residues

(Continued)

Table 2
(Continued)

Ref.	Descriptor symbols[a] (number)	Explanation
	H	Presence of a string of six hydrophobic amino acids including alanine
	6HA	The presence of a string of four amino acids
	First_Residue	First residue flag: first residue is isoleucine: 1, string begins with valine or leucine: 0 , otherwise: 2
Cherkasov, 2005	(77)	(Electronegativity-based)
	EO_Equalized*	Iteratively equalized electronegativity of a molecule
	Average_EO_Pos*	Arithmetic mean of electronegativities of atoms with positive partial charge
	Average_EO_Neg*	Arithmetic mean of electronegativities of atoms with negative partial charge (Hardness-based)
	Global_Hardness	Molecular hardness - reversed softness of a molecule
	Sum_Hardness*	Sum of hardnesses of atoms of a molecule
	Sum_Pos_Hardness	Sum of hardnesses of atoms with positive partial charge
	Sum_Neg_Hardness*	Sum of hardnesses of atoms with negative partial charge
	Average_Hardness*	Arithmetic mean of hardnesses of all atoms of a molecule
	Average_Pos_Hardness	Arithmetic mean of hardnesses of atoms with positive partial charge
	Average_Neg_Hardness*	Arithmetic mean of hardnesses of atoms with negative partial charge
	Smallest_Pos_Hardness*	Smallest atomic hardness among values for positively charged atoms
	Smallest_Neg_Hardness*	Smallest atomic hardness among values for negatively charged atoms
	Largest_Pos_Hardness*	Largest atomic hardness among values for positively charged atoms
	Largest_Neg_Hardness*	Largest atomic hardness among values for negatively charged atoms
	Hardness_of_Most_Pos*	Atomic hardness of an atom with the most positive charge
	Hardness_of_Most_Neg*	Atomic hardness of an atom with the most negative charge

	(Softness based)
Global_Softness	Molecular softness – sum of constituent atomic softnesses
Total_Pos_Softness	Sum of softnesses of atoms with positive partial charge
Total_Neg_Softness*	Sum of softnesses of atoms with negative partial charge
Average_Softness	Arithmetic mean of softnesses of all atoms of a molecule
Average_Pos_Softness	Arithmetic mean of softnesses of atoms with positive partial charge
Average_Neg_Softness*	Arithmetic mean of softnesses of atoms with negative partial charge
Smallest_Pos_Softness	Smallest atomic softness among values for positively charged atoms
Smallest_Neg_Softness	Smallest atomic softness among values for negatively charged atoms
Largest_Pos_Softness	Largest atomic softness among values for positively charged atoms
Largest_Neg_Softness	Largest atomic softness among values for positively charged atoms
Softness_of_Most_Pos	Atomic softness of an atom with the most positive charge
Softness_of_Most_Neg	Atomic softness of an atom with the most negative charge
	(Charge-based)
Total_Charge	Sum of absolute values of partial charges on all atoms of a molecule
Total_Charge_Formal	Sum of charges on all atoms of a molecule (formal charge of a molecule)
Average_Pos_Charge*	Arithmetic mean of positive partial charges on atoms of a molecule
Average_Neg_Charge*	Arithmetic mean of negative partial charges on atoms of a molecule
Most_Pos_Charge	Largest partial charge among values for positively charged atoms
Most_Neg_Charge	Largest partial charge among values for negatively charged atoms
	(Sigma based)
Total_Sigma_mol_i*	Sum of inductive parameters sigma (molecule$\rightarrow$ atom) for all atoms within a molecule
Total_Abs_Sigma_mol_i	Sum of absolute values of group inductive parameters sigma (molecule$\rightarrow$atom) for all atoms within a molecule

(Continued)

Table 2
(Continued)

Ref.	Descriptor symbols[a] (number)	Explanation
	Most_Pos_Sigma_mol_i*	Largest positive group inductive parameter sigma (molecule→atom) for atoms in a molecule
	Most_Neg_Sigma_mol_i*	Largest (by absolute value) negative group inductive parameter sigma (molecule→atom) for atoms in a molecule
	Most_Pos_Sigma_i_mol	Largest positive atomic inductive parameter sigma (atom→molecule) for atoms in a molecule
	Most_Neg_Sigma_i_mol	Largest negative atomic inductive parameter sigma (atom→molecule) for atoms in a molecule
	Sum_Pos_Sigma_mol_i*	Sum of all positive group inductive parameters sigma (molecule →atom) within a molecule
	Sum_Neg_Sigma_mol_i*	Sum of all negative group inductive parameters sigma (molecule →atom) within a molecule
		(Steric based)
	Largest_Rs_mol_i	Largest value of steric influence Rs(molecule→atom) in a molecule
	Smallest_Rs_mol_i*	Smallest value of group steric influence Rs(molecule→atom) in a molecule
	Largest_Rs_i_mol*	Largest value of atomic steric influence Rs(atom→molecule) in a molecule
	Smallest_Rs_i_mol	Smallest value of atomic steric influence Rs(atom→molecule) in a molecule
	Most_Pos_Rs_mol_i	Steric influence Rs(molecule→atom) ON the most positively charged atom in a molecule
	Most_Neg_Rs_mol_i*	Steric influence Rs(molecule→atom) ON the most negatively charged atom in a molecule
	Most_Pos_Rs_i_mol	Steric influence Rs(atom→molecule) OF the most positively charged atom to the rest of a molecule
	Most_Neg_Rs_i_mol*	Steric influence Rs(atom→molecule) OF the most negatively charged atom to the rest of a molecule
		(Other descriptors provided by MOE [85])
	a_acc*	Number of hydrogen bond acceptor atoms
	a_don*	Number of hydrogen bond donor atoms

ASA*	Water accessible surface area
ASA_H*	Water accessible surface area of all hydrophobic atoms
ASA_P*	Water accessible surface area of all polar atoms
ASA-*	Water accessible surface area of all atoms with negative partial charge
ASA+*	Water accessible surface area of all atoms with positive partial charge
FCharge*	Total charge of the molecule
b_1rotN	Number of rotatable single bonds
logP(o/w)*	Log of the octanol/water partition coefficient
logS*	Log of the aqueous solubility
mr	Molecular refractivity
PC-*	Total negative partial charge
PC+*	Total positive partial charge
RPC-	Relative negative partial charge
RPC+*	Relative positive partial charge
TPSA **	Polar surface area
vdw_area*	Van der Waals surface area calculated using a connection table approximation
vdw_vol	Van der Waals volume calculated using a connection table approximation
vol	Van der Waals volume calculated using a grid approximation
VSA	Van der Waals surface area using polyhedral representation
vsa_acc*	Approximation to the sum of VDW surface areas of pure hydrogen bond acceptors
vsa_acid*	Approximation to the sum of VDW surface areas of acidic atoms
vsa_base	Approximation to the sum of VDW surface areas of basic atoms
vsa_don	Approximation to the sum of VDW surface areas of pure hydrogen bond donors
vsa_hyd*	Approximation to the sum of VDW surface areas of hydrophobic atoms
Weight*	Molecular weight

[a]Three descriptors (z1, x2, z3) for each position in the peptide that is varied.
*These descriptors were used in the classification analysis described in the text for Set A and Set B.

Table 3
Summary of Results of QSAR Studies on Antimicrobial Activity

Ref.	Models	Number of descriptors (set size)[a]	Data description	Method[b]	Data size	Accuracy of predictions[c]	Predicted to observed correlation(R^2)[d]
Strom et al., 2001 (73)	Principal component analysis (PCA)	12	Lactoferricin from bovine (LFB)	LOO	20	Plot of predicted to observed MICs shows good correlation	nd
	Projections to latent structures (PLS)		Lactoferricin from murine (LFM) 18 derivatives of LFM that varied at up to 4 positions.				
Lejon et al., 2001 (71)	Projections to latent structures (PLS)	12	Lactoferricin from bovine (LFB) Lactoferricin from murine (LFM), plus 18 derivatives of LFM that varied at up to 4 positions (same data as Strom et al., 2001)	Single training set	20	Good correlation visible in plot	nd
Lejon et al., 2004 (72)	Projections to latent structures (PLS)	15	LFB derivatives	Single training set	11	Prediction of 2 peptides E. coli and S. aureus outside training set was poor	0.957 (E. coli) 0.924 (S. aureus)

Reference	Method		Peptide set	Validation	n	Comments	Result
	Projections to latent structures (PLS)	39	LFB derivatives	Single training set	23	Prediction of 2 peptides *E.coli* and *S. aureus* outside training set was poor	0.966 (*E. coli*) 0.905 (*S. aureus*)
Jenssen et al., 2005 (74)	PCA on all descriptors and biological response	42	LFB, LFM, LFC (goat), LFH (human) derivatives	Single training set	50	Prediction of 1 peptide for *S. aureus* was good	0.79 (*E. coli*) 0.75 (*S. aureus*)
	Projections to latent structures	75	Lactoferricin derivatives	Single training set	7	—	0.8989 (HSV-1) 0.8276 (HSV-2)
Frecer et al., 2004 (76)	Multivariate linear regression	3	Cyclic peptides similar to protegrin	LOO	7	—	0.98
Frecer, 2006 (77)	Multiple linear regression chosen using genetic function approximation (GFA) allowing up to 5 descriptors in the final models	2 (25)	Potegrin 1 derivatives	LOO	97	—	0.604
Ostberg and Kaznessis, 2005 (78)	Multiple linear regression using the most statistically significant combinations of descriptors	5 (18)	Protegrin 1 derivatives	Single training set	55	—	0.69

(Continued)

Table 3
(Continued)

Ref.	Models	Number of descriptors (set size)[a]	Data description	Method[b]	Data size	Accuracy of predictions[c]	Predicted to observed correlation(R^2)[d]
Hilpert et al..2006 (36)	Binary classification algorithm	51	Scambled variants of Bac2A	Single training set	49	Prediction of active/inactive was 74% accurate on training data	nd
Cherkasov.2005a (79) for descriptors only	Artificial neural networks (ANN)₃	44 (77)	Random peptides chosen according to two amino acid frequency distributions Sets A and B contained 933 and 500 peptides, respectively (*see* text for details, unpublished data)	Training and validation within one set, independent testing on second set	1433	Set A models predicted activity with up to 83% accuracy on Set B Set B models predicted up to 43% accuracy on Set A (*see* text for details)	nd

[a]Where present, number in parentheses is the number of available descriptors from which the descriptors used in the model were chosen.
[b]Method indicates the method of evaluating the model: LOO indicates leave-one-one cross-validation, single training set indicates that the same dataset used to build the model was used to determine accuracy.
[c]Describes accuracy where cross-correlation squared parameter was not calculated; nd = not determined.
[d]Cross-correlation parameter squared; nd = not determined.

29 physicochemical properties for groups of related peptides that were varied at certain positions. These properties included calculated properties such as molecular weight and empirical values such as HPLC retention times at nine different conditions of pH and eluent mixture. The first three principal components were named "z values" and were obtained from these 29 properties. They represent a mixture of these 29 properties in fixed proportions, the so-called "loadings". These loadings represent the importance of each of the 29 properties on the z value. These z values have the important property that they describe variations in the data independently of one another (they are orthogonal in a mathematical sense). The value of each of the three z values is related approximately to hydrophilicity/hydrophobicity, volume, and electronic properties. The result is that the properties of highly similar peptides can be represented quite well by the three z values in place of the original 29 descriptors used by Hellberg et al. *(75)* or the 20 used by Strom et al. *(73)*. Lejon et al. *(71)* found that similar results could be obtained for the same data set of 20 lactoferricin and lactoferricin derivates using the z values to describe the effect of amino acid substitution on peptide helicity, HPLC retention times, and antibacterial activity. Further studies on an expanded set of peptides *(72)* again demonstrated good predictive ability using z values for peptide analogues where only a few amino acid substitutions were made. However, as noted by the authors, the predictions became lower in accuracy when more than one or two substitutions were made in a single peptide; this indicates the limitation of the approach for more general antibacterial prediction.

3.2.2. QSAR of Protegrin Analogues and De Novo Peptides

In another approach, Frecer et al. *(76)* used a de novo design strategy to produce synthetic peptides with structural similarity to cyclic β-sheet defense peptides such as protegrin. The peptides were designed based on assumptions of mechanisms of antibacterial activity and the importance of charge, amphipathicity and lipophilicity (hydrophobicity). Peptides were designed using a combination of molecular modeling, molecular dynamics, and docking methods to have a structure containing tandemly repeated cationic and nonpolar amino acids that forms an amphipathic β-hairpin that binds to lipid A (an important surface structure on gram-negative bacteria). A total of seven peptides were constructed and synthesized for QSAR analysis. Only three descriptors were used to model antibacterial, hemolytic, and cytotoxicity: total charge, an amphipathicity index, and a lipophilicity index. Antimicrobial activity was found to give good correlation with charge and amphipathicity, while hemolytic activity was largely determined by lipophilicity, and cytotoxic effects on monocytes were due mainly to charge and amphipathicity index. This suggested a strategy of substituting amino acids in the hydrophobic face with less lipophilic ones to decrease hemolytic effect without significantly affecting antimicrobial or cytotoxic activity.

In a second paper, Frecer *(77)* performed QSAR analysis on 97 protegrin derivatives of 14 amino acids in length, whose activity was already published. In this study he calculated 14 descriptors including features such as charge, overall lipophilicity, separate lipophilicity of polar and nonpolar faces of the molecule, molecular surface areas for polar and nonpolar faces, total numbers of lipophilic and aromatic residues, and numbers of hydrogen bond donors and receivers. In addition, 10 amphipathicity measures were calculated from these involving ratios of, for example, charge to overall lipophilicity. A genetic function approximation (GFA) was used to generate linear equations involving up to five descriptors to describe antibacterial and hemolytic activity. Models were evaluated for lack-of-fit score and the best equation found. They found only moderate predictive power with antibacterial activity due mostly to charge and amphipathicity (ratio of charge to lipophilicity of nonpolar face). Also, hemolytic activity was found to be due to lipophilicity of the nonpolar face for this set of peptides with moderate correlation.

Ostberg and Kaznessis *(78)* examined protegrin and analogues using QSAR descriptors such as charge, molecular weight, as well as molecular structural properties such as volume, density, globularity, energy components, and solvent accessible surface area (SASA). The data set in this study consisted of 62 protegrin and analogues and the multivariate linear regression produced moderate correlation between predicted and actual activity: antibacterial activity was found using five descriptors, four descriptors for cytotoxicity, and four descriptors for hemolysis.

3.2.3. QSAR of Scrambled Bactenecin-Derived Peptides

A linear variant of the bovine cationic peptide bactenecin, Bac2A, has been used in studies of positional importance of amino acids. Hilpert et al. *(36)* examined the effect of scrambling the amino acid sequence of Bac2A and investigated the activity of the resulting peptides. A QSAR analysis was performed on a total of 49 peptides using 18 descriptors based largely on positions of arginines, distributions of hydrophobic amino acids, and water-accessible surface. A binary classification algorithm was used to create a decision tree to classify peptides that are active or inactive, with an accuracy of 74% trained on the full set of peptides.

3.3. Limitations of Current Studies

There are several limitations of existing QSAR modeling of antibacterial activity. The primary limitation concerns the size of the data sets. Despite the use of z values to reduce the number the variables used for analysis of lacto-ferricin derivatives, the number of peptides (the number of equations) is still small compared to the number of variables. While techniques such as PLS or PCA can be used to produce models based on this amount of data, other powerful modeling techniques will not work. In addition to having more confidence in parameter estimation, having much more data than number of variables also allows more definitive determination of model performance using techniques such as cross-validation whereby a representative fraction of the data is set aside for testing after models have been built on the remaining data.

Another limitation of published QSAR studies into antibacterial activity of peptides is due to the types of models used. Both PCA and PLS produce models that are linear with respect to the QSAR descriptors. Often the choice to use simpler linear models is made deliberately (e.g., as stated *[76,77]*) since the resulting models allow straightforward interpretation for predicting improvements in peptide activity. However, more complex models such as ANNs are capable of modeling nonlinear relationships as well, where descriptors interact with one another in a nonadditive manner. The main disadvantages of using more complex models such as ANNs are the cryptic nature of the models produced (contributions of individual descriptors to activity are not clear) and the requirement for larger amounts of data, since the number of parameters used to determine the model grows more quickly than the number of QSAR descriptors on which it is based.

3.4. Inductive QSAR Descriptors

3.4.1. Introduction

In addition to the modeling techniques, QSAR descriptors used to date in modeling antibacterial peptides have relied on sets of peptides with a high degree of similarity. The QSAR descriptors for the lactoferricin studies have measured differences in activity between peptides that differ in only a few amino acids, for example. However, a set of QSAR descriptors has been developed recently that allows for full 3D-sensitive properties of peptides: the inductive QSAR descriptors (reviewed in ref. *80*), which we feel may dramatically improve the prediction of activity of antimicrobial peptides by allowing models to be constructed that are valid for a large range of peptides rather than in narrow range of nearly identical peptides.

Previously, inductive QSAR descriptors have been successfully applied to a number of molecular modeling studies, including quantification of antibacterial activity of organic compounds *(89)*, calculation of partial charges in small molecules and proteins *(81)*, and in comparative docking analysis as well as in *in silico* lead discovery *(82)*. Inductive QSAR descriptors have been used in models that can distinguish between antimicrobial compounds, conventional drugs, and drug-like substances and have been shown to give up to 97% accurate separation of the three types of molecular activities on an extensive set of 2686 chemical structures *(83)*. These descriptors have been used in different types of models for classification of compounds, from ANNs, *k*-nearest neighbors, linear discriminative analysis, and multiple linear regression. It has been found that ANNs result generally in more accurate predictions, followed closely by k-nearest neighbors methods *(84)*.

3.4.2. Results

The use of QSAR descriptors for prediction and understanding of the activity of antimicrobial peptides has previously been limited to comparisons between peptides that differ in only a small number of amino acids. This has been primarily due to the cost and difficulty of producing large numbers of peptides as well as the cost of assaying their activity. However, with the recent advance in high-throughput peptide synthesis technique in combination with rapid assay of activity with the luminescence-based assay, robust amounts of data have begun to be available *(35)*.

Historically QSAR modeling of antibacterial peptides has mainly utilized descriptors that are designed to model differences in properties of similar peptides, or used ones such as charge, amphipathicity, and lipophilicity, whose relationship is known a priori from amino acid substitution studies. Where larger sets of QSAR descriptors have been used for the protegrin and analogues, the models have been limited to linear regression, resulting in lower predictive ability. We believe that the use of inductive QSAR descriptors within a complex modeling methodology such as ANNs may yield dramatic improvements in the near future in prediction of antimicrobial activity of highly dissimilar peptides. While ANNs do not directly explain the contributions of each QSAR descriptor, it is clear from previous studies that relevant 3D structures are being captured. Use of trained ANNs allows the rapid in silico screening of larger numbers of potential peptides that may be synthesized as candidates for drug leads without a detailed understanding of the contributions of individual molecular properties.

To be successful, this approach requires much larger sets of peptides with a wide range of activity and amino acid sequence diversity, as can be provided by the recent advances in high-throughput peptide synthesis and antibacterial assay *(31)*. We have recently performed such a study utilizing two sets of synthetic peptides containing 933 peptides (Set A) and 500 peptides (Set B). We describe the results briefly here. The two sets of peptides were synthesized and measured using the luminescence assay for activity against *P. aeruginosa* yielding a relative IC_{50} (as a fraction of the IC_{50} of Bac2A). These peptides showed a wide range of activity, from inactive to much more active than the control peptide Bac2A. We then calculated the set of QSAR descriptors (descriptors due to Cherksov et al., indicated in **Table 2**) for each peptide using molecular modeling software *(85)*. (Because of computational time constraints and uncertainty concerning the appropriate water or lipid environment of the active peptide, the 3D structure of the peptides was not calculated; rather an initial, α-helical structure was used.) Where descriptor values correlated with another descriptor at greater than 0.95 (absolute value), one descriptor was dropped so that the descriptor set did not cause problems in model building. A total of 44 descriptors were identified for modeling (indicated in **Table 2** with *).

We wished to determine if inductive QSAR combined with an advanced machine learning technology such as ANNs is able to distinguish peptides with high activity versus those without high activity, as such a capability would be powerful for an *in silico* search for completely novel antibiotic peptides. (The structure of an ANN is shown in **Fig. 2**.) We therefore classified each peptide as more active (<0.75 IC_{50} of Bac2A) or not more active (>0.75 IC_{50} of Bac2A)

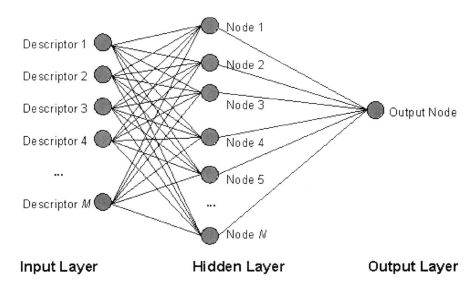

Fig. 2. Structure of an artificial neural network. The network consists of three layers: the input layer, the hidden layer, and the output layer. The input nodes take the values of the normalized QSAR descriptors. Each node in the hidden layer takes the weighted sum of the input nodes (represented as lines) and transforms the sum into an output value. The output node takes the weighted sum of these hidden node values and transforms the sum into an output value between 0 and 1.

than control. Given the large amount of data available, a rigorous methodology was feasible. We performed a 10-fold stratified, cross-validation where each of the two sets of peptides were randomly divided into 10 groups, requiring the number of active peptides in each section to be approximately the same in each group. By consecutively leaving one group of 10% out for use in validation and combining the other 9 groups into a single training group, 10 groups were constructed for training the models. ANNs were constructed using simulation software *(86)*, by training on each of the 10 training groups and using the left-out validation group to signal the end of training and preventing overfitting. (Overfitting occurs when there are too few data for the complexity of the model: the model tends then to "memorize" the data and while it describes the training data very well, it performs poorly when used to predict data outside the training set.)

The performance of any model built on data is best measured by accessing the predictions of the model in situations that did not exist in the training data. Given the large amounts of peptide activity data we had available, we chose to assess the model predictions on the set of data that was not involved in any way in the construction of the models: we used the Set A models to predict the activity of Set B peptides, and Set B models to predict activity of Set A peptides. Rather than attempt to predict IC_{50}, each ANN produced a single number between 0 and 1 to indicate the likelihood that the peptide was more active than Bac2A. An example of the performance of one selected ANN model built on 90% of Set A peptides for predicting Set B peptide activity is shown in **Fig. 3**. The positive predictive value (PPV, the fraction of peptides that are active out of all peptides predicted active) is quite low for ANN output threshold values up to 0.9; but the PPV increases dramatically to 1.0 for threshold values greater than 0.9. For threshold values less than 0.9, most peptides predicted to be active are in fact inactive (PPV < 0.5). However, for high threshold values (<0.999) nearly all peptides predicted to be active are in fact active (PPV > 0.5), but few peptides are predicted to be active (only 2% of peptides at threshold of 0.999). For such high threshold values, many active peptides are incorrectly classified as inactive. However, this may not be a problem for an in silico screening of large numbers of peptides where high confidence is needed that peptides predicted to be active are worth the expense of experimental investigation. Such screening programs may not be concerned that many active peptides are incorrectly classified as inactive, only that those predicted active are most likely active.

This is illustrated by examining the activities of test peptides ranked according to the ANN outputs (values near 1 indicating likelihood to be more active peptides, and values close to 0 indicating likelihood to not be more active peptides). The results are summarized in **Table 4**. Here, the proportion of

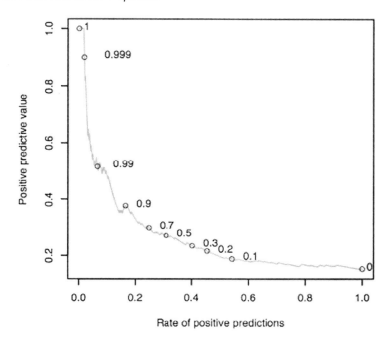

Fig. 3. Predictive performance for active peptides by an ANN model. The positive predictive value (PPV, number of true positives/(true positives + false positives)) is shown versus the rate of positive predictions ((true positives + false positives)/(all predictions)) for varying ANN output threshold. ANN outputs above the threshold are positive predictions (i.e., predict peptide is active). For example, the threshold value of 0.5 on the curve corresponds to about 30% of the peptides are predicted to be active (positive) from the position on the *x*-axis, and the PPV is 0.27 from the *y*-axis. For the threshold value of 0.999, the PPV is 0.90, but only 2% of peptides are predicted active.

peptides is shown that match the prediction for several positions of the peptides in the list. The data indicate good predictive accuracy for the topmost and bottommost peptides in the ranked list. The models built on the larger, more diverse set of peptides (Set A) were much better able to predict the activity of the other, smaller set (Set B). The enhancement ratio (the effectiveness of using the ANN-inductive QSAR modeling compared to without) is an impressive 12.6 (83.0/6.6) for models built on Set A predicting Set B, and 4.7 (43.0/9.2) for Set B models predicting Set A in the top section of peptides.

These results show that the new availability of high-quantity and -quality antibacterial peptide data allows more rigorous treatment and evaluation using machine learning techniques. In addition, the diversity of peptide sequences is dramatically improved—no longer is it necessary to start with a prototype

Table 4
Accuracy of Inductive QSAR-ANN Models for Predicting High-Activity Peptides

Section[a]	Accuracy	
	Set A models[b][mean (SD) %]	Set B models[b][mean (SD) %]
Top 10	83.0 *(11)*	43.0 (9.5)
Top 25	55.6 (3.0)	44.8 (9.2)
Top 50	45.8 (3.6)	41.8 (3.3)
Top 100	33.3 (2.1)	35.0 (3.3)
Bottom 100	92.4 (2.5)	96.8 (2.1)
Bottom 50	96.4 (1.6)	97.8 (2.2)
Bottom 25	95.6 (1.3)	97.6 (2.8)
Bottom 10	92.0 (4.2)	97.0 (4.8)

Set A models were trained on Set A peptides and tested using Set B peptides. Set B models were trained on Set B peptides and tested using Set A peptides.

[a]The section column indicates the selection of the ranked peptides that were used in calculating accuracy; for example, Top 10 indicates the peptides ranked by model output in the top 10 by the ANN model were considered.

[b]Numbers indicate the percentage of peptides in the section that were predicted correctly: for the top sections, these were correctly predicted more active (relative IC_{50} was > 1.0); for the bottom sections, these are are the percentage of peptides correctly predicted less active (relative IC_{50} was < 1.0).

peptide of known activity in order to screen for potentially improved peptides. These techniques, such as artificial neural network modeling combined with atomic-resolution inductive QSAR methodology, are therefore expected to allow in silico screening of very large numbers of antibacterial peptides that may lead to novel therapeutics in an efficient and effective manner.

References

1. Hancock, R.E.W. and Lehrer, R. (1998) Cationic peptides: a new source of antibiotics. *Trends Biotechnol.* **16**, 82–88.
2. Finking, R. and Marahiel, M.A. (2004) Biosynthesis of nonribosomal peptides. *Annu. Rev. Microbiol.* **58**, 453–488.
3. Garcia-Olmedo, F., Molina, A., Alamillo J.M., and Rodriguez-Palenzuela P. (1998) Plant defense peptides. *Biopolymers* **47**, 479–491.
4. Kawabata, S., Beisel, H.G., Huber, R., et al. (2001) Role of tachylectins in host defense of the Japanese horseshoe crab Tachypleus tridentatus. *Adv. Exp. Med. Biol.* **484**, 195–202.
5. Bulet, P., Stocklin, R. and Menin, L. (2004) Anti-microbial peptides: from invertebrates to vertebrates. *Immunol. Rev.* **198**, 169–184.

6. Iwanaga, S. and Kawabata, S. (1998) Evolution and phylogeny of defense molecules associated with innate immunity in horseshoe crab. *Front. Biosci.* **3**, D973–D984.

7. Imler, J.L. and Bulet, P. (2005) Antimicrobial peptides in Drosophila: structures, activities and gene regulation. *Chem. Immunol. Allergy* **86**, 1–21.

8. Cannon, J.P., Haire, R.N. and Litman G.W. (2002) Identification of diversified genes that contain immunoglobulin-like variable regions in a protochordate. *Nat. Immunol.* **3**, 1200–1207.

9. Litman, G.W., Anderson, M.K. and Rast, J.P. (1999) Evolution of antigen binding receptors. *Annu. Rev. Immunol.* **17**, 109–147.

10. Nochi, T. and Kiyono, H. (2006) Innate immunity in the mucosal immune system. *Curr. Pharm. Des.* **12**, 4203–4213.

11. Yang, D., Biragyn, A., Hoover, D.M., Lubkowski, J. and Oppenheim, J.J. (2004) Multiple roles of antimicrobial defensins, cathelicidins, and eosinophil-derived neurotoxin in host defense. *Annu. Rev. Immunol.* **22**, 181–215.

12. Li, J., Xu, X., Xu, C., et al. (2007) Anti-infection peptidomics of amphibian skin. *Mol. Cell. Proteomics.* Epub ahead of print, Jan 31, 2007.

13. Yang, D., Biragyn, A., Kwak, L.W. and Oppenheim, J.J. (2002) Mammalian defensins in immunity: more than just microbicidal. *Trends Immunol.* **23**, 291–296.

14. Jenssen, H., Hamill, P. and Hancock, R.E.W. (2006) Peptide antimicrobial agents. *Clin. Microbiol. Rev.* **19**, 491–511.

15. Papo, N. and Shai, Y. (2005) Host defense peptides as new weapons in cancer treatment. *Cell. Mol. Life Sci.* **62**, 784–790.

16. Gallo, R.L., Ono, M., Povsic, T., et al. (1994) Syndecans, cell surface heparan sulfate proteoglycans, are induced by a proline-rich antimicrobial peptide from wounds. *Proc. Natl. Acad. Sci. USA* **91**, 11035–11039.

17. Hancock, R.E.W. and Rozek, A. (2002) Role of membranes in the activities of antimicrobial cationic peptides. *FEMS Microbiol. Lett.* **206**, 143–149.

18. Mani, R., Cady, S.D., Tang, M., Waring, A.J., Lehrer, R.I. and Hong, M. (2006) Membrane-dependent oligomeric structure and pore formation of a beta-hairpin antimicrobial peptide in lipid bilayers from solid-state NMR. *Proc. Natl. Acad. Sci. USA* **103**, 16242–16247.

19. Hwang, P.M. and Vogel, H.J. (1998) Structure-function relationships of antimicrobial peptides. *Biochem. Cell. Biol.* **76**, 235–246.

20. Gazit, E., Miller, I.R., Biggin, P.C., Sansom, M.S.P., and Shai, Y. (1996) Structure and orientation of the mammalian antibacterial peptide cecropin P1 within phospholipid membranes. *J. Mol. Biol.* **258**, 860–870.

21. Brogden, K.A. (2005) Antimicrobial peptides: pore formers or metabolic inhibitors in bacteria? *Nat. Rev. Microbiol.* **3**, 238–250.

22. Paschke, M. (2006) Phage display systems and their applications. *Appl. Microbiol. Biotechnol.* **70**, 2–11.

23. Westerlund-Wikstrom, B. (2000) Peptide display on bacterial flagella: principles and applications. *Int. J. Med. Microbiol.* **290**, 223–230.

24. Yan, X. and Xu, Z. (2006) Ribosome-display technology: applications for directed evolution of functional proteins. *Drug Discov. Today* **11**, 911–916.

25. Pini, A., Giuliani, A., Falciani, C., et al. (2005) Antimicrobial activity of novel dendrimeric peptides obtained by phage display selection and rational modification. *Antimicrob. Agents Chemother.* **49**, 2665–2672.

26. Xie, Q., Matsunaga, S., Wen, Z., et al. (2006) In vitro system for high-throughput screening of random peptide libraries for antimicrobial peptides that recognize bacterial membranes. *J. Pept. Sci.* **12**, 643–652.

27. Houghten, R.A. (1985) General method for the rapid solid-phase synthesis of large numbers of peptides: Specificity of antigen-antibody interaction at the level of individual amino acids. *Proc. Natl. Acad. Sci. USA* **82**, 5131–5135.

28. Pellois, J.P., Zhou, X., Srivannavit, O., Zhou, T., Gulari, E. and Gao, X. (2002) Individually addressable parallel peptide synthesis on microchips. *Nat. Biotechnol.* **20**, 922–926.

29. Geysen, H.M., Meloen, R.H. and Barteling, S.J. (1984) Use of peptide synthesis to probe viral antigens for epitopes to a resolution of a single amino acid. *Proc. Natl. Acad. Sci. USA* **81**, 3998–4002.

30. Frank, R. (1992) Spot synthesis: an easy technique for positionally addressable, parallel chemical synthesis on a membrane support. *Tetrahedron* **48**, 9217–9232.

31. Hilpert, K. and Hancock, R.E.W. (2007) Use of luminescent bacteria for rapid screening and characterization of short cationic antimicrobial peptides synthesized on cellulose by peptide array technology. *Nature Protocols*, **3**, 1652–1660.

32. Kamradt, T. and Volkmer-Engert, R. (2004) Cross-reactivity of T lymphocytes in infection and autoimmunity. *Mol. Divers.* **8**, 271–280.

33. Hilpert, K., Winkler, F.H.D. and Hancock, R.E.W. (2007) Peptide arrays on cellulose support: SPOT synthesis - a time and cost efficient method for synthesis of large numbers of peptides in a parallel and addressable fashion. *Nature Protocols*, **2**, 1333–1349.

34. Lewenza, S., Falsafi, R.K., Winsor, G., et al. (2005) Construction of a mini-Tn5-luxCDABE mutant library in Pseudomonas aeruginosa PAO1: a tool for identifying differentially regulated genes. *Genome Res.* **15**, 583–589.

35. Hilpert, K., Volkmer-Engert, R., Walter, T. and Hancock, R.E.W. (2005) High-throughput generation of small antibacterial peptides with improved activity. *Nat. Biotechnol.* **23**, 1008–1012.

36. Hilpert, K., Elliott, M.R., Volkmer-Engert, R., et al. (2006) Sequence requirements and an optimization strategy for short antimicrobial peptides. *Chem. Biol.* **13**, 1101–1107.

37. Gennaro, R., Skerlavaj, B., and Romeo, D. (1989) Purification, composition and activity of two bactenecins, antibacterial peptides of bovine neutrophils. *Infect. Immun.* **57**, 3142–3146.

38. Scocchi, M., Romeo, D., and Zanetti, M. (1994) Molecular cloning of Bac7, a praline- and arginine-rich antimicrobial peptide from bovine neutrophils. *FEBS Lett.* **352**, 197–200.

39. Wu, M. and Hancock, R.E.W. (1999) Improved derivatives of bactenecin, a cyclic dodecameric antimicrobial cationic peptide. *Antimicrob. Agents Chemother.* **43**, 1274–1276.

40. Moore, A.J., Beazley, W.D., Bibby, M.C. and Devine, D.A. (1996) Antimicrobial activity of cecropins. *J Antimicrob. Chemother.* **37**, 1077–1089.

41. Sipos, D., Andersson, M., and Ehrenberk, A. (1992) The structure of the mammalian antibacterial peptides cecropin P1 in solution, determined by proton-NMR. *Eur. J. Biochem.* **209**, 163–169.

42. Turner, J., Cho, Y., Dinh, N.N., Waring, A.J., and Lehrer, R.I. (1998) Activities of LL-37, a cathelin-associated antimicrobial peptide of human neutrophils. *Antimicrob. Agents Chemother.* **42**, 2206–2214.

43. Zimmermann, G.R., Legault, P., Selsted, M.E. and Pardi, A. (1995) Solution structure of bovine neutrophil beta-defensin-12: the peptide fold of the beta-defensins is identical to that of the classical defensins. *Biochemistry* **34**, 13663–13671.

44. Mandal, M., Jagannadham, M.V. and Nagaraj, R. (2001) Antibacterial activities and conformations of bovine β-defensin BNBD-12 and analogs: structural and disulfide bridge requirements for activity. *Peptides* **23**, 413–418.

45. Lee, D.L., and Hodges, R.S. (2003) Structure-activity relationships of de novo designed cyclic antimicrobial peptides based on Gramicidin S. *Biopolymers* **71**, 28–48.

46. Wadhwani, P., Afonin, S., Ieronima, M., Buerck, J., and Ulrich, A.S. (2006) Optimizad protocol for síntesis of cyclic Gramicidin S: starting amino acid is key to high yield. *J. Org. Chem* . **71**, 55–61.

47. Ryge, T.S., Doisy, X., Ifrah, D., Olsen, J.E., and Hansen, P.R. (2004) New indolicidin analogues with potent antibacterial activity. *J. Pept. Res.* **64**, 171–185.

48. Selsted, M.E., Tang, Y.-Q., Morris, W.L., et al. (1993) Purification, primary structures, and antibacterial activities of b-defensins, a new family of antimicrobial peptides from bovine neutrophils. *J. Biol. Chem.* **268**, 6641–6648.

49. Samuelsen ,O., Haukland, H.H., Ulvatne, H. and Vorland, L.H. (2004) Anticomplement effects of lactoferrin-derived peptides. *FEMS Immunol. Med. Microbiol.* **41**, 141–148.

50. Chen, H.-L., Yen, C.-C., Lu, C.-Y., Yu, C.-H., and Chen, C.-M. (2006) Synthetic porcine lactoferricin with a 20 residue peptide exhibits antimicrobial activity against *Escherichia coli*, *Staphylococcus aureus*, and *Candida albicans*. *J. Agric. Food Chem.* **54**, 3277–3282.

51. Dorschner, R.A., Pestonjamasp, V.K., Tamakuwala, S., et al. (2001) Cutaneous injury induces the release of cathelicidin anti-microbial peptides active against group A streptococcus. *J. Invest. Dermatol.* **117**, 91–97.

52. Gudmundsson, G.M., Lidholm, D.-A., Åsling, B., Gan, R., and Boman, H.G. (1991) The cecropin locus. Cloning and expression of a gene cluster encoding three antibacterial peptides in *Hyalaphora cecropia*. *J. Biol. Chem.* **266**, 11510–11517.

53. Gesell, J., Zasloff, M. and Opella, S.J. (1997) Two-dimensional 1H NMR experiments show that the 23-residue magainin antibiotic peptide is an alpha-helix in dodecylphosphocholine micelles, sodium dodecylsulfate micelles, and trifluoroethanol/water solution. *J. Biomol. NMR* **9**, 127–135.

54. Wei, G., Campagna, A.N. and Bobek, L.A. (2006) Effects of MUC7 peptides on the growth of bacteria and on *Streptococcus mutans* biofilm. *J. Antimicrob. Chemother.* **57**, 1100–1109.

55. Giacometti, A., Cirioni, O., Barchiesi, F., Del Prete, M.S. and Scalise G. (1999). Antimicrobial activity of polycationic peptides. *Peptide* **20**, 1265–1273.

56. Giacometti, A., Cirioni, O., Barchiesi, F., Fortuna, M., and Scalise, G. (1999) In-vitro activity of cationic peptides alone and in combination with clinically used antimicrobial agents against *Pseudomonas aeruginosa. J. Antimicrob. Chemother.* **44**, 641–645.

57. Olasupo, N.A., Fitzgerald, D.J., Gasson, M.J., and Narbad, A. (2003) Activity of natural antimicrobial compounds against *Escherichia coli* and *Salmonella enterica* serovar Typhimurium. *Lett. Appl. Microbiol.* **36**, 448–451.

58. van den Hooven, H.W., Doeland, C.C.M., van de Kamp, M., Konings, R.N.H., Hilbers, C.W., and van de Ven, F.J.M. (1996) Three-dimensional structure of the lantibiotic nisin in the presence of membrane-mimetic micelles of dodecylphospho-choline and of dodecylsulphate. *Eur. J. Biochem.* **235**, 382–393.

59. Clausell, A., Rabanal, F., Garcia-Subirats, M., Alsina, M.A., and Cajal, Y. (2005) Synthesis and membrane action of polymyxin B analogues. *Luminescence.* **20**, 117–123.

60. Li, C., Lewis, M.R., Gilbert, A.B., et al. (1999) Antimicrobial activities of amine- and guanidine- functionalized cholic acid derivatives. *Antimicrob. Agents Chemother.* **43**, 1347–1349.

61. Regna, P.P., Solomons, I.A., Forscher, B.K., and Timreck, A.E. (1949) Chemical studies on polymyxin B. *J. Clin. Invest.* **28**, 1022–1027.

62. Agerberth, B., Lee, J.Y., Bergman, T., et al. (1991) Amino acid sequence of PR-39, isolation from pig intestine of a new member of the family of Pro, Arg, rich antibac-terial peptides. *Eur. J. Biochem.* **202**, 849–854.

63. Shi, J., Ross, C.R., Chengappa, M.M., Sylte, M.J., McVey, D.S., and Blecha, F. (1996) Antibacterial activity of a synthetic peptide (PR-26) derived from PR-39, a praline-arginine-rich neutrophil antimicrobial peptide. *Antimicrob. Agents Chemother.* **40**, 115–121.

64. Aumelas, A., Mangoni, M., Roumestand, C., et al. (1996) Synthesis and solution structure of the antimicrobial peptide protegrin-1. *Eur. J. Biochem.* **237**, 575–583.

65. Fahrner, R.L, Dieckmann, T., Harwig, S.S.L., Lehrer, R.I., Eisenberg, D., and Feigon, J. (1996) Solution structure of protegrin-1, a broad-spectrum antimicrobial peptide from porcine leukocytes. *Chem. Biol.* (London) **3**, 543–550.

66. Steinberg, D.A., Hurst, M.A., Jufii, C.A., et al. (1997) Protegrin-1: a broad-spectrum, rapidly microbicidal peptide with in vivo activity. *Antimicrob. Agents Chemother.* **41**, 1738–1742.

67. Kawano, K., Yoneya, T., Miyata, T., et al. (1990) Antimicrobial peptide, tachy-plesin I, isolated from hemocytes of the horseshoe crab (*Tachypleus tridentatus*). *J. Biol. Chem.* **265**, 15365–15367.

68. Nakamura, T., Furunaka, H., Miyata, T., Tokunaga, F., Muta, T., and Iwanaga, S. (1988) Tachyplesin, a class of antimicrobial peptide from the hemocytes of the horshoe crab (*Tachypleus tridentatus*). *J. Biol. Chem.* **263**, 16709–16713.

69. Perkins, R., Fang, H., Tong, W., and Welsh, W.J. (2003) Quantitative structure-activity relationship methods: perspectives on drug discovery and toxicology. *Environ. Toxicol. Chem.* **22**, 1666–1679.

70. Weaver, D.C. (2004). Applying data mining techniques to library design, lead generation and lead optimization. *Curr. Opin. Chem. Biol.* **8**, 264–270.
71. Lejon, T., Strom, M.B., and Svendsen, J.S. (2001) Antibiotic activity of pentadecapeptides modelled from amino acid descriptors. *J. Pept. Sci.* **7**, 74–81.
72. Lejon, T., Stiberg, T., Strom, M.B., and Svendsen, J.S. (2004) Prediction of antibiotic activity and synthesis of new pentadecapeptides based on lactoferricins. *J. Pept. Sci.* **10**, 329–335.
73. Strom, M.B., Stensen, W., Svendsen, J.S., and Rekdal, O. (2001) Increased antibacterial activity of 15-residue murine lactoferricin derivatives. *J. Pept. Res.* **57**, 127–139.
74. Jenssen, H., Gutteberg, T.J., and Lejon, T. (2005) Modeling of anti-HSV activity of lactoferricin analogues using amino acid descriptors. *J. Pept. Sci.* **11**, 97–103.
75. Hellberg, S., Sjostrom, M., Skagerberg, B., and Wold, S. (1987) Peptide quantitative structure-activity relationships, a multivariate approach. *J. Med.Chem.* **30**, 1126–1135.
76. Frecer, V., Ho, B., and Ding, J.L. (2004) De novo design of potent antimicrobial peptides. *Antimicrob. Agents Chemother.* **48**, 3349–3357.
77. Frece, V. (2006) QSAR analysis of antimicrobial and haemolytic effects of cyclic cationic antimicrobial peptides derived from protegrin-1. *Bioorgan. Medic. Chem.* **14**, 6065–6074.
78. Ostberg, N., and Kaznessis, Y. (2004) Protegrin structure–activity relationships: using homology models of synthetic sequences to determine structural characteristics important for activity. *Peptides* **26**, 197–206.
79. Cherkasov, A. (2005) Inductive QSAR descriptors. Distinguishing compounds with antibacterial activity by artificial neural networks. *Int. J. Mol. Sci.* **6**, 63–86.
80. Cherkasov, A. (2005) 'Inductive' descriptors. 10 successful years in QSAR. *Curr. Computer-Aided Drug Design* **1**, 21–42.
81. Cherkasov, A. (2003) Inductive electronegativity scale. Iterative calculation of inductive partial charges. *J. Chem. Inf. Comp. Sci.* **43**, 2039–2047.
82. Cherkasov, A., Shi, Z., Fallahi, M., and Hammond, G.L. (2005) Successful in silico discovery of novel non-steroidal ligands for human sex hormone binding globulin. *J. Med. Chem.* **48**, 3203–3213.
83. Karakoc, E., Sahinalp, S.C., and Cherkasov, A. (2006) Comparative QSAR- and fragments distribution analysis of drugs, druglikes, metabolic substances, and antimicrobial compounds. *J. Chem. Inf. Model.* **46**, 2167–2182.
84. Karakoc, E., Cherkasov, A., and Sahinalp, S.C. (2006) Distance based algorithms for small biomolecule classification and structural similarity search. *Bioinformatics* **15**, 243–251.
85. *Molecular Operational Environment.* (2004) Chemical Computation Group Inc., Montreal, Canada.
86. SNNS: Stuttgart Neural Network Simulator, version 4.2, from University of Tübingen, Stuttgart, Germany. Available at http://www-ra.informatik.uni-tuebingen.de/SNNS/.

9

Investigating the Mode of Action of Proline-Rich Antimicrobial Peptides Using a Genetic Approach: A Tool to Identify New Bacterial Targets Amenable to the Design of Novel Antibiotics

Marco Scocchi, Maura Mattiuzzo, Monica Benincasa, Nikolinka Antcheva, Alessandro Tossi, and Renato Gennaro

Summary

The proline-rich antimicrobial peptides (PRPs) are considered to act by crossing bacterial membranes without altering them and then binding to, and functionally modifying, one or more specific targets. This implies that they can be used as molecular hooks to identify the intracellular or membrane proteins that are involved in their mechanism of action and that may be subsequently used as targets for the design of novel antibiotics with mechanisms different from those now in use. The targets can be identified by using peptide-based affinity columns or via the genetic approach described here. This approach depends on chemical mutagenesis of a PRP-susceptible bacterial strain to select mutants that are either more resistant or more susceptible to the relevant peptide. The genes conferring the mutated phenotype can then be isolated and identified by subcloning and sequencing. In this manner, we have currently identified several genes that are involved in the mechanism of action of these peptides, including peptide-transport systems or potential resistance factors, which can be used or taken into account in drug design efforts.

Key Words: Antimicrobial peptide; proline-rich peptide; Bac7; antibiotic resistant mutant; chemical mutagenesis; fluorescence quenching; cell uptake.

1. Introduction

Host defense peptides are ancient effectors of innate immunity found in all organisms, which have evolved several different mechanisms of action *(1–3)*. Many of these peptides display amphipathic, cationic scaffolds that allow

From: *Methods in Molecular Biology, vol. 494: Peptide-Based Drug Design*
Edited by: L. Otvos, DOI: 10.1007/978-1-59745-419-3_9, © Humana Press, New York, NY

membrane interaction, leading to subsequent membrane damage and/or peptide translocation into the cytoplasm where they can interact with molecular targets. In most cases, this membrane interaction is directed to anionic phospholipids, driven by electrostatic effects *(4)*. In some instances, however, as for example bacterial lantibiotics or some plant defensins, specific membrane components are involved (respectively, lipid II and sphingomyelins) *(5)*. In any case, direct interaction with the membrane is a principal factor in the mode of action, subsequently leading to membrane compromising or peptide translocation.

The proline-rich antimicrobial peptides (PRPs) are thought to act by a specific two-stage mechanism *(6–8)*: *(i)* they have the ability to cross biological membranes without altering them, and so penetrate the bacterial cytoplasm in which *(ii)* they bind to, and functionally modify, one or more specific targets *(9)*. In this respect, one such target has been identified as the molecular chaperone DnaK for the insect PRP pyrrhocoricin *(10,11)*. It has also been proposed that the binding of the porcine proline-rich PR-39 to DNA may be involved in its killing mechanism *(12)*.

A unique characteristic of both insect and mammalian PRPs is that the all-D enantiomers of the natural peptides are poorly, or not at all, active *(6)*. This implies that either or both the internalization and target inactivation mechanisms require a stereospecific molecular interaction. Studies to identify the mode of action of PRPs must thus take this into account. One procedure is that of using peptide-based affinity columns to which bacterial lysates are applied *(10)*. In principle, only tightly binding species (the putative molecular target) should be retained on the column. In practice, it is difficult to detect other than cytoplasmic targets and only if these are abundantly expressed. These shortcomings can be avoided by combining affinity-binding and genetic methods *(13)*. These are based on selection of bacterial isolates that are more resistant or more susceptible to PRPs than the parental strain, following isolation and identification of the genes underlying the altered susceptibility. These methodologies can be quite potent as they can lead to the identification of *(i)* proteins involved in membrane translocation of peptides, *(ii)* target proteins that mediate the antibacterial activity of PRPs, or *(iii)* proteins involved in PRP clearance (pumps, proteases, etc.) that may be responsible for resistance to peptide. The following chapter describes a method to select *Escherichia coli* clones resistant to the bovine PRP Bac7 that led to the identification of a membrane protein (SbmA) that may be involved in peptide internalization *(14)* and of a protease (OpdB) that could be implicated in PRP resistance.

Clearly, identification of resistance mechanisms and of potential intracellular targets all lead to information which would be very useful for the development of novel anti-infective drugs by peptide-based design, once the structure of these targets is known. In addition, identification of transport proteins that may translocate peptides inside the bacterial cells paves the way for the use of this route to make these cells susceptible to drugs that normally do not cross the bacterial membrane. These drugs could penetrate the cytoplasm as cargo linked to the appropriate peptide shuttle.

2. Materials

Unless specifically indicated, all materials should be stored according to the manufacturers' instructions.

2.1. Set-Up of the Selection Method

1. Luria-Bertani (LB) agar, Mueller-Hinton (MH) agar or M9 agar (all from Difco Laboratories). The choice of the medium depends on the peptide used. If the peptide's activity is salt-sensitive, use of a low-salt minimal medium such as M9 is more appropriate.
2. Stock solution of each of the proline-rich peptides (PRP) to be tested in MilliQ water slightly acidified with acetic acid (peptide concentration from 1 to 10 mg/mL).
3. Overnight culture of the wild-type strain *E. coli* HB101 in the medium of choice. Other bacterial strains may also be used as required.
4. Determination of the number of Colony Forming Units/mL (CFUs) via turbidity at 600 nm. It is appropriate to develop a calibration curve; in our hands an optical density of 0.3 at 600 nm corresponds to approximately 4.6×10^7 CFU/mL for the *E. coli* HB101 strain.
5. Incubator set to 37°C.

2.2. Chemical Mutagenesis of Bacteria

1. N-methyl-N'-nitro-N-nitrosoguanidine (MNNG) stock solution (1 mg/mL) in acetone (Sigma).
2. Rifampicin (Sigma), stock solution: 50 mg/mL in methanol. Final concentration in the agar plates: 100 μg/mL.
3. Sodium citrate buffer, 0.1 M, pH 5.5 (Buffer 1).
4. Sodium phosphate buffer, 0.1 M, pH 7.0 (Buffer 2).

2.3. Construction of the Genomic Libraries

1. pUC18 plasmid.
2. BamHI (10 U/μL), Sau3AI (10 U/μL), and their corresponding 10X reaction buffers (Promega).
3. Calf Intestinal Alkaline Phosphatase (CIAP) (20 U/μL) and its 10X reaction buffer (Promega).
4. Agarose, gel-electrophoresis grade (Invitrogen).

5. 50X Tris-Acetate-EDTA (TAE buffer): 2 M TRIS base, 5.2% (w/v) acetic acid and 50 mM Na$_2$-EDTA, pH 8.3.
6. 0.5 μg/mL ethidium bromide (Sigma) incorporated into the agarose gel.
7. T4 DNA Ligase 10 U/μL and its 10X reaction buffer (USB, GE Healthcare).
8. SOB medium: 20 g/L bacto-tryptone (Difco Laboratories), 5 g/L bacto-yeast extract (Difco Laboratories), 0.5 g/L NaCl, 16.6 g/L KCl.
9. SOC medium: SOB medium with the addition of 20 mM sterile glucose and 10 mM sterile MgCl$_2$.
10. Agar medium (*see* 2.1) containing 100 μg/mL ampicillin (ampicillin stock solution: 100 mg/mL in MilliQ water, store at −20°C).
11. GFX™ Gel Purification Kit (GE Healthcare Biosciences).
12. Incubator set at 16°C.
13. Agarose gel electrophoresis apparatus.
14. Cuvettes for electroporation (Biorad).
15. Electroporation apparatus.

2.4. Identification of the Resistance Associated Genes

1. GFX™ Micro Plasmid Prep Kit (GE Healthcare).
2. EcoRI (10 U/μL), HindIII (10 U/μL), and their corresponding 10X reaction buffers (Promega).
3. pUC18 universal primers M13 uni (−43) and M13 rev (−49).

2.5. Characterization of PRP Resistance

1. Buffered high-salt solution: 10 mM Na–phosphate, 400 mM NaCl, 10 mM MgCl$_2$, pH 7.5 (Buffer 3).
2. Agar medium, liquid broth (*see* **Subheading 2.1**).

2.6. Synthesis of Cys-Tagged and Fluorescently Labeled Peptides

1. Fluorophore: BODIPY®FL N-(2-aminoethyl)maleimide (Invitrogen); excitation max at 505 nm, emission max at 513 nm. BODIPY®FL solution prepared fresh at 1 mg/ml in acetonitrile (AcCN).
2. Peptides with a cysteine residue added to the C-terminus to provide a free sulphydrile group (for peptide synthesis protocols *see* **ref. 15**).
3. Phosphate-buffered saline (PBS).
4. Peptide stock solution 0.1 mg/mL in PBS containing 20% AcCN.
5. Analytical and semi-preparative reversed-phase columns (e.g. Waters Symmetry and Delta-Pak™ C$_{18}$; Millipore).
6. RP-HPLC and mass spectrometry instruments.

2.7. Flow Cytometry

1. Mueller-Hinton broth (*see* **Subheading 2.1**).
2. Buffer 3 (*see* **Subheading 2.5**).

3. Trypan Blue (Sigma-Aldrich), stock solution of 5 mg/mL in MilliQ water filtered using a 0.22 μm membrane filter (Millipore).
4. Propidium iodide (Sigma-Aldrich), stock solution of 1 mg/mL in MilliQ water filtered using a 0.22 μm membrane filter (Millipore).
5. Flow cytometer (e.g., Cytomics FC 500 instrument; Beckman-Coulter, Inc., Fullerton, CA) equipped with an argon laser (488 nm, 5 mW) and a photomultiplier tube fluorescence detector for green (525 nm) or red (610 nm) filtered light. All detectors set on logarithmic amplification. Optical and electronic noise are eliminated by setting an electronic gating threshold on the forward scattering detector, while the flow rate is kept at a data rate below 300 events/second to avoid cell coincidence. At least 10,000 events are acquired for each sample.

3. Methods

The mode of action of the PRPs is determined by their interaction with membrane and/or intracellular molecular targets that mediate their internalization and bactericidal mechanism of action. Both types of molecular targets can be identified using a method based on the isolation of PRP-resistant mutant clones. These are subsequently genetically analyzed to identify the gene(s) responsible for the resistant phenotype, thus allowing the identification of proteins that may be involved in the peptide's mechanism of action and in subsequent drug design to develop antibiotic molecules with novel mechanism of action with respect to those on market. The first step can be accomplished by mutagenizing a peptide-susceptible bacterial species (in our case the *E. coli* HB101 strain) and by isolating a collection of peptide-resistant mutants (P-RES) able to grow on solid medium in the presence of bactericidal concentrations of the selected peptide. In this respect, it is important to set-up the appropriate selection conditions allowing the growth of the resistant-mutant clones while inhibiting the growth of those that remain susceptible. The second step is carried out by transforming the wild-type HB101 strain with plasmid libraries, each constructed with the genomic DNA of the selected P-RES clones (*see* **Note 1**). The selection of those clones acquiring a peptide-resistant phenotype following transformation should allow the isolation of the plasmid-carried recombinant fragments responsible for resistance and thus allow the identification of the gene(s) responsible for the altered susceptibility. The process is schematically outlined in **Fig. 1**.

In our specific study, the selection of Bac7-resistant mutants led to the identification of several genes providing an increased resistance phenotype (*see* **Table 1**). Among these are a protease capable of hydrolyzing PRPs and a mutated gene encoding for SbmA, a membrane protein that is likely a subunit of an ABC transporter system in numerous Gram-negative species (*see* **Subheading 3.8**). As PRPs at active concentrations penetrate the cells without

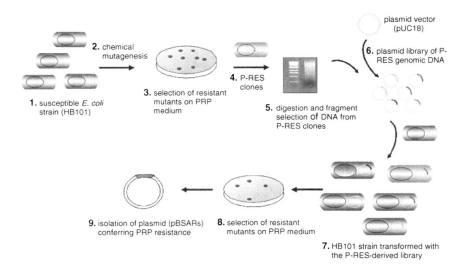

Fig. 1. Schematic representation of the method used to identify resistance-conferring genes for PRP. MNNG was used as a chemical mutagen (*see* **Subheading 3.2**) and the selection process was performed in a medium containing Bac7(1-35) at 10X MIC (*see* **Subheading 3.1**).

Table 1
MIC Values for Bac7 Fragments of 20 PRP-Resistant Clones (HMM1 to HMM20)
Compared to Those of the Parental Wild-Type Strain HB101

Strain	MIC (μM)			
	LB medium		MH medium	
	Bac7(1–35)	Bac7(1–16)	Bac7(1–35)	Bac7(1–16)
E. coli HB101	4	4	0,5	1
E. coli HMM1 to HMM20	>32	>32	2–4	4–8

apparent membrane damage, we hypothesized that the SbmA protein could be involved in the transport of Bac7 within susceptible bacterial cells. A robust demonstration of this hypothesis could be achieved for instance by reconstitution of the transport system in an artificial membrane (e.g., liposomes, and subsequent demonstration of peptide's penetration in the vesicles). However, this would require the identification of the other partners of the likely multi-subunit ABC transporter, followed by their expression and isolation. In a first instance, to obtain a rapid and less laborious confirmation of the putative involvement of

SbmA in peptide's translocation, we devised a simple cytofluorimetric method (*see* **Subheadings 3.6 and 3.7**) that couples the use of fluorescently-labelled peptides with an impermeant quencher that can bleach the fluorescence of the peptide molecules outside or on the surface of bacteria but not of those translocated into the cytoplasm.

3.1. Set-Up of the PRP-Resistant Selection Conditions

1. Melt the agar medium (*see* **Subheading 2.1**) and let it cool in a water bath.
2. When the medium reaches a temperature of about 55°C, divide it into aliquots and add in each of them increasing concentrations of the PRP (from 1 to 10 μM in the case of Bac7-derived fragments). Stir or shake thoroughly to get a homogenous peptide distribution.
3. Pour approximately 15–20 mL into a 10 cm Petri dish and let the agar solidify at room temperature.
4. Plate an overnight suspension of the wild-type HB101 strain.
5. Check the growth of bacteria on the PRP-containing plates.
6. Determine the minimum concentration of the peptide able to completely inhibit the growth of the wild-type strain. In our case, to inhibit the growth of the HB101 strain, an active fragment of Bac7, Bac7(1-35), had to be added to the plate at 10 μM.

3.2. Generation of E. coli Mutants Resistant to PRP

The following protocol for the isolation of mutants showing an increasing resistance to PRPs is based on the method of J.H. Miller *(16)*.

1. Collect 120 mL of a mid-log phase culture of *E. coli* HB101 (~1 × 10⁸ CFU/ml), wash twice in Buffer 1 by centrifugation at 1000*g* for 10 min, resuspend the final pellet in 2 mL of Buffer 1 and place it on ice.
2. Add 50 μL of MNNG stock solution to 1.95 mL of the cell suspension, and 50 μL of Buffer 1 only to a control sample. Incubate all the samples at 37°C, with shaking, for different times.
3. At the end of each incubation time, wash the sample with 5 mL of the Buffer 2 to remove the mutagen, and resuspend the treated cells in 2 mL of the same buffer. Plate the 10^{-4} and 10^{-5} dilutions of each mutagenized sample and the 10^{-5} and 10^{-6} dilutions of the control on agar medium to check the viability of the treated cells. Dilute 1:20 each of the mutagenized samples in broth medium and incubate overnight at 37°C with mild-shaking.
4. Plate the 10^{-6} dilution of each of the overnight cultures, including the control, on agar plates to monitor the presence of viable cells. To look for rifampicin-resistant (Rif^r) mutants, plate undiluted and a 10^{-1} dilution of the control and a 10^{-2} dilution for each mutagenized culture on agar plates with rifampicin (100 μg/mL).

To look for PRP-resistant mutants, plate an undiluted sample of each mutagenized culture on agar medium containing the PRP at the selected concentration (*see* **Subheading 3.1**). Incubate all the plates at 37°C and store the cultures at 4°C.

5. Score all plates and calculate the titer of viable cells, the titer of the Rif' mutants, and the titer of the peptide-resistant mutants (*see* **Note 2**).

3.3. Construction of the Genomic DNA Libraries

1. Prepare a few µg of BamHI-linearized and dephosphorylated pUC18 vector using standard methods *(17)*.
2. To prepare the DNA fragments, chose a P-RES clone selected as described in section 3.2, and inoculate a single colony in 5 mL of LB broth. It may be convenient to assay the extent of resistance of the clones by the method described in **Subheading 3.5.**
3. Extract the genomic DNA from an overnight culture using a standard method *(18)*.
4. Incubate 60 µg of the genomic DNA with 0.01 U/µg of Sau3AI at 37°C. Check the size of the partial digested fragments by 0.8% agarose gel electrophoresis in TAE buffer and continue the reaction until fragments of 2-4 kb will be visible (30-45 min).
5. Separate the products on 0.8% agarose gel electrophoresis, excise the DNA fragments of a length comprised between 2 and 4 kb and purify them from the gel (*see* **Note 3**).
6. After preparing the fragments and the vector, mix them in a microfuge tube for 5 min at 55°C. Add 30 ng of plasmid vector and the DNA inserts to a molar ratio of 1:10. Cool the mixture on ice and add 1 µL of 10X ligation buffer and 1-2 units of T4 DNA ligase to a final volume of 10 µL.
7. Incubate the reaction mixtures overnight at 16°C.
8. Pipette 50 µL of electrocompetent HB101 cells *(19)*, with an efficiency of transformation of at least 10^7 CFU/µg of DNA, into an ice-cold sterile microfuge tube and place it on ice.
9. Add a volume of 1-5 µL of the ligation mixture.
10. Transfer the sample to an ice-cold cuvette taking care to dry the outside of the cuvette. Place the cuvette in the electroporation device and deliver to the cells an electrical pulse of 1.8 kV for approximately 5 msec (use a capacitance of 10 µF and a resistance of 600 ohm).
11. Remove the electroporated cells as quickly as possible and incubate them in a microcentrifuge tube with 1 mL of SOC medium at 37°C for 1 h. Plate different volumes of cells (90–300 µL) on agar medium supplemented with ampicillin and the relevant peptide in order to select PRP-resistant mutant clones (*see* **Note 4**).
12. In parallel, plate 50 µL of the transformed cells on plates supplemented with ampicillin as a transformation control.
13. Incubate the plates overnight at 37°C.

3.4. Identification of Genes Responsible for Resistant Phenotypes

1. Select the clones able to grow on medium supplemented with an inhibitory concentration of PRP and inoculate them in 5 mL of LB broth supplemented with ampicillin.
2. Incubate overnight at 37°C.
3. Extract the recombinant plasmids (pBSARs) *(17)* and double digest 5 μL (about 1 μg) of each of them with 1-5 U of EcoRI and HindIII restriction enzymes (*see* **Note 5**).
4. Separate the products on 0.8% agarose gel to verify the presence of the inserts corresponding to the genomic fragments and determine their length.
5. Sequence the plasmid at the 5' and 3' ends of the inserts using the universal primers located on the pUC18 vector.
6. Compare the sequences with a DNA sequence databank to identify the genomic fragment responsible of the resistant phenotype. In this respect, the on-line server Colibri (http://genolist.pasteur.fr/Colibri/), containing a database dedicated to the analysis of the *E. coli* genome, may be of great help.

3.5. Characterization of Mutants

The degree of resistance of the mutant clones to the relevant PRP may be assayed by a number of methods. The broth microdiluition susceptibility test, a reference method for determination of the Minimum Inhibitory Concentration (MIC), may be performed according to the guidelines of the National Committee for Clinical Laboratory Standard (NCCLS) as described *(20)*. An example of the increase in the MIC values for Bac7(1-35) in mutagenized HB101 is shown in **Table 1**. Another suitable and simple method, here described, is the count of bacterial viable cells after incubation with the peptide.

1. Add 50 μL of a log-phase bacterial suspension, diluted to approximately 1.5×10^6 cells/mL, in a microfuge tube and then add 100-x μL of liquid medium (e.g., MH broth), where x corresponds to the volume of peptide to be assayed (x = 0 in the control tubes). Use the PRP at a bactericidal concentration for the parental wild-type strain (HB101). Different PRP peptide concentrations may be used (e.g. 5X MIC).
2. Incubate the samples at 37°C for different times (1-4 h).
3. Serially dilute aliquots of the samples in Buffer 3.
4. Plate in duplicate on agar medium 50 μL of the 10^{-3} and 10^{-4} dilutions of the control or the 10^{-1} to 10^{-4} dilutions of the peptide-treated samples.
5. Incubate overnight at 37°C to allow colony counts.

3.6. Synthesis of Cys-Tagged and Fluorescently Labeled Peptides

For all the cell internalization assays, PRPs have to be labeled with a fluorescent dye as a tracer. One such dye, BODIPY, functionalized with the

2-aminoethyl maleimide group (*see* **Note 6**), may be easily linked to the free sulfhydril group of a cysteine residue added to this purpose to the C-terminus of the peptide (*see* **Note 7**). This fluorophore is relatively nonpolar and electrically neutral, so it does not affect the peptide's charge.

1. Add drop-wise and under a dimmed light five molar equivalents of BODIPY®FL N-(2-aminoethyl) maleimide to 20 mL of peptide stock solution in PBS under nitrogen bubbling. This avoids precipitation of the dye in the aqueous solution.
2. After 5 h incubation with stirring, add a new aliquot of the peptide (2 mg) and leave overnight under stirring in the dark.
3. Monitor the reaction periodically with RP-HPLC. The retention time of the BODIPY-linked peptide increases, as does the absorption at 280 nm.
4. Upon completion (about 24 h), add a 20-fold excess of cysteine to scavenge the excess of thiol-reactive reagent.
5. After 30 min incubation, filter the solution and purify the labelled peptide by semi-preparative RP-HPLC.
6. The reaction product is then analyzed by ESI-MS (The binding of BODIPY®FL causes a mass increase of 414 Da).

3.7. Characterization of the Mutant Phenotype by Flow Cytometry

1. Dilute mid-log phase bacteria to 1×10^6 cells/mL in Mueller-Hinton broth.
2. Incubate the bacterial suspension with the fluorescently-labelled peptides at 37°C for different times.
3. After incubation, centrifuge the treated cells at 5.000g for 5 min and resuspend the pellet in Buffer 3.
4. Repeat step 3 at least three times.
5. To discriminate between the fluorescence due to binding of the BODIPY-tagged peptide on the surface of the bacterial cells, that is accessible to quenching, from that given by the internalized peptide, that is not, add the impermeant quencher Trypan Blue at a final concentration of 1 mg/mL and incubate for 10 min at room temperature.
6. Keep the samples on ice until they are analyzed by flow cytometry at λ_{ex} of 488 nm and λ_{em} of 525 nm.
7. Check the integrity of the bacterial membrane by measuring the percentage of propidium iodide uptake, by adding propidium iodide to an aliquot of the peptide-treated bacteria and to controls at a final concentration of 10 μg/mL.
9. Analyze the cells in the flow cytometer at λ_{ex} of 530 nm and λ_{em} of 620 nm after incubation for 4 min at 37°C.
10. Take into account only samples in which PI-positive cells are less than 3%.
11. Perform data analysis with the WinMDI software (Dr. J. Trotter, Scripps Research Institute, La Jolla, CA). Significance of differences among groups can be assessed by using the Instat program (GraphPad Software Inc., San Diego, CA) and performed by the paired t test. Values of $P<0.05$ are considered statistically significant.

3.8. Examples of PRP Resistance-Conferring Genes Identified Using the Genetic Approach

A first screening, aimed at discovering genes conferring resistance via over-expression of wild-type genes or *trans*-dominant mutations, allowed the identification of several recombinant plasmids (pBSARs) containing *E. coli* DNA fragments of a P-RES clone leading to a resistant phenotype toward PRPs (**Table 2**). Interestingly, this phenotype concerned PRPs of both mammalian and insect origin, indicating that they may have a common mechanism and providing a fascinating example of convergent evolution. The DNA fragments originated from 5 different regions of the *E. coli* genome. In five clones, the only complete gene always present was *sbmA*, which encodes for an inner membrane protein. Five additional clones always contained one complete gene, termed *opdB*, which encodes for the oligopeptidase B, a member of the prolyl oligopeptidase family of serine peptidases. A single clone encoded for the LacI repressor and for MhpR, a transcriptional activator of the 3-hydroxyphenylpropionate degradation pathway, while others harbored sequences of unknown function (*see* **Table 2** for the genes present in the recombinant clones determining a resistant phenotype) *(14)*. All these clones were able to grow at concentrations of Bac7-derived peptides that were from 4- to 8-fold higher than the susceptible parental strain. In addition, when each of these plasmids was used to transform the parent *E. coli* strain, transformants acquired a resistant phenotype, showing that the harbored genomic regions were by themselves sufficient to confer this phenotype.

Here, the *opdB* and *sbmA* genes only will be shortly discussed as an example. The former gene encodes for the oligopeptidase B and is unmutated in all the five DNA fragments identified, indicating that its over-expression determines the resistant phenotype. This enzyme exhibits a trypsin-like specificity, its substrates

Table 2
List of Recombinant Plasmids (pBSARs) Leading to a Resistant Phenotype Toward Bac7 and of the Genes Contained in Each Insert

Plasmid	Insert size (bp)	Included genes
pBSAR-1/4/7/8	2700	*opdB; yobC*
pBSAR-2	2700	*yjeO; yjeN; yjeM*
pBSAR-3	2500	*yfcB; yfcN*
pBSAR-5/13	3200	*ampH; sbmA; yaiW(partial)*
pBSAR-6	5000	*yaiZ; yaiY; yaiW; ampH sbmA*
pBSAR-9	2800	*lacI; mhpR*
pBSAR-10/12	3500	*sbmA; yaiW; yaiY; yaiZ*
pBSAR-11	4000	*yobC; opdB*

are restricted to low-molecular mass peptides, and is broadly distributed among unicellular eukaryotes and Gram-negative bacteria. Although its function is still unknown, this oligopeptidase has emerged as an important virulence factor and potential therapeutic target in infectious diseases caused by *Trypanosoma* species. Actually, the OpdB proteinase may find basic cleavage sites in Bac7, and Bac7-derived fragments are substrates of OpdB, at least *in vitro*, as shown by mass analysis of the Bac7(1-35)-derived fragments after incubation in the presence of recombinant enzyme (unpublished results). In this case, rather then identifying the target for potential drug design, our method has identified a potential resistance factor to PRPs which must be taken into account during a further drug design process in the case PRPs are used as a lead compound.

At variance with *OpdB*, the *sbmA* gene found in the plasmids leading to a resistant phenotype was mutated at a single position (nucleotide 826). The *E. coli sbmA* gene encodes a 406-amino acid inner membrane protein predicted to have seven transmembrane-spanning domains. Orthologs of this gene have been found in numerous Gram-negative species, but not in Gram-positive bacteria. It was first identified as a gene whose product is implicated in the uptake of the microcins B17 and J25 *(21,22)* antibiotic peptides of bacterial origin, and of the glycopeptide antibiotic bleomycin. Strains lacking this gene are in fact completely resistant to microcins *(23)*.

Despite the fact that the precise function of SbmA is still unknown, there are several indications that suggest it may mediate the transport of different types of compounds allowing for their internalization. In fact, SbmA is classified as a subunit of a membrane transporter belonging to the ABC (ATP binding cassette) superfamily, although its natural substrate(s) for transport, the partner transmembrane domain and the cytoplasmic nucleotide binding domain are at present unknown. Our mutagenesis results indicate that SbmA is involved in the mode of action of Bac7 as well as of other PRPs *(14)*. Several findings support this conclusion: *(i)* the *sbmA826* mutation leads to an increased resistance specifically to PRPs and not to other types of antimicrobial peptides (e.g., α-helical peptides); *(ii)* *sbmA* null mutants also show a Bac7-resistant phenotype; *(iii)* introduction of the wild-type *sbmA* gene into *sbmA*-mutated strains restores the sensitivity to Bac7; *(iv)* the level of expression of SbmA correlates directly with bacterial susceptibility to Bac7(1-35); *(v)* the cell-associated fluorescence, after addition of an impermeant quencher, is much lower in the mutated than in the parental strain when a fluorescently-labelled Bac7-derived peptide is used *(14)*.

It is further worth noting that the observed level of Bac7-resistance is equivalent for the point mutated and null mutants, thus indicating that the point mutation is sufficient to completely abolish its function regarding Bac7 resistance. Significantly, this mutation leads to a Glu to Lys substitution at position 276 at a site predicted to be located between two transmembrane helices, and which is within the most conserved region of the protein in different Gram-negative species. This corresponds to the short hydrophilic EAA sequence motif that is highly conserved in ABC import systems, and is localized in a cytoplasmic loop that constitutes an important interaction site with the cognate ABC-ATPase subunit *(24)*. Taking everything into account, the hypothesis that SbmA may be an uptake system carrying compounds into the bacterial cells is quite solid, and PRPs appear to use it to be internalized into bacterial cells *(21,25)*.

In conclusion, our genetic approach has allowed us to demonstrate that the entrance of PRPs into Gram-negative bacterial cells is not simply mediated by the intrinsic capacity of these peptides to cross the membrane lipid bilayers, but that a specific protein–mediated transport mechanism is necessary. Our data clearly indicate that SbmA is involved in the transport of peptides and that it likely represents an important part of this process, as its inactivation leads to an increased resistance to PRPs. This type of bacterial transporters may thus represent a Trojan horse for translocation of specific peptides into bacterial cells. This also suggests possible novel strategies for using PRPs or their analogs as shuttle systems for the internalization of non-permeant drugs with anti-infective potential into bacteria.

4. Notes

1. By this approach, positive clones are expected to arise from overexpression of wild-type genes conferring resistance or from *trans*-dominant mutations giving a mutant (resistant) phenotype *(14)*. To identify loss-of-function recessive mutations the method should be reversed: P-RES resistant clones should be transformed with a genomic library constructed with the DNA of the wild-type HB101, and clones with restored peptide susceptibility analyzed.

2. Isogenicity between the wild type strain and the mutated colonies may be checked by ERIC-PCR based on the enterobacterial repetitive intergenic consensus (ERIC) sequences *(26)*. Genomic DNA is extracted and amplified by PCR with the specific primer 5′-AAGTAAGTGACTGGGGTGAGCG-3′ for the conserved ERIC sequence. Amplified PCR products are separated by electrophoresis on 2% agarose gel and the band patterns compared with those of the parental HB101 strain. A nonisogenic *E. coli* strain may be used as a control.

3. To obtain highly purified DNA samples, an additional extraction with phenol/chloroform followed by one with chloroform may be performed. The DNA is then precipitated by adding NaCl to a final concentration of 0.3 M and 2.5 volumes of 100% ethanol. After centrifugation, the pellet is washed twice with 70% ethanol, dried on air and resuspended in sterile water. This additional procedure greatly increases the efficiency of transformation.

4. The plasmid library can be subjected to a round of amplification using the *E. coli* XL-1 Blue strain before being introduced into the final strain. In this case, the

ligation products (*see* **Subheading 3.3**) are used to transform aliquots of the competent XL-1 Blue cells, an alpha-complementation *E. coli* strain (*17*). The use of this strain allows distinguishing between recombinant clones with an insert (white colonies) and clones with an empty vector (blue colonies) and assessing the percentage of the recombinant clones. 50 μL of transformed cells are plated on agar medium supplemented with ampicillin, with 20 μg/mL X-gal (the substrate of the β-galactosidase) and with the inducer IPTG (0.1 mg/mL). The remaining cells are added to 10-50 mL of LB medium supplemented with ampicillin and grown overnight at 37°C. The plasmid DNA is extracted from 3 mL of the overnight culture and used to transform the PRP-susceptible strain.

5. Other combinations of restriction enzymes may be used to excise the cloned fragment form the plasmid, each cleaving at a different side of the cloning site. However, avoid using BamHI, as the restriction site may not have been reconstructed in the recombinant fragments after ligation.

6. Several BODIPY fluorophores with different excitation and emission maxima are available, and each can have alternative linkage chemistries. It is important to note that the linkage chemistry affects the fluorescence intensity. One cannot therefore quantitatively compare peptides labelled using different linkage chemistries. Once having selected a fluorophore/ linkage chemistry it is better to stick with it.

7. It is better to label the PRPs at the C-terminus. In our experience, if the peptide is labelled to the N-terminus we often noticed a significant decrease in antimicrobial activity. Before using labelled PRPs, it is in any case advisable to check whether this chemical modification has altered its antimicrobial activity compared with that of the unmodified molecule by both MIC and growth inhibition assays (*see* **Subheading 3.5**). In our experience the addition of the BODIPY to the C-terminus of Bac7(1-35) did not modify its antibacterial properties.

Acknowledgments

This study was supported by grants from the Italian Ministry for University and Research (PRIN 2005) and by FVG Region LR11/2003 Grant 200502027001.

References

1. Brogden, K. A. (2005) Antimicrobial peptides: pore formers or metabolic inhibitors in bacteria? *Natl. Rev. Microbiol.* **3**, 238–250.

2. Brown, K. L. and Hancock, R. E. (2006) Cationic host defense (antimicrobial) peptides. *Curr. Opin. Immunol.* **18**, 24–30.

3. Hancock, R. E. and Sahl, H. G. (2006) Antimicrobial and host-defense peptides as new anti-infective therapeutic strategies. *Natl. Biotechnol.* **24**, 1551–1557.

4. Shai, Y. (2002) Mode of action of membrane active antimicrobial peptides. *Biopolymers* **66**, 236–248.

5. Pag, U. and Sahl, H. G. (2002) Multiple activities in lantibiotics – models for the design of novel antibiotics? *Curr. Pharm. Des.* **8**, 815–833.

6. Gennaro, R., Zanetti, M., Benincasa, M., Podda, E. and Miani, M. (2002) Pro-rich antimicrobial peptides from animals: structure, biological functions and mechanism of action. *Curr. Pharm. Des.* **8**, 763–778.

7. Otvos, L., Jr. (2002) The short proline-rich antibacterial peptide family. *Cell Mol. Life Sci.* **59**, 1138–1150.

8. Podda, E., Benincasa, M., Pacor, S., Micali, F., Mattiuzzo, M., Gennaro, R. and Scocchi, M. (2006) Dual mode of action of Bac7, a proline-rich antibacterial peptide. *Biochim. Biophys. Acta* **1760**, 1732–1740.

9. Cudic, M. and Otvos, L., Jr. (2002) Intracellular targets of antibacterial peptides. *Curr. Drug Targets* **3**, 101–106.

10. Otvos, L., Jr., Rogers, M. E., Consolvo, P. J., Condie, B. A., Lovas, S., Bulet, P. and Blaszczyk-Thurin, M. (2000) Interaction between heat shock proteins and antimicrobial peptides. *Biochemistry* **39**, 14150–14159.

11. Kragol, G., Lovas, S., Varadi, G., Condie, B. A., Hoffmann, R. and Otvos, L., Jr. (2001) The antibacterial peptide pyrrhocoricin inhibits the ATPase actions of DnaK and prevents chaperone-assisted protein folding. *Biochemistry* **40**, 3016–3026.

12. Boman, H. G., Agerberth, B. and Boman, A. (1993) Mechanisms of action on Escherichia coli of cecropin P1 and PR-39, two antibacterial peptides from pig intestine. *Infect. Immun.* **61**, 2978–2984.

13. Shi, Y., Cromie, M. J., Hsu, F. F., Turk, J. and Groisman, E. A. (2004) PhoP-regulated Salmonella resistance to the antimicrobial peptides magainin 2 and polymyxin B. *Mol. Microbiol.* **53**, 229–241.

14. Mattiuzzo, M., Bandiera, A., Gennaro, R., Benincasa, M., Pacor, S., Antcheva, N. and Scocchi, M. (2007) Role of the *Escherichia coli* SbmA in the antimicrobial activity of proline-rich peptides. *Mol. Microbiol.* **66**, 151–163.

15. Tossi, A., Scocchi, M., Zanetti, M., Gennaro, R., Storici, P. and Romeo, D. (1997) An approach combining rapid cDNA amplification and chemical synthesis for the identification of novel, cathelicidin-derived, antimicrobial peptides. *Methods Mol. Biol.* **78**, 133–150.

16. Miller, J. H. (1992) *A Short Course in Bacterial Genetics. A Laboratory Manual and Handbook for Escherichia coli and Related Bacteria. Volume 1.* Cold Spring Harbor Laboratory Press, New York.

17. Sambrook, J. and Russell, D. W. (2001) *Molecular Cloning: a Laboratory Manual*, 3rd ed. Cold Spring Harbor Laboratory Press, New York

18. Chen, W. P. and Kuo, T. T. (1993) A simple and rapid method for the preparation of gram-negative bacterial genomic DNA. *Nucleic Acids Res.* **21**, 2260.

19. Dower, W. J., Miller, J. F. and Ragsdale, C. W. (1988) High efficiency transformation of *E. coli* by high voltage electroporation. *Nucleic Acids Res.* **16**, 6127–6145.

20. Benincasa, M., Scocchi, M., Podda, E., Skerlavaj, B., Dolzani, L. and Gennaro, R. (2004) Antimicrobial activity of Bac7 fragments against drug-resistant clinical isolates. *Peptides* **25**, 2055–2061.

21. Lavina, M., Pugsley, A. P. and Moreno, F. (1986) Identification, mapping, cloning and characterization of a gene (sbmA) required for microcin B17 action on *Escherichia coli* K12. *J. Gen. Microbiol.* **132**, 1685–1693.

22. Salomon, R. A. and Farias, R. N. (1995) The peptide antibiotic microcin 25 is imported through the TonB pathway and the SbmA protein. *J. Bacteriol.* **177**, 3323–3325.

23. Yorgey, P., Lee, J., Kordel, J., Vivas, E., Warner, P., Jebaratnam, D. and Kolter, R. (1994) Posttranslational modifications in microcin B17 define an additional class of DNA gyrase inhibitor. *Proc. Natl. Acad. Sci. U S A* **91**, 4519–4523.

24. Locher, K. P., Lee, A. T. and Rees, D. C. (2002) The E. coli BtuCD structure: a framework for ABC transporter architecture and mechanism. *Science* **296**, 1091–1098.

25. de Cristobal, R. E., Solbiati, J. O., Zenoff, A. M., Vincent, P. A., Salomon, R. A., Yuzenkova, J., Severinov, K. and Farias, R. N. (2006) Microcin J25 uptake: His5 of the MccJ25 lariat ring is involved in interaction with the inner membrane MccJ25 transporter protein SbmA. *J. Bacteriol.* **188**, 3324–3328.

26. Meacham, K. J., Zhang, L., Foxman, B., Bauer, R. J. and Marrs, C. F. (2003) Evaluation of genotyping large numbers of *Escherichia coli* isolates by enterobacterial repetitive intergenic consensus-PCR. *J. Clin. Microbiol.* **41**, 5224–5226.

10

Serum Stability of Peptides

Håvard Jenssen and Stein Ivar Aspmo

Summary

Hospitals worldwide have lately reported a worrying increase in the number of isolated drug-resistant pathogenic microbes. This has to some extent fueled at least academic interest in design and development of new lead components for novel drug design. Much of this interest has been focused on antimicrobial peptides and peptides in general, primarily due to their natural occurrence and low toxicity. However, issues have been raised regarding the stability of peptide therapeutics for systemic use. The focus of this chapter is assays for measuring peptide stability in the presence of serum, both *in vitro* and *in vivo*.

Key Words: Peptide degradation; peptide-serum stability; peptide half-life.

1. Introduction

Bacterial resistance has increased dramatically *(1)*, presenting a huge global health threat and a challenge for developers of antimicrobial drugs *(2)*. Though small molecules still dominate drug discovery, small proteins and peptides have recently been recognized as suitable leads in several fields of drug design.

Numerous investigations have been undertaken, screening large numbers of naturally occurring peptides, chemical peptide libraries and genetic/recombinant libraries, resulting in the discovery of a rather limited number of bioactive peptides. Some peptides have demonstrated positive effect on controlling diabetes, e.g., glucagons like peptide-1 and analogues of this *(3)*, deletion peptides of insulin *(4)*, and a deletion peptide (V_{437}LGGGVALL RVIPALDSLTPANED$_{460}$) of heat shock protein 60 *(5,6)*. A modified deletion peptide (G_{256}EBGIAGNKGDQGPKGEBGPA$_{276}$) (modifications highlighted) from type II collagen has demonstrated suppression of autoimmune arthritis *(7)*.

From: *Methods in Molecular Biology, vol. 494: Peptide-Based Drug Design*
Edited by: L. Otvos, DOI: 10.1007/978-1-59745-419-3_10, © Humana Press, New York, NY

Some other peptides, e.g., dipeptidyl peptidase IV *(8)*, endopeptidase 3.4.24.16 inhibitor *(9)*, and a copolymer of tyrosine, glutamic acid, alanine, and lysine *(10)*, have also demonstrated immunomodulatory activity. Thus analogues of these may some day play an important role in treatment of autoimmune and inflammatory diseases. It should also be noted that a vast number of naturally derived peptides also has demonstrated antimicrobial activity toward a wide range of pathogens *(11)*, resulting in possible templates for further optimization studies clearly demonstrating the potential importance of peptides as pharmacological tools.

However, the majority of positive leads identified so far have been isolated from the natural reservoir of peptides. A possible explanation for this may be that these peptides have undergone natural selection over billions of years, resulting in enhanced *in vivo* stability. Another explanation may be that the majority of random peptide sequences from chemical and recombinant libraries do not fold into stable structures *(12)*; however, this may not nessesarily render these peptides more susceptible to proteolytic degradation.

Peptides have several advantages over small molecular drugs, e.g., they have higher affinity and specificity to interact with its target, while their toxicity profile remains low. However, several issues have prevented small proteins and peptides from becoming a mainstream template in drug design. This is primarily due to rapid renal clearance and low *in vivo* stability as a result of protease degradation, giving them a rather short half-life. Bioavailability and limited access to intracellular space has also been debated. However, peptide-based antimicrobials are with no doubt suitable for topical applications *(13)*, offering decreased potential of resistance induction *(14)* compared to other antimicrobials.

Synthetic peptides often lack the conformational stability required for a successful drug; therefore determination of peptide stability in serum constitutes a powerful and important screening assay for the elimination of unstable peptides in the pipeline of drug development (*see* **Note 1**). Peptide stability in serum can rather easily be determined by reverse phase–high-performance liquid chromatography (RP-HPLC) and mass spectroscopy (MS) from both *in vitro* and *in vivo* studies.

2. Materials

2.1. Solid Phase Peptide Synthesis and Purification

1. PAL-PEG resin, Fmoc amino acids and coupling reagent *N,N´*-diisopropylethylamine (DIPEA) and *O*-benzotriazole-*N,N,N´,N´*-tetramethyl-uronium-hexafluoro-phosphate (HBTU) (Advanced ChemTech, Louisville, KY).

2. Cleaving reagent trifluoroacetic acid (TFA; Shanghai Fluoride Chemicals, Shanghai, China).
3. Acetonitrile (PerSeptive Biosystems, Framingham, MA), also known by the synonyms methyl cyanide, cyanomethane, ethanenitrile, and ethyl nitrile, is rather toxic and precautions should be taken.
4. Automatic peptide synthesizer, Milligen 9050 PepSynthesizer (Milligen, Milford, MA).
5. RP-HPLC (Waters Corporation, Milford, MA).
6. Electron-spray interface VG QUATTRO quadruple mass spectrometer (VG Instruments Inc., Altrincham, UK).

2.2. In Vitro Peptide Stability in Serum/Reaction Kinetics

1. RPMI medium 1640 (Gibco, Invitrogen, Carlsbad, CA) supplemented with 25% (v/v) of human serum (Fisher BioReagents, Fisher Scientific, Pittsburg, PA) (*see* **Note 2**).
2. Peptide stock solution: 1–10 mg/mL dissolved in pure dimethyl sulfoxide (DMSO; Sigma, St. Louis, MO).
3. Trichloroacetic acid (TCA; Sigma, St. Louis, MO) (aq) 6% or 15% (w/v), or 96% ethanol.
4. TFA (Shanghai Fluoride Chemicals, Shanghai, China).
5. RP-HPLC buffer A: 0.08% TFA in water, buffer B: 0.08% TFA in acetonitrile.

2.3. In Vivo Peptide Stability Assay/Pharmacokinetics in Mice

1. Peptide dissolved in sterile phosphate-buffered saline (PBS) pH 7.2.
2. Balb/c mice (Charles River Laboratories, Wilmington, MA).
3. TCA (Sigma, St. Louis, MO) (aq) 15% (w/v).

3. Methods

Today peptide synthesis is a mainstream operation in many big laboratories. Several companies have in addition specialized in providing custom-ordered peptides, e.g., large-scale peptides (made in solution), peptide libraries (made on cellulose membranes), or most commonly small peptide quantities (made on resin, solid phase synthesis). Although peptides are considered rather stabile in many test assays, concerns have been raised regarding their stability in presence of human serum.

In a general serum stability assay, the peptide is subjected to human serum (*see* **Note 3**) at realistic temperature conditions and incubated for various time intervals, comparable to traditional protease assays. The reactions are stopped by TCA or ethanol, precipitating larger serum proteins while leaving peptides and peptide derivatives soluble for further analysis. To obtain reliable and repro-ducible results, peptide degradation is performed under conditions where the peptide concentration is not the rate-limiting concentration, but rather where the reaction speed is linearly dependent on the serum concentration (e.g., at 25% serum), i.e., the enzyme concentration is rate limiting, not the substrate concentration.

Peptide analysis is performed on the supernatant with either mass spectrometry, e.g., MALDI-TOF *(15)*, or with RP-HPLC under stability-specific chromatography conditions. It should be noted that mass spectrometry is rarely a valid quantitative measure without utilizing isotopic internal standards, while RP-HPLC analysis of peptides is directly quantitative with a UV detector. However, mass spectrometry of the degradation products will provide insight into if or where the peptide is cleaved and/or if other modifications to the peptide have occurred in the serum (glycosylations, phosporylations, or degly-cosylations, dephosphorylations, etc.). It should also be taken into account that several factors may cause misleading stability results for the peptides (*see* **Note 4**).

In vivo testing of peptide stability is obviously of more relevance than *in vitro* testing. However, a better term would probably be peptide pharmacokinetics (in mice), given that these measurements basically are also done *in vitro*.

3.1. Solid Phase Peptide Synthesis and Purification

1. Solid phase peptide synthesis has been described earlier in great detail *(16)*. But, in brief, a standard way of doing this is by using Fmoc-chemistry (9-fluorenylmethoxycarbonyl), a baselabile protection group that easily can be removed during peptide chain elongation.
2. The peptides can be synthesized automatically using Fmoc amino acids and PAL-PEG resin on a Milligen 9050 PepSynthesizer *(17)*.
3. The first amino acid is coupled onto the resin in the presence of DIPEA, using HBTU as a coupling reagent.
4. The next amino acid in the sequence is coupled to the active N-terminal end of the first amino acid, and the process can be repeated over and over until the entire peptide is made.
5. When the peptide is finished it can be cleaved from the solid support (resin) using TFA. This acid-based cleavage process also results in deprotection of the amino acid side chains.
6. The peptide is then precipitated out in ether and washed extensively with additional ether prior to drying.
7. The dried peptide is then weighed out and dissolved in water containing 1% acetonitrile, to a concentration of 20 mg/mL of peptide.
8. The peptides are purified by preparative RP-HPLC on a C18 column with a loading capacity of 20 mg.

9. Eluted fractions are analyzed by electron-spray interface on a VG QUATTRO quadruple mass spectrometer, freeze-dried, and dissolved in distilled, pyrogen-free water immediately prior to use.

3.2. In Vitro Peptide Stability in Serum/Reaction Kinetics

1. One mL of RPMI supplemented with 25% (v/v) of human serum are allocated into a 1.5 Eppendorf tube and temperature-equilibrated at 37± 1°C for 15 min before adding 5 µL of peptide stock solution to make a final peptide concentration of 50 µg/mL. Dilution of the serum will render the proteolytic enzymes the limiting factor; enable a linear degradation of the peptides.
2. The initial time is recorded, and at known time intervals (*see* **Note 5**) 100 µL of the reaction solution is removed and added to 200 µL of 6% aqueous TCA (or 96% ethanol) for precipitation of serum proteins. In some reaction mixtures TCA concentrations up to 15% may be required for proper serum protein precipitation (*see* **Note 6**).
3. The cloudy reaction sample is cooled (4°C) for 15 min and then spun at 18,000 g (Eppendorf centrifuge) for 2 min to pellet the precipitated serum proteins.
4. The reaction supernatant is then analyzed using RP-HPLC (**Fig. 1**) on, for example, a 5-µm 25 × 0.4-cm Vydac C-18 column. A linear gradient from 100% buffer A, to 50–50% of buffer A and buffer B, is used over 30 min. A flow rate of

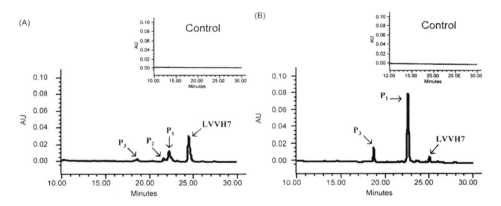

Fig. 1. Example of UV absorbence profiles (280 nm) of RP-HPLC chromatograms of human LVV-hemorphin-7 (LVVH7) products after 10 min of incubation with (**A**) rat brain homogenate and (**B**) rat brain microsomal fraction at 37°C. The figures show how degradation is more prominent in the microsomal fraction of brain tissue, i.e., how the original LVVH7 peak is reduced, while degradation products 1 and 3, P1 and P3, increases in **B** compared to **A**. Control chromatograms with homogenate and microsomal fraction with no added human LVVH7 are represented in small windows (modified from **ref. 27**).

1 mL/min can be sufficient, and absorbance is detected at 214 nm (or 280 nm if aromatic amino acids are present in the peptide), absorbance unit full scale (AUFS) may be set to 0.2, and the analysis is done at room temperature (*see* **Note 7**).

5. Kinetic analysis can be carried out by least-squares analysis of the logarithm of the integration peak area versus time. Correction for small, interfering serum peaks that co-elute with peptide peaks (subtraction of background) is sometimes necessary.

6. If available, a LC system where it is possible to split the flow from the column into a UV detector and an electrospray ion spray mass spectrometer (ESI-MS), or, if available, a MALDI spotting system, is highly recommended. It is then possible to get both quantitative and qualitative results of the degradation simultaneously.

3.3. *In Vivo Peptide Stability Assay/Pharmacokinetics in Mice*

1. The peptide is dissolved in 200 μL sterile PBS to obtain a final peptide concentration of 5 mg/kg, individual for each mouse.

2. Injection is done subcutaneously under the shoulder blade into healthy Balb/c mice, and a total of four mice are assigned to each time point.

3. Blood samples are taken from the eye vein in the mice at six different time points, the first being right after peptide administration, then after 5, 15, 45, 120, and 240 min (*see* **Note 8**).

4. The blood samples are typically about 100 μL in total, and the blood cells are removed by centrifugation, 5 min at 16,000 g.

5. The supernatant (plasma) is transferred to a clean tube, and the serum proteins are precipitated by mixing with 20 μL 15% TCA per 100 μL plasma (*see* **Note 6**).

6. Removal of the precipitated serum proteins and analysis of the proteolytic degradation of the injected peptide in the different plasma samples are done in accordance with **steps 3–6** in **Subheading 3.2.**

4. Notes

1. There are numerous ways of modifying a peptide in an attempt to increase its stability and half-life in serum *(18)*. Modification may obviously but not always change the peptide's biological activity:

   - The most common modifications are probably N-terminal acetylation or glycosylation or C-terminal amidation *(19)*; the latter most certainly increases the activity of short antimicrobial peptides.
   - Substitution of terminal amino acids from L-amino acids to D- will make the peptide more resistant to proteolytic degradation, but both ends needs to be changed to achieve increased stability *(20)*.
   - Peptides may be glycosylated *(20,21)*, sulfated, or phosphorylated on the tyrosine residues *(22)* to increase their stability.

- Conjugation of the peptide with polyethylene glycol (PEGylation) is quite powerful *(23)*, increasing the peptides' water solubility and reducing their immunogenecity *(24)*.
- Conjugation of the peptide to albumin *(25)* or the Fc domain of, for instance, human gamma immunoglobulin *(26)* may extend the half-life of the peptide.

2. Human blood samples can rather easy be converted to human serum with the use of BD Vacutainer® SST centrifugation tubes (Becton Dickinson Diagnostics, Franklin Lakes, NJ). If serum purification is considered over custom orders, the reader should consider pooling serum from several individuals. The serum should also be heat inactivated at 56°C for 30 min. Inactivation may cause precipitation; thus sterile filtration of the serum is recommended.

3. Obviously, tissue homogenates, or even fractionated tissue homogenates *(27)*, other than human blood serum can be tested using the described method, but for pharmacological testing, human blood serum is usually the most relevant medium and is therefore used as the example in this chapter.

4. Several factors are known to inflict on the peptides serum stability, causing misleading results:

   - Nonhuman plasma is known to have different levels of certain peptidases compared to human serum and may in turn give misleading (too high or too low) stability results *(28,29)*.
   - Plasma from patients with pancreatitis or other disease states may change the plasma peptidase activity—for example, certain viral diseases *(30)*.
   - Serum from subjects/animals of different ages may also affect the level of certain peptidases *(31)*.
   - Plasma prepared with strong chelators such as ethylene diamine tetraacetic acid (EDTA) may act as peptidase inhibitors for metalloproteases and peptidases using divalent cations as cofactors *(28)*.
   - Immunoassay methods may not be stability specific *(32)*.
   - Radiotracer methods where the probe degradation occurs through a label-specific pathway are not suitable *(33)*.

5. Measurement of the proteolytic serum activity should ideally be done at different time intervals. However, studies have demonstrated that an adequate time point for detection of peptide degradation in sterile serum is 1 h. This time point gave almost full degradation of the unstable control peptide, with comparable levels of the degradation products with both RP-HPLC and mass spectroscopy *(15)*.

6. A higher concentration of TCA will push the equilibrium of charged or polar substances with some size, even peptides, further from soluble toward precipitation, thus reducing the amount of detectable peptide in the sample *(34)*.

7. Amphipathic peptides may stick to amphipathic surfaces, i.e., cells, media components, or chemical instruments, thus making their quantitative analysis rather difficult *(35)*.

8. Retro-orbital bleeding is suggested as blood samples for measuring peptides *in vivo* stability/pharmacokinetics in mice should be as sterile as possible to

ensure maximum activity of proteolytic enzymes and minimal interference with the assay *(35)*. The Institutional Animal Care and Use Committee (IACUC) has raised concernes with retro-orbital bleeding, and several institutions prefer tail vein bleeding instead. The small size of the Balb/c mice make tail vein bleeding almost impossible, but the mice may be replaced with bigger animals; e.g., an outbred strain CD-1® mouse (Charles River Laboratories Inc., Wilmington, MA) may serve as a good replacement for this type of study.

References

1. Overbye, K.M., and Barrett, J.F. (2005) Antibiotics: where did we go wrong? *Drug Discov. Today* **10**, 45–52.
2. Levy, S.B., and Marshall, B. (2004) Antibacterial resistance worldwide: causes, challenges and responses. *Nat Med.* **10**(12 Suppl.), S122–129.
3. Gallwitz, B. (2005) New therapeutic strategies for the treatment of type 2 diabetes mellitus based on incretins. *Rev Diabet Stud.* **2**, 61–69.
4. Martinez, N.R., Augstein, P., Moustakas, A.K., et al. (2003) Disabling an integral CTL epitope allows suppression of autoimmune diabetes by intranasal proinsulin peptide. *J. Clin. Invest.* **111**, 1365–1371.
5. Cohen, I.R. (2002) Peptide therapy for type I diabetes: the immunological homunculus and the rationale for vaccination. *Diabetologia* **45**, 1468–1474.
6. Raz, I., Elias, D., Avron, A., Tamir, M., Metzger, M., and Cohen, I.R. (2001) Beta-cell function in new-onset type 1 diabetes and immunomodulation with a heat-shock protein peptide (DiaPep277): a randomised, double-blind, phase II trial. *Lancet* **24**, 1749–1753.
7. Myers, L.K., Sakurai, Y., Tang, B., et al. (2002) Peptide-induced suppression of collagen-induced arthritis in HLA-DR1 transgenic mice. *Arthritis Rheum.* **46**, 3369–3377.
8. Lorey, S., Stockel-Maschek, A., Faust, J., et al. (2003) Different modes of dipeptidyl peptidase IV (CD26) inhibition by oligopeptides derived from the N-terminus of HIV-1 Tat indicate at least two inhibitor binding sites. *Eur. J. Biochem.* **270**, 2147–2156.
9. Vincent, B., Jiracek, J., Noble, F., et al. (1997) Effect of a novel selective and potent phosphinic peptide inhibitor of endopeptidase 3.4.24.16 on neurotensin-induced analgesia and neuronal inactivation. *Br. J. Pharmacol.* **121**, 705–710.
10. Fridkis-Hareli, M., Santambrogio, L., Stern, J.N., Fugger, L., Brosnan, C., and Strominger, J.L. (2002) Novel synthetic amino acid copolymers that inhibit autoantigen-specific T cell responses and suppress experimental autoimmune encephalomyelitis. *J. Clin. Invest.* **109**, 1635–1643.
11. Jenssen, H., Hamill, P., and Hancock, R.E.W. (2006) Peptide antimicrobial agents. *Clin. Microbiol. Rev.* **19**, 491–511.
12. Watt, P.M. (2006) Screening for peptide drugs from the natural repertoire of biodiverse protein folds. *Nat. Biotechnol.* **24**, 177–183.

13. Bush, K., Macielag, M., and Weidner-Wells, M. (2004) Taking inventory: antibacterial agents currently at or beyond phase 1. *Curr. Opin. Microbiol.* **7**, 466–476.
14. Ge, Y., MacDonald, D.L., Holroyd, K.J., Thornsberry, C., Wexler, H., and Zasloff, M. (1999) *In vitro* antibacterial properties of pexiganan, an analog of magainin. *Antimicrob. Agents Chemother.* **43**, 782–788.
15. Hoffmann, R., Bulet, P., Urge, L., and Otvos, L., Jr. (1999) Range of activity and metabolic stability of synthetic antibacterial glycopeptides from insects. *Biochim. Biophys. Acta* **1426**, 459–467.
16. Howl, J., ed. (2005) *Peptide Synthesis and Applications.* Totowa, NJ: Humana Press.
17. Rekdal, O., Andersen, J., Vorland, L.H., and Svendsen, J.S. (1999) Construction and synthesis of lactoferricin derivatives with enhanced antibacterial activity. *J. Pept. Sci.* **5**, 32–45.
18. Sato, A.K., Viswanathan, M., Kent, R.B., and Wood, C.R. (2006) Therapeutic peptides: technological advances driving peptides into development. *Curr. Opin. Biotechnol.* **17**, 638–642.
19. Landon, L.A., Zou, J., and Deutscher, S.L. (2004) Is phage display technology on target for developing peptide-based cancer drugs? *Curr. Drug Discov. Technol.* **1**, 113–132.
20. Powell, M.F., Stewart, T., Otvos, L, Jr., et al. (1993) Peptide stability in drug development. II. Effect of single amino acid substitution and glycosylation on peptide reactivity in human serum. *Pharm. Res.* **10**, 1268–1273.
21. Otvos, L., Jr., Urge, L., Xiang, Z.Q., et al. (1994) Glycosylation of synthetic T helper cell epitopic peptides influences their antigenic potency and conformation in a sugar location-specific manner. *Biochim. Biophys. Acta* **1224,** 68–76.
22. Otvos, L., Jr., Cappelletto, B., Varga, I., et al. (1996) The effects of post-translational side-chain modifications on the stimulatory activity, serum stability and conformation of synthetic peptides carrying T helper cell epitopes. *Biochim. Biophys. Acta* **1313**, 11–19.
23. Veronese, F.M., and Pasut, G. (2005) PEGylation, successful approach to drug delivery. *Drug Discov. Today* **10**, 1451–1458.
24. Werle, M., and Bernkop-Schnurch, A. (2006) Strategies to improve plasma half life time of peptide and protein drugs. *Amino Acids* **30**, 351–367.
25. Leger, R., Thibaudeau, K., Robitaille, M., et al. (2004) Identification of CJC-1131-albumin bioconjugate as a stable and bioactive GLP-1(7–36) analog. *Bioorg. Med. Chem. Lett.* **14**, 4395–4398.
26. Dumont, J.A., Low, S.C., Peters, R.T., and Bitonti, A.J. (2006) Monomeric Fc fusions: impact on pharmacokinetic and biological activity of protein therapeutics. *BioDrugs* **20**, 151–160.
27. Murillo, L., Piot, J.M., Coitoux, C., and Fruitier-Arnaudin, I. (2006) Brain processing of hemorphin-7 peptides in various subcellular fractions from rats. *Peptides* **27**, 3331–3340.
28. McDermott, J.R., Smith, A.I., Biggins, J.A., Hardy, J.A., Dodd, P.R., and Edwardson, J.A. (1981) Degradation of luteinizing hormone-releasing hormone by serum and plasma *in vitro*. *Regul. Pept.* **2**, 69–79.

29. Walter, R., Neidle, A., and Marks, N. (1975) Significant differences in the degradation of pro-leu-gly-NH2 by human serum and that of other species (38484). *Proc. Soc. Exp. Biol. Med.* **148**, 98–103.

30. Springer, C.J., Eberlein, G.A., Eysselein, V.E., Schaeffer, M., Goebell, H., and Calam, J. (1991) Accelerated *in vitro* degradation of CCK-58 in blood and plasma of patients with acute pancreatitis. *Clin Chim Acta.* **198**, 245–253.

31. White, N., Griffiths, E.C., Jeffcoate, S.L., Milner, R.D.G., Preece, M.A. (1980) Age-related changes in the degradation of thyrotrophin releasing hormone by human and rat serum. *J. Endocrin.* **86**, 397–402.

32. Frohman, L.A., Downs, T.R., Williams, T.C., Heimer, E.P., Pan, Y.C., and Felix, A.M. (1986) Rapid enzymatic degradation of growth hormone-releasing hormone by plasma *in vitro* and *in vivo* to a biologically inactive product cleaved at the NH2 terminus. *J. Clin. Invest.* **78**, 906–913.

33. Wroblewski, V.J. (1991) Mechanism of deiodination of 125I-human growth hormone *in vivo*. Relevance to the study of protein disposition. *Biochem. Pharmacol.* **42**, 889–897.

34. Cudic, M., Lockatell, C.V., Johnson, D.E., and Otvos, L., Jr. (2003) *In vitro* and *in vivo* activity of an antibacterial peptide analog against uropathogens. *Peptides* **24**, 807–820.

35. Otvos, L., Jr., Snyder, C., Condie, B., Bulet, P., and Wade, J.D. (2005) Chimeric antimicrobial peptides exhibit multiple modes of action. *Int. J. Pept. Res. Ther.* **11**, 29–42.

11

Preparation of Glycosylated Amino Acids Suitable for Fmoc Solid-Phase Assembly

Mare Cudic and Gayle D. Burstein

Summary

Many biological interactions and functions are mediated by glycans, consequently leading to the emerging importance of carbohydrate and glycoconjugate chemistry in the design of novel drug therapeutics. Despite the challenges that carbohydrate moieties bring into the synthesis of glycopeptides and glycoproteins, considerable progress has been made during recent decades. Glycopeptides carrying many simple glycans have been chemically synthesized, enzymatic approaches have been utilized to introduce more complex glycans, and most recently native chemical ligation has enabled synthesis of glycoproteins from well-designed peptide and glycopeptide building blocks. Currently, general synthetic methodology for glycopeptides relies on preformed glycosylated amino acids for the stepwise solid-phase peptide synthesis. The formation of glycosidic bonds is of fundamental importance in the assembly of glycopeptides. As such, every glycosylation has to be regarded as a unique problem, demanding considerable systematic research. In this chapter we will summarize the most common chemical methods for the stereoselective synthesis of *N*- and *O*-glycosylated amino acids. The particular emphasis will be given to the preparation of building blocks for use in solid-phase glycopeptide synthesis based on the 9-fluorenylmethoxycarbonyl (Fmoc) protective group strategy.

Key Words: Glycosylation; *N*- and *O*-glycosylated amino acids; solid-phase synthesis; glycopeptide; Fmoc approach.

1. Introduction

Glycopeptides are a rapidly growing family of molecules which contain a carbohydrate domain and a peptide domain. They are the product of a post- or

From: *Methods in Molecular Biology, vol. 494: Peptide-Based Drug Design*
Edited by: L. Otvos, DOI: 10.1007/978-1-59745-419-3_11, © Humana Press, New York, NY

co-translational modification of proteins. The glycans not only affect the structures, the chemical, physical, and biochemical properties of proteins, but also
their functions *(1,2)*. Glycoproteins play an important role in various biological
events such as cell adhesion, differentiation, and proliferation *(3–5)*, as well
as in numerous pathological processes ranging from viral and bacterial infections *(6–11)* to cancer metastasis *(12–15)*. Moreover, the structure of glycans,
which decorate the cell surface of higher organism, changes with onset of cancer
(14,16) and inflammation *(17–19)*. The most commonly observed glycoproteins can be classified into two principal groups: *N*- and *O*-glycosides. In *N*-
linked glycoproteins, the glycans are attached through their reducing termini
(*N*-acetylglucosamine) to the side chain amide group of an asparagine residue of
the Asn-Xaa-Ser/Thr consensus sequence where Xaa is any amino acid except
proline. *N*-Linked glycoproteins have a common pentasaccharide core that can
be altered further by various glycosyltransferases *(20)* (**Fig. 1**).

In *O*-linked glycoproteins, the glycans are usually attached to the side
chain of serine and threonine. The most common core of *O*-linked glycans
is an α-GalNAc (**Fig. 2**). The typical example is mucin type glycoproteins in which the sequential attachment of monosaccharides leads to
several different core structures, core1-6 *(21,22)*. A second important group
of *O*-glycoproteins, the proteoglycans, contain β-D-xylosyl residue linked to
serine *(23)*. Recently it has been found that serine and threonine residues in
nuclear pore proteins, transcription factors, and cytoskeletal proteins may carry
β-linked D-*N*-acetylglucosamine residues *(24)*. The collagens, which are structural proteins, contain hydroxylysine, which is frequently glycosylated with
β-D-galactosyl or an α-D-glucosyl-(1→2)-β-D-galactosyl moieties *(25)*. Some

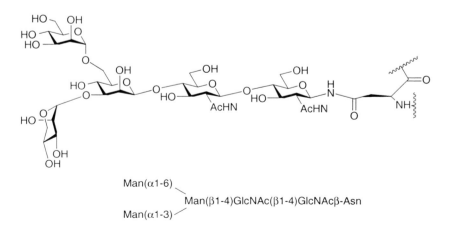

Fig. 1. Pentasaccharide core structure of *N*-linked glycoproteins.

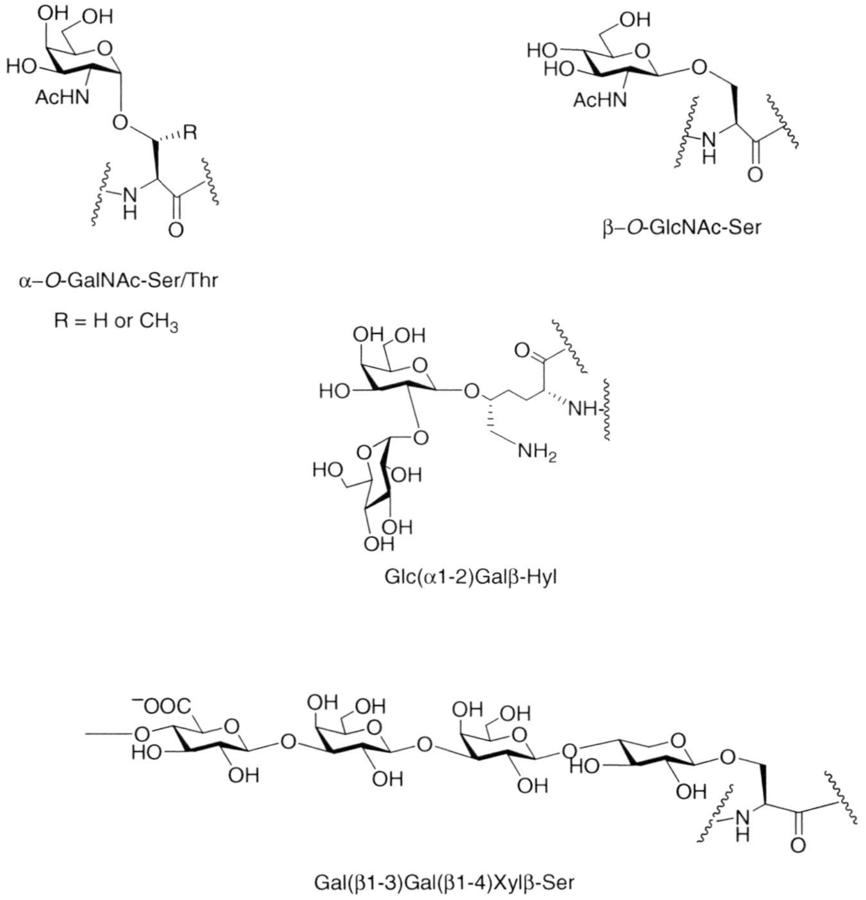

α–*O*-GalNAc-Ser/Thr

R = H or CH$_3$

β–*O*-GlcNAc-Ser

Glc(α1-2)Galβ-Hyl

Gal(β1-3)Gal(β1-4)Xylβ-Ser

Fig. 2. The most common saccharide cores of *O*-linked glycans.

uncommon glycosyl-peptide linkages found in nature include C-glycosides, S-glycosides, and carbohydrates linked via a phosphodiester bridge.*(26–29)*.

Isolation of well-defined glycoproteins or glycopeptides from natural sources is difficult or even impossible because of the microheterogenicity, a unique property of natural glycoproteins. Therefore, in order to understand the role of glycosylation of proteins, we rely on the synthesis of these compounds, either by chemical, enzyme, or recombinant approach *(30)*. Despite the challenges that carbohydrate moieties bring into the synthesis of glycopeptides and glyco-proteins, considerable progress has been made during recent decades. Routine procedures can now be employed for the chemical synthesis of glycopeptides carrying many simple glycans *(1,31–34)*. Enzymatic approaches have been utilized to introduce more complex glycans into synthetic glycopeptides *(35–37)*. The advent of native chemical ligation now makes it possible to synthesize glycoproteins from well-designed peptide and glycopeptide building blocks *(38,39)*. These advances in the field will further improve medical applications of these important and diverse biological molecules.

In this chapter we will summarize chemical methods for the stereoselective attachment of carbohydrates to amino acids, with particular emphasis on the preparation of building blocks for use in solid-phase glycopeptide synthesis based on the 9-fluorenylmethoxycarbonyl (Fmoc) protective group strategy.

2. Materials

1. Solvents for glycosylation reactions were purchased from Acros Organics, extra dry with molecular sieves, containing less than 50 ppm of water. Solvents used for "flash" chromatography, extractions, and crystallization are HPLC grade and can be purchased from Fisher Scientific.
2. Thin-layer chromatography silica gel 60 F_{254} (Whatman) plates were obtained from Fisher Scientific.
3. Silica gel (60 Å, 230-400 Mesh) for flash chromatography was purchased from Fisher Scientific.
4. Sodium hydrogen carbonate, sodium carbonate, ammonium hydrogen carbonate, anhydrous sodium sulfate, and sodium azide were purchased from Fisher Scientific.
5. Acetic acid, hydrochloric acid, and acetic anhydride were obtained from Fisher Scientific.
6. Hydrobromic acid (33% v/v) in acetic acid was obtained from Acros Organics.
7. Mercury acetate was purchased from Sigma-Aldrich.
8. Peracetylated sugars were purchased from Sigma or Senn Chemicals.
9. Glycosyl acceptors: Fmoc-protected amino acids were obtained from Novabiochem; D/L-hydroxylysine dihydrochloride monohydrate was obtained from Acros Organics.
10. Acid-washed molecular sieves (AW-300), activated molecular sieves (4 Å), and Celite (Celite 521) were obtained from Sigma-Aldrich.
11. Silver trifluoromethanesulfonate, methylsilyl trifluoromethanesulfonate, dimethylmethylthiosulfonium-trifluoromethane sulfonate (DMTST), and boron trifluoride diethyl etherate were obtained from Sigma-Aldrich.
12. Triethylamine, *N,N*-diisopropylethylamine, and 2,6-di-*tert*-butyl-4-methylpyridine were purchased from Acros Organics.
13. Trichloroacetonitrile and thiophenol were obtained from Acros Organics.
14. Palladium on carbon (Pd-C) was obtained from Acros Organics.
15. Pentafluorophenol was purchased from Acros Organics.
16. DCC or DIC were purchased from Novabiochem.

3. Methods

A crucial step in the chemical synthesis of glycopeptides is the incorporation of the saccharide into the peptide. Two approaches can be considered to accomplish this: the direct glycosylation of a properly protected full-length peptide and the use of a preformed glycosylated amino acid building block for the stepwise synthesis of the peptide backbone (**Scheme 1**).

An advantage of the direct glycosylation method is that the route is more convergent, permitting fast access to glycopeptides differing in glycan structure. Condensations between glycosyl amines and peptides containing aspartic or glutamic acid residues by direct approach have been accomplished in syntheses of *N*-linked glycopeptides (*40–42*). However, glycosylamines often undergo formation of intramolecular aspartimides (*43*). Direct *O*-glycosylation is often plagued by low yields due to the low reactivity of the side-chain hydroxyls and the low solubility of the peptides under conditions commonly employed for chemical glycosylation (*44–46*). Currently the most general synthetic methodology for glycopeptides relies on preformed glycosylated amino acids for the stepwise solid-phase peptide synthesis (*1,31–34*). Consequently, efficient synthetic routes to glycosylated amino acids for use as building blocks in solid-phase synthesis are of central importance in research dealing with preparation of glycopeptides.

3.1. Synthesis of Glycosylated Amino Acids

The formation of glycosidic bonds is of fundamental importance in the assembly of glycopeptides. As a consequence of immense structural variety, the development of a universal glycosylation reaction has failed. As such, every

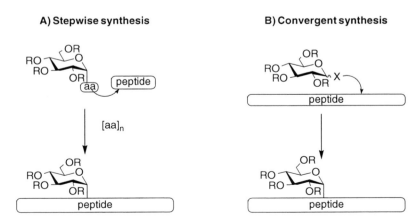

Scheme. 1. Synthetic strategies for glycopeptide preparations.

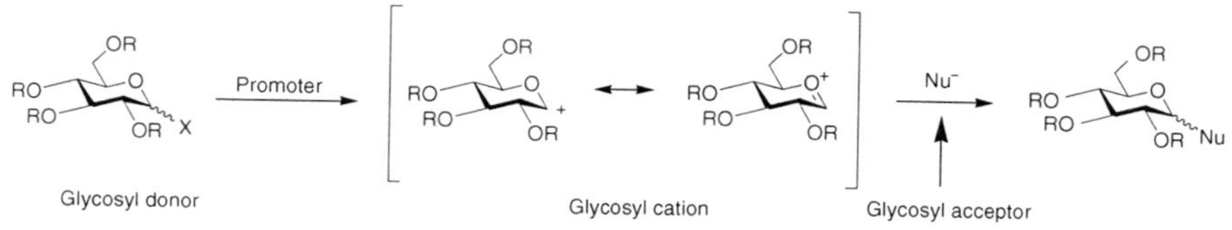

Scheme. 2. The formation of glycosidic bonds.

glycosylation has to be regarded as a unique problem demanding considerable systematic research.

In the most common chemical approach a glycosyl donor, i.e., a saccharide with a leaving group at the anomeric center is activated with a promoter to give a glycosyl cation susceptible to nucleophilic attack by a glycosyl acceptor (**Scheme 2**). In order to be efficient, a glycosylation reaction must occur with high stereoselectivity and in the case of multiple acceptor sites also with high regioselectivity, and provide the desired glycoside in high yield. The problem of regioselectivity can be controlled by the use of protective groups. However, glycosidic bonds are labile towards strong acids *(47)* and O-linked glycopeptides may undergo side reactions, such as β-elimination on treatment with strong bases *(48)*. Thus, protecting groups for glycosylated amino acids have to be carefully chosen. Extensive reviews covering protective groups in general and protective groups common in carbohydrate chemistry have appeared and the topic will not be discussed in any detail *(49–52)*.

3.2. O-Glycosylated Amino Acids

The reactivity of both the glycosyl donor and acceptor is another factor strongly affected by their respective protecting groups. Benzyl, allyl, isopropylidene, benzylidene, and silyl protected saccharides are more reactive than saccharides protected with electron-withdrawing groups such as acetyl and benzoyl groups. The stereochemical result of a glycosylation reaction is dependent on whether or not the glycosylation is proceeding with neighboring group participation of the substituent at C-2. A neighboring group at C-2 will lead to a 1,2-*trans* glycoside (**Fig. 3**). The initially formed oxocarbenium ion is in the equilibrium with the more stable, charge delocalized acyloxonium ion formed by the participation of the acyl group. Nucleophilic ring opening of the acyloxonium ion gives the desired 1,2-*trans* glycoside, whereas attack on the acyl carbon results in orthoester formation. Under the usually acidic glycosylation conditions, the orthoester rearranges to the more stable glycoside via the acyloxonium ion. The formation of orthoesters can become the main reaction when neutral or basic reaction conditions are applied. The formation of stable orthoester is more common with an acetyl group than with a sterically hindered

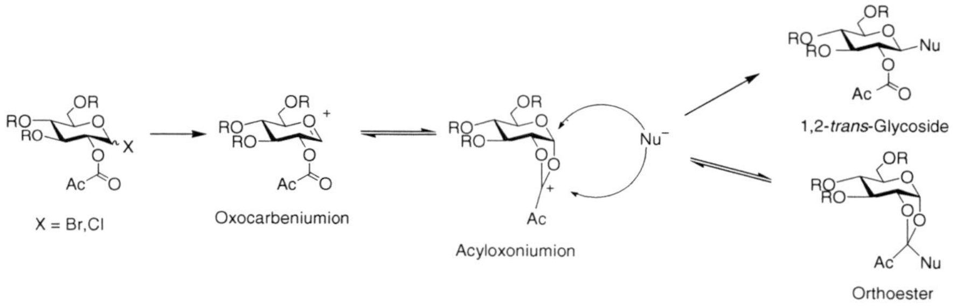

Fig. 3. Glycosidation pathways with a participating group in the 2-position.

pivaloyl group or a benzoyl group, which can delocalize the positive charge more efficiently.

Glycosylation without a neighboring effect will result in the formation of both 1,2-*trans* as well as 1,2-*cis* glycosides. The in situ anomerization procedure is the most common method for the synthesis of 1,2-*cis* glycosides. The soluble catalyst facilitates the rapid interconversion between the α- and β-glycosyl bromides *via* contact ion pairs (**Fig. 4**). The α-bromide is stabilized by the anomeric effect, and the more reactive β-bromide is present only in low concentrations. The higher reactivity of the β-bromide and the higher energy barrier for attack on the α-bromide leads to the selective formation of the 1,2-*cis* glycoside from the β-bromide. It is important to use a solvent of low polarity in order to prevent the dissociation of the ion pairs with concomitant loss of stereoselectivity.

The most widely used approaches in the synthesis of the *O*-glycosylated amino acids, Koenigs-Knorr, activation of anomeric acetate, trichloroacetimidate, and thioglycosides will be explained in detail.

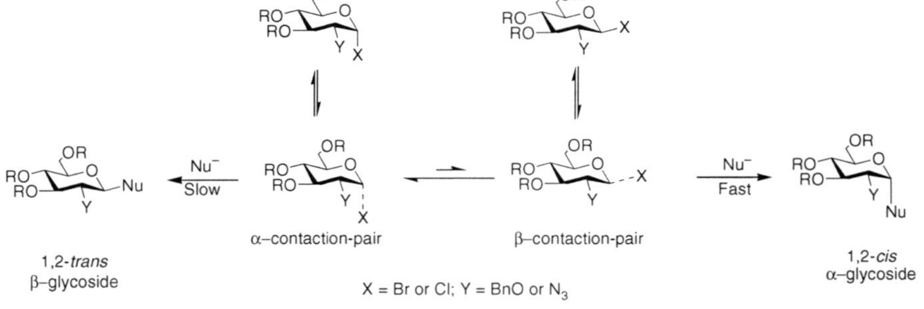

Fig. 4. Glycosidation pathways with a nonparticipating group in the 2-position.

3.2.1. Koenigs-Knorr Coupling

The Koenigs-Knorr method was, for a long time, the only available method for glycosylation and still represents the most widely used procedure for the synthesis of 1,2-*trans* glycosides in the chemistry of carbohydrate derivatives *(53)*. The α-glycosyl halide generated in the activation step *(54)* can be readily further activated in the glycosylation step by promoters, i.e., heavy metal salts, resulting in an irreversible glycosyl transfer to the acceptor. Insoluble silver salts, such as Ag_2CO_3, as well as soluble ones, such as AgOTf and $AgClO_3$, have been employed in the synthesis of glycosides.

The Koenigs-Knorr method has been used for the formation of 1,2-*trans* and 1,2-*cis* glycosidic bond. 1,2-*trans*-O-Linked glycosylated amino acids, β-D-Glc-(1→O)-Ser *(55)*, β-D-Gal-(1→O)-Hyl *(56,57)* (**Fig. 5**), α-D-Man-(1→O)-Ser/Thr *(58,59)*, and β-D-GlcNAc-(1→O)-Ser/Thr *(60)* are often prepared using a participating group at C-2 of the glycosyl donor. In the case of GlcNAc, it is desirable to replace the *N*-acetyl group with an electron-withdrawing group like *N*-allyloxycarbonyl (Aloc) *(61)*, *N*-trichloroethoxycarbonyl (Troc) *(62–65)*, or *N*-dithiasuccinoyl (Dts) *(66,67)* in order to decrease side product (oxazoline) formation. 1,2-*cis*-O-Linked glycosylated amino acids, α-D-GlcNAc-(1→O)-Ser/Thr, have been prepared by utilizing the nonparticipating azido group at C-2 of the GalNAc donor *(68)*. Often, 1-bromo *(69,70)* (**Fig. 6**) and 1-chlorosugars

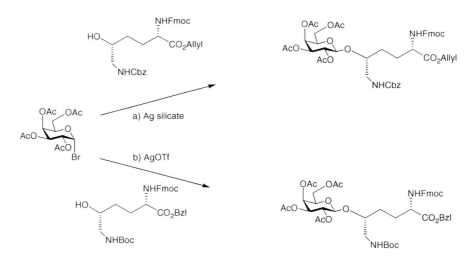

Fig. 5. Synthesis of 1,2-*trans* glycosides using a participating group at C-2: for example, synthesis of glycosylated hydroxylysine derivative (β-D-Gal-(1→O)-Hyl) using (a) insoluble promoter-silver silicate *(56)*; (b) soluble promoter-silver trifluoromethanesulfonate *(57)*.

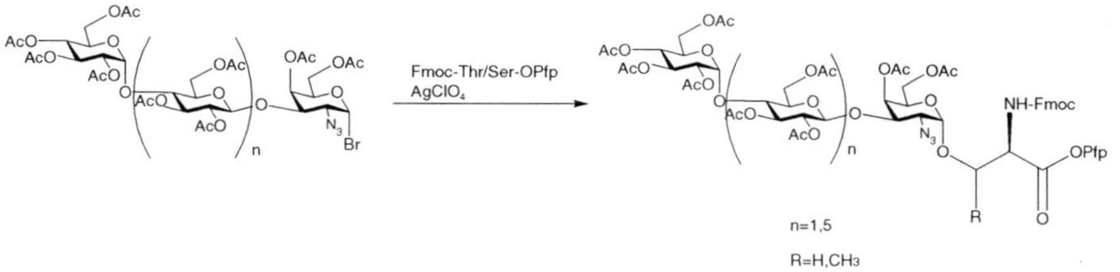

Fig. 6. Synthesis of 1,2-*cis* glycosides using a nonparticipating group at C-2: for example, synthesis of glycosylated GalNAc derivative (α-D-GalNAc-(1→O)-Ser/Thr) using AgClO$_4$ promoter at - 45°C *(70)*.

(71–74) have been employed in Koenigs-Knorr–type glycosylations of Fmoc-protected serine and threonine esters. In order to obtain the desired α-D-GlcNAc-(1→O)-Ser/Thr building block, the azido group has to be reduced and subsequently acetylated. Alternatively, reductive acetylation can be performed in one step with thioacetic acid *(69–71)*.

3.2.1.1. PREPARATION OF GLYCOSYL HALIDES

1. Dissolve peracetylated sugar (5 mol) in HOAc (5 mL/1 g) and cool to 0°C.
2. Add HBr in HOAc (33% v/v, 2 mL) dropwise over the period of 10 min.
3. Leave reaction mixture overnight in refrigerator at +4°C.
4. Add ice into the reaction mixture, extract with chloroform, wash chloroform extract twice with water.
5. Evaporate, and crystallize the product from diethylether (*see* **Note 1**).

3.2.1.2. KOENIGS-KNORR REACTION

1. Dissolve the glycosyl acceptor (1 eq) in dichloromethane and stir with activated molecular sieves (4 Å) under an inert atmosphere at 20°C (*see* **Note 2**).
2. Add silver salt (2 eq) and base (2 eq), protect from light, and cool suspension to −15°C (*see* **Note 3**).
3. At −15°C add glycosyl donor (2 eq) (from **Subheading 3.2.1.1.**), stir for 30 min and then stir overnight at room temperature (*see* **Note 4**).
4. Filter reaction mixture over Celite and wash the Celite layer well with CH$_2$Cl$_2$.
5. Combine filtrates, evaporate and purify by flash chromatography.

3.2.2. *Activation of Anomeric Acetate*

This breakthrough method uses 1,2-*trans* peracetates in the presence of Lewis acids as a promoter to glycosylate N^α-Fmoc amino acids with an unprotected α-carboxyl group. The deprotection of that glycosylated amino acid carboxyl group prior to incorporation into a peptide can be avoided *(62)*. In addition, the advantage of this method is that the starting materials are readily available

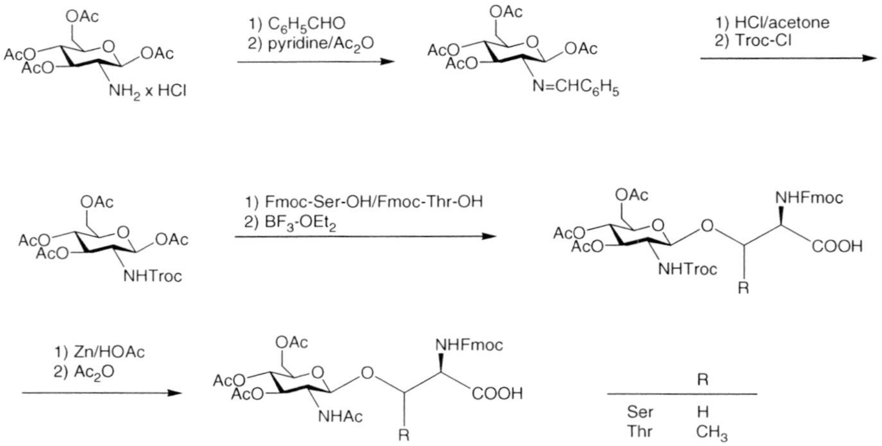

Fig. 7. Synthesis of Fmoc-Ser/Thr-OH building blocks containing glycosylated *O*-β-GlcNAc *(62)*.

and the method does not require extensive experience in synthetic carbohydrate chemistry. 1,2-*trans*-1-*O*-acetylated saccharides have been used for preparation of β-*O*-glycosides in the D-galactose and glucose series **(Fig. 7)** and for the synthesis of α-D-mannosides *(62)* **(Fig. 8)**.

3.2.2.1. Synthesis of β-Peracetylated Sugars (with Pyridine as Catalyst)

1. Add acetic anhydride (10 eq, 2 eq per OH group) into dry pyridine (approx 10 mL pyridine for 1 g of sugar) and cool down the reaction mixture to 0°C (*see* **Note 5**).
2. Add sugar (1 eq) and stir the solution until it dissolves completely at 0°C.
3. Let solution stand overnight at room temperature.

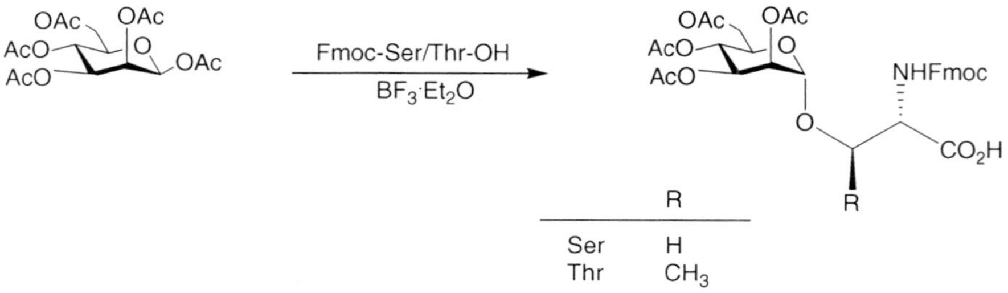

Fig. 8. Synthesis of Fmoc-Ser-OH *(62)* and Fmoc-Thr-OH *(75)* building blocks containing glycosylated 1-*O*-α-Man.

4. Pour solution over ice and water then extract the desired peracetylated sugar into chloroform.
5. Wash chloroform extract with water, sodium hydrogen carbonate, and again with water. Dry chloroform solution with anhydrous sodium sulfate, evaporate and isolate β-anomer by crystallization or perform flash chromatography to separate α- from β-anomer (*see* **Note 6**).

3.2.2.2. CONVERSION OF THE α-PERACETYLATED SUGARS INTO CORRESPONDING β-ANOMERS

1. Perform **steps 1–5** from **Subheading 3.2.1.1.** with α-peracetylated sugars.
2. Dissolve resulting glycosyl halide (1 eq) in acetic acid (approx 10 mL pyridine for 1 g of sugar).
3. Add mercuric acetate (2 eq) at room temperature and stir reaction mixture for 1 h.
4. Dilute solution with chloroform, wash with water, dry with anhydrous sodium sulfate, evaporate, and isolate β-anomer of peracetylated sugar by crystallization (diisopropylether).

3.2.2.3. ACTIVATION OF ANOMERIC ACETATE

1. Dissolve the glycosyl acceptor, N^α-Fmoc amino acid (1.2 eq) in dichloromethane (or ACN) under an inert atmosphere at room temperature (*see* **Note 2**).
2. Add Lewis acid, $BF_3 \cdot Et_2O$ (3 eq).
3. Add glycosyl donor, 1,2-*trans*-peracetylated monosaccharide (1 eq) and stir overnight at room temperature (*see* **Note 7**).
4. Dilute reaction mixture with dichloromethane and wash with 1 *M* aqueous HCl and water.
5. Combine organic filtrates, dry over Na_2SO_4, evaporate, and purify by flash chromatography.

3.2.3. Trichloroacetimidate Method

The trichloroacetimidate method has been developed into a widely applicable method and is often superior to the Koenigs-Knorr method. The general significance of *O*-glycosyl trichloroacetimidates lies in their ability to act as strong glycosyl donors under relatively mild acid catalysis. Neighboring group participation will give rise to 1,2-*trans* glycosides. When nonparticipating protecting groups are selected, α-trichloroacetimidates will yield β-glycosides and β-trichloroacetimidate α-glycosides. This method has been used in the synthesis of the mucin-type glycopeptides *(53,76–81)* containing α-D-GlcNAc-(1→O)-Ser/Thr types of glycosidic linkage (**Fig. 9**).

3.2.3.1. SYNTHESIS OF TRICHLOROACETIMIDATE

1. Dissolve the reducing sugar derivative (1 eq) in dry dichloromethane.
2. Add trichloroacetonitrile (2.9 eq) and base (1.1 eq) (*see* **Note 8**).

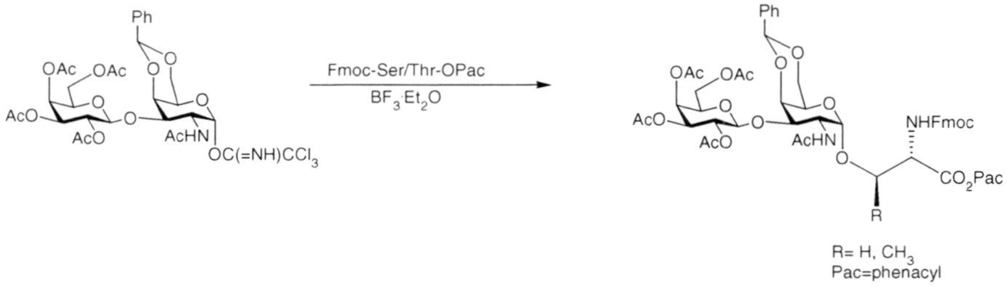

Fig. 9. Synthesis of the T antigen building block by α-glycosylation of Ser and Thr with GalNAc donor, which retains the 2-acetamido functionality at C-2 *(79)*.

3. Evaporate solution and purify the reaction mixture by flash chromatography (*see* **Note 9**).

3.2.3.2. TRICHLOROACETIMIDATE METHOD

1. Dissolve the glycosyl acceptor (1 eq) in dichloromethane (or acetonitrile, inert solvent) and stir with activated molecular sieves (4 Å) under an inert atmosphere at 20°C (*see* **Note 2**).
2. Add glycosyl donor-trichloroacetimidate (2 eq) and stir for 15 min at −15°C (*see* **Note 10**).
3. Add slowly trimethylsilyl trifluoromethanesulfonate (TMSOTf) (0.1 eq) and stir reaction mixture at −15°C for 1 h and at room temperature overnight (*see* **Note 11**).
4. Dilute reaction mixture with CH_2Cl_2 and wash with saturated aqueous $NaHCO_3$.
5. Combine filtrates, evaporate, and purify by flash chromatography.

3.2.4. Thioglycosides

Thioglycosides *(82)* can be activated by a wide variety of promotors, with DMTST being regularly used now and preferred over volatile and extremely toxic methyl-triflate. A great variety of methods for the preparation of alkyl- and aryl-1-thioglycosides of aldoses have been described *(52)*. Commonly, fully acetylated hexopyranoses react with thiols such as thiophenol or thioethanol in the presence of Lewis acid such as $BF_3 \cdot Et_2O$ to give predominantly 1,2-*trans* products. Thioglycoside method has been used in the synthesis of the mucin-type glycopeptides *(83,84)* containing α-D-GalNAc-(1→O)-Ser/Thr type of glycosidic linkage, 1,2-*trans*-O-linked glycosylated amino acids, β-D-Gal-(1→O)-Hyl *(57)*, and β-D-Xyl-(1→O)-Ser (**Fig. 10**) *(85)*.

3.2.4.1. SYNTHESIS OF THIOGLYCOSIDES

1. Dissolve the peracetylated sugar derivative (1 eq) in dry CH_2Cl_2 under an inert atmosphere at room temperature (*see* **Note 2**).

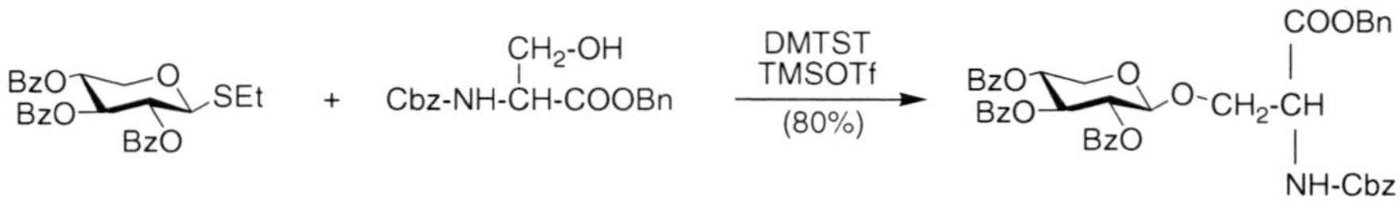

Fig. 10. Synthesis of β-D-Xyl-(1→O)-Ser derivative using thioglycoside derived from xylose (*85*).

2. Add thiophenol (1.5 eq), followed by Lewis acid, $BF_3 \cdot Et_2O$ (3 eq), and stir overnight at room temperature.
3. Wash reaction mixture two times with saturated aqueous sodium bicarbonate ($NaHCO_3$) and then with water.
4. Combine filtrates, evaporate and purify by flash chromatography.

3.2.4.2. THIOGLYCOSIDE METHOD

1. Dissolve the glycosyl donor (1 eq) and acceptor (1 eq) in dichloromethane and stir with activated molecular sieves (4 Å) under an inert atmosphere at room temperature (*see* **Note 2**).
2. Add DMTST (1.5 eq) and stir overnight at room temperature (*see* **Note 12**).
3. At the same temperature add *N,N*-diisopropylethylamine (1.5 eq) to neutralize the acid.
4. Filter reaction mixture over Celite and wash the Celite well with CH_2Cl_2.
5. Wash the organic filtrate with saturated aqueous $NaHCO_3$.
6. Combine filtrates, evaporate and purify by flash chromatography.

3.3. N-Glycosylated Amino Acids

N-Glycosidic linkage is commonly formed between a glycosyl amine and an activated aspartic acid derivative instead of glycosylation of asparagine (**Scheme 3**). Therefore, the first step toward the synthesis of *N*-glycosylated amino acids is the preparation of glycosyl amines. Two of the most commonly used procedures for the synthesis of glycosyl amines include the reduction of glycosyl azides and the treatment of reducing sugars with saturated ammonium

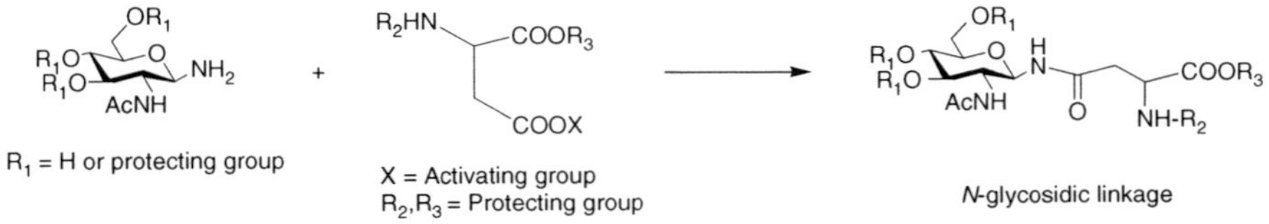

Scheme. 3. Formation of *N*-glycosidic linkage.

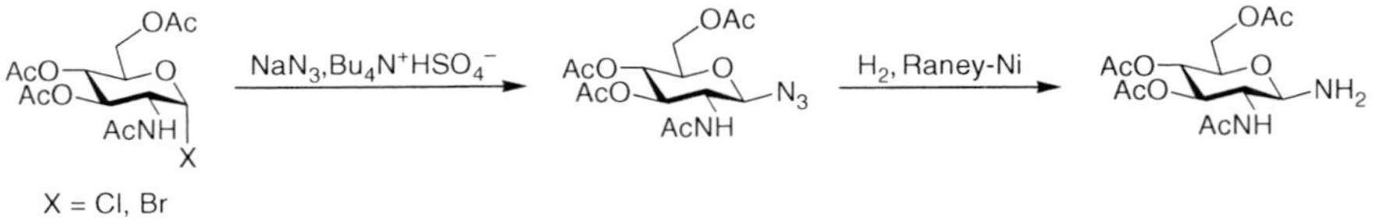

Fig. 11. Synthesis of glycosyl amine by reduction of glycosyl azide *(88,95)*.

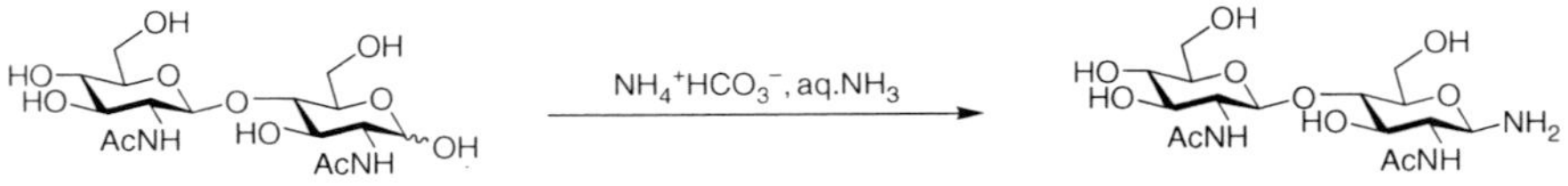

Fig. 12. Synthesis of glycosyl amine by amination of reducing sugar (chitobiose) *(97)*.

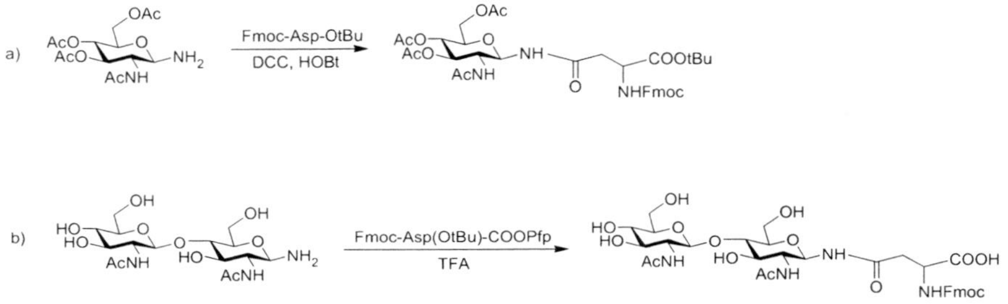

Fig. 13. Synthesis of β-linked asparagines glycosides: (a) dicyclohexylcarbodiimide (DCC) and 1-hydroxybenzotriazole (HOBt) coupling *(98)*, (b) pentafluorophenyl (OPfp) ester of aspartic acid derivative coupling *(97)*.

hydrogen carbonate. Glycosyl azides are most often obtained from the corresponding glycosyl halides *(86–88)* and then reduced to glycosyl amines by several available approaches *(87,89–95)* (**Fig. 11**).

The second route to glycosyl amine involves the introduction of the amino group at the reducing end of an unprotected carbohydrate in the presence of saturated ammonium bicarbonate *(96)* (**Fig. 12**).

Formation of an amide bond between a glycosyl amine and aspartic acid derivative or an aspartic acid–containing peptide can be than achieved by methods used for the formation of a peptide bond (**Fig. 13**) *(40,97–99)*.

3.3.1. Synthesis of Glycosylamine from Glycosyl Azides

1. Prepare glycosyl halide according to **steps 1–5** from **Subheading 3.2.1.1.**
2. Dissolve glycosyl halide (1 eq) in DMF and add NaN₃ (1.5 eq) and stir overnight at 80°C.

3. Dilute the reaction mixture with water, extract the glycosyl azide with chloroform, and evaporate the solvent *in vacuo*.
4. Dissolve formed glycosyl azide in methanol (containing 1 eq of HCl), add Pd-C (20 mol %), and stir at room temperature under a hydrogen atmosphere, Pd-C (20 mol %) until complete by TLC (usually overnight).
5. Filter reaction mixture through Celite, wash with methanol, and concentrate *in vacuo*.
6. Purify product by flash chromatography (*see* **Note 13**).

3.3.2. Synthesis of Glycosylamine from Reducing Sugars

1. Dissolve reducing sugar in saturated aqueous NH_4HCO_3 (approx 20 mL per 0.5 g of sugar) and stir for 6 d at 30°C (*see* **Note 14**).
2. Dilute reaction mixture with water (10 mL) and concentrate *in vacuo* to half of its original volume in order to remove excess of NH_4HCO_3. Repeat this step seven or eight times (*see* **Note 15**).
3. In the last step evaporate all water and dry the product in desiccator (*see* **Note 16**).

3.3.3. Formation of N-Glycoside Bond

1. Dissolve Fmoc-Asp(OtBu)-OH (1 eq) and pentafluorophenol (1 eq) in DMF (3 mL for 1 mmol).
2. Add DCC or DIC (2 eq) and stir at room temperature for 30 min (*see* **Note 17**).
3. Dissolve glycosyl amine (1 eq) in DMF-H_2O (2:1, v/v, 2 mL) and add to the reaction mixture. Stir overnight.
4. Remove solvent *in vacuo* and wash the residue with cold ether and cold water.
5. Purify product by flash chromatography.

4. Notes

1. If crystallization doesn't occur at room temperature, leave it in the freezer overnight.
2. All glycosylation reactions are performed under argon atmosphere and in the presence of molecular sieves (4 Å). Solvents used were extra dry with molecular sieves, containing less than 50 ppm of water.
3. The Koenigs-Knorr reaction will yield exclusively 1,2-*trans* glycosides, especially when insoluble silver salts are used as promoters.
4. In the case of GlcNAc, it is desirable to replace the *N*-acetyl with an electron-withdrawing group like *N*-allyloxycarbonyl (Aloc), *N*-trichloroethoxycarbonyl (Troc), or *N*-dithiasuccinoyl (Dts) in order to decrease side product (oxazoline) formation.
5. The hydroxyl groups of a carbohydrate will react readily with acetic anhydride in the presence of an acidic catalyst or a basic catalyst such as sodium acetate or pyridine *(54)*. The acidic catalyst may be zinc chloride, hydrogen

chloride, sulfuric acid, trifluoroacetic anhydride, perchloric acid, or an acidic ion-exchange resin. The complete acetylation of nonreducing sugars can be effected by any method. In the case of a reducing sugar, the anomeric configuration obtained depends on the catalyst used in acetylation and the temperature. Acetylation of α-D- or β-D-glucopyranose with acetic anhydride and pyridine at 0°C occurs without appreciable anomerization. The formation of the acetylated β-D-anomer is favored at higher temperatures with sodium acetate as a catalyst.

6. This step is recommended since α-anomer can be converted to β-anomer.
7. When anomeric acetate moiety is GlcNAc it is desirable to protect the *N*-acetyl group *(100)*.
8. Both of the anomeric glycosyl trichloroacetimidates can be obtained in pure form, depending on the base (NaH, K_2CO_3, or DBU) used for deprotonation of the reducing sugar).
9. When using NaH, always filtrate the reaction mixture over the Celite before the evaporation step.
10. The high reactivity of glycosyl trichloroacetimidate can lead to side reactions or even decomposition of the donor before reaction with the acceptor. To improve yield and the stereocontrol of the reaction, the so-called inverse glycosylation procedure is often used. In this case the glycosyl acceptor and the catalyst are dissolved together followed by the addition of the glycosyl donor.
11. The amount of Lewis acid employed varies from case to case. In many cases a few drops of diluted solution of TMSOTf is sufficient to complete the reaction, in other cases repeated additions of the catalyst are required. This might be particularly necessary when orthoesters, formed as intermediates, are to be isomerized to the respective glycosides.
12. Methyl triflate is volatile and extremely toxic; therefore the use of DMTST is recommended. Other promotes such as *N*-iodosuccinimide (NIS) can be used.
13. The use of Pd/C *(87,89)*, neutral Raney Ni (W-2) *(91,95)*, and Lyndlar's catalyst *(90)* have been reported in literature. Conditions used for the reduction of glycosyl azide have to be carefully optimized in order to suppress anomerization and the formation of undesired α-glycosides. Use of basic conditions and procedures that avoid noble metals leads to decreased α-glycosides formation *(94)*.
14. Temperature and time of reaction may vary, as well as the final yields.
15. Glycosyl amines are not very stable; therefore the water bath temperature should not exceed 30°C. Always prepare them freshly before further conversion to glycopeptides.
16. The attempts to separate the starting reducing sugars from the glycosyl amines by using Amberlit 15 (H⁺) ion-exchange resin was not successful with all sugars *(96,97)*.
17. DCC and DIC are known allergens and should be handled with care.

References

1. Seitz, O. (2000) Glycopeptide synthesis and the effects of glycosylation on protein structure and activity. *Chem. BioChem.* **1**, 214–246.
2. Lis, H. and Sharon, N. (1993) Protein glycosylation—structural and functional aspects. *Eur. J. Biochem.* **218**, 1–27.
3. Dwek, R. (1999) Glycobiology: toward understanding the function of sugars. *Chem. Rev.* **96**, 683–720.
4. Varki, A. (1993) Biological roles of oligosaccharides: All the theories are correct. *Glycobiology* **3**, 97–130.
5. Bertozzi, C. and Kiessling, L. (2001) Chemical glycobiology. *Science* **291**, 2357–2364.
6. Wyatt, R. and Sodroski, J. (1998) The HIV-1 envelope glycoproteins: fusogens, antigens, and immunogens. *Science* **280**, 1884–1888.
7. Geyer, H. and Geyer, R. (2000) Glycobiology of viruses. *Carbohydr. Chem. Biol.* **4**, 821–838.
8. Troy, F. (1992) Polysialylation: from bacteria to brain. *Glycobiology* **2**, 5–23.
9. Rostand, K. and Esko, J. (1997) Microbial adherence to and invasion through proteoglycans. *Infect. Immun.* **65**, 1–8.
10. Moncada, D., Kammanadiminti, S., and Chadee, K. (2003) Mucin and toll-like receptors in host defense against intestinal parasites. *Trends Parasitol.* **19**, 305–311.
11. Van Kooyk, Y., Engering, A., Lkkerkerker, A.N., Ludwig, I.S., and Geijtenbeck, T.B. (2004) Pathogens use carbohydrates to escape immunity induced by dendritic cells. *Curr. Opin. Immun.* **16**, 488–493.
12. Takano, R., Muchmore, E., and Dennis, J. (1994) Sialyation and malignant potential in tumor cell glycosylation mutants. *Glycobiology* **4**, 665–674.
13. Muramatsu, T. (1993) Carbohydrate signals in metastasis and prognosis of human carcinoma. *Glycobiology* **3**, 294–296.
14. Kim, Y. and Varki, A. (1997) Perspectives on the significance of altered glycosylation of glycoproteins in cancer. *Glycoconjugate J.* **14**, 569–576.
15. Dennis, J., Granovsky, M., and Warren, C.E. (1999) Glycoprotein glycosylation and cancer progression. *Biochim. Biophys. Acta* **1473**, 21–34.
16. Hakomori, S. (1989) Aberrant glycosylation in tumors and tumor-associated carbohydrate antigens. *Adv. Cancer Res.* **52**, 257–331.
17. Lowe, J. (2001) Glycosylation, immunity and autoimmunity. *Cell* **104**, 809–812.
18. Gleeson, P. (1994) Glycoconjugates in autoimmunity. *Biochim. Biophys. Acta* **1197**, 237–255.
19. Rudd, P., Elliott, T., Cresswell, P., Wilson, I., and Dwek, R. (2001) Glycosylation and the immune system. *Science* **291**, 2370–2376.
20. Kornfeld, R. and Kornfeld, S. (1985) Assembly of asparagine-linked oligosaccharides. *Annu. Rev. Biochem.* **54**, 631–664.
21. Carraway, K. and Hull, S. (1991) Cell surface mucin-type glycoproteins and mucin-like domains. *Glycobiology* **1**, 131–138.

22. Hanisch, F.-G. and Muller, S. (2000) Muc1: the polymorphic apperance of human mucin. *Glycobiology* **10**, 439–449.

23. Kjellen, L. and Lindahl, U. (1991) Proteoglycans: structure and interactions. *Annu. Rev. Biochem.* **60**, 443–475.

24. Zachara, N. and Hart, G. (2002) The emerging significance of *O*-GlcNAc in cellular regulation. *Chem. Rev.* **203**, 431–438.

25. Kivirikko, K. and Myllyla, R. (1982) Post-translational enzymes in the biosynthesis of collagen: Intracellular enzymes. *Methods Enzymol.* **82**, 245–304.

26. Vliegenthart, J. and Casset, F. (1998) Novel forms of protein glycosylation. *Curr. Opin. Struct. Biol.* **8**, 565–571.

27. Lote, C. and Weiss, J. (1971) Identification of digalactosylcysteine in a glycopeptide isolated from urine by a new preparative technique. *FEBS Lett.* **16**, 81–85.

28. Gonzalez de Peredo, A., Klein, D., Macek, B., Hess, D., Peter-Katalinic, J., and Hofsteenge, J. (2002) C-mannosylation and O-fucosylation of thrombospodin type I repeats. *Mol. Cell Proteomics* **1**, 11–18.

29. Haynes, P.A. (1998) Phosphoglycosylation: a new stuctural class of glycosylation? *Glycobiology* **8**, 1–5.

30. Davis, B. (2002) Synthesis of glycoproteins. *Chem. Rev.* **102**, 579–601.

31. Kunz, H. (1987) Synthesis of glycopeptides, partial structures of biological recognition components. *Agnew. Chem. Int. Ed.* **26**, 294–308.

32. Meldal, M., Glycopeptide synthesis, in *Neoglycoconjugates: Preparation and Applications* (Lee, Y. and Lee, R.T., eds.). Academic Press, San Diego, 1994, pp. 145–198.

33. Kihlberg, J. and Elofsson, M. (1997) Solid-phase synthesis of glycopeptides: immunological studies with T cell simulating glycopeptides. *Current Med. Chem.* **4**, 79–110.

34. Herzner, H., Reipen, T., Schultz, M., and Kunz, H. (2000) Synthesis of glycopeptides containing carbohydrates and peptide recognition motifs. *Chem. Rev.* **100**, 4495–4537.

35. Koeller, K.M. and Wong, C.H. (2001) Enzymes for chemical synthesis. *Nature* **409**, 232–239.

36. Flitsch, S.L., Macmillan, D., Bill, R.M., Sage, K.A., and Fern, D. (2000) Selective in vitro glycosylation of recombinant proteins: semi-synthesis of novel homogeneous glycoforms of human erythropoietin. *Chem. Biol.* **4**, 619–625.

37. Palcic, M. (1999) Biocatalytic synthesis of oligosaccharides. *Curr. Opin. Biotechnol.* **10**, 616–624.

38. Wu, B., Chen, J., Warren, D., Chen, G., Hua, Z., and Danishefsky, S. (2006) Building complex glycopeptides: Development of cysteine-free native chemical ligation protocol. *Agnew. Chem. Int. Ed.* **45**, 4116–4125.

39. Brik, A., Yang, Y.-Y., Ficht, S., and Wong, C.H. (2006) Sugar-assisted glycopeptide ligation. *J. Am. Chem. Soc.* **128**, 5626–5627.

40. Cohen-Anisfeld, S. and Lansbury, P., Jr. (1993) A practical, convergent method for glycopeptide synthesis. *J. Am. Chem. Soc.* **115**, 10531–10537.

41. Vetter, D., Tumelty, D., Singh, S., and Gallop, M. (1995) A versatile solid-phase synthesis of *N*-linked glycopeptides. *Angew. Chem. Int. Ed.* **34**, 60–62.
42. Offer, J., Quibell, M., and Johnson, T. (1996) On-resin solid-phase synthesis of asparagine-*N*-linked glycopeptides: use of *N*-(2-acetoxy-4-methoxybenzyl) (AcHmb) aspartyl amide-bond protection to prevent unwanted aspartimide formation. *J. Am. Chem. Soc. Perkin Trans I.*, 175–182.
43. Bodanszky, M. and Natarajan, S. (1975) Side reactions in peptide-synthesis. 2. Formation of succinimide derivatives from aspartyl residues. *J. Org. Chem.* **40**, 2495–2499.
44. Hollosi, M., Kollat, E., Laczko, I., Medzihradsky, K., Thurin, J., and Otvos, L.J. (1991) Solid-phase synthesis of glycopeptides: glycosylation of resin-bound serine-peptides by 3,4,6-tri-*O*-acetyl-D-glucose-oxazoline. *Tetrahedron Lett.* **32**, 1531–1534.
45. Andrews, D. and Seale, P. (1993) Solid-phase synthesis of *O*-mannosylated peptides: two strategies compared. *Int. J. Pept. Protein Res.* **42**, 165–170.
46. Paulsen, H., Schleyer, A., Mathieux, N., Meldal, M., and Bock, K. (1997) New solid-phase oligosaccharide synthesis on glycopeptides bound to a solid phase. *J. Am. Chem. Soc. Perkin Trans I.*, 281–293.
47. Mort, A. and Lamport, D.T. (1977) Anhydrous hydrogen fluoride deglycosylates glycoproteins. *Anal. Biochem.* **82**, 289–309.
48. Wakabayashi, K. and Pigman, W. (1974) Synthesis of some glycodipeptides containing hydroxyamino acids, and their stabilities to acids and bases. *Carbohydr. Res.* **35**, 3–14.
49. Greene, T.W. and Wuts, P., *Protective Groups in Organic Synthesis*. John Wiley, New York, 1991.
50. Mogemark, M. and Kihlberg, J. (2006) Glycopeptides. *Org. Chem. Sugars* 755–801.
51. Arsequell, G. and Valencia, G. (1997) *O*-Glycosyl α-amino acids as building blocks for glycopeptide synthesis. *Tetrahedron: Asymmetry* **8**, 2839–2876.
52. Lindhorst, T.K., *Essentials of Carbohydrate Chemistry and Biochemistry*. Wiley-VCH Verlag, Weinheim, 2003, pp. 39–78.
53. Schmidt, R. (1986) New methods for the synthesis of glycosides and oligosaccharides - Are there alternatives to the Koenigs-Knorr method? *Angew. Chem. Int. Ed.* **25**, 212–235.
54. Wolfrom, M.L. and Thompson, A. (1963) Acetylation. *Meth. Carbohydr. Chem.*, 211–215.
55. Reimer, K., Meldal, M., Kusumoto, S., Fukase, K., and Bock, K. (1993) Small-scale solid-phase *O*-glycopeptide synthesis of linear and cyclized hexapeptides from blood-clotting factor IX containing O-(α-D-Xyl-(1–3)-α-D-Xyl-(1–3)-β-D-Glc)-L-Ser. *J. Am. Chem. Soc. Perkin Trans I.*, 925–932.
56. Holm, B., Broddefalk, J., Flodell, S., Wellner, E., and Kihlberg, J. (2000) An improbed synthesis of a galactosylated hydroxylysine building block and its use in solid-phase glycopeptide synthesis. *Tetrahedron Lett.* **56**, 1579–1586.

57. Cudic, M., Lauer-Fields, J.L., and Fields, G.B. (2005) Improved synthesis of 5-hydroxylysine (Hyl) derivatives. *J. Pept. Res.* **65**, 272–283.
58. Vegad, H., Gray, C., Somers, P., and Dutta, A. (1997) Glycosylation of Fmoc amino acids: preparation of mono- and di-glycosylated derivatives and their incorporation into Arg-Gly-Asp(RGD)-containing glycopeptides. *J. Am. Chem. Soc. Perkin Trans I.*, 1429–1441.
59. Jansson, A., Meldal, M., and Bock, K. (1992) Solid-phase synthesis and characterization of *O*-dimannosylated heptadecapeptide analogues of human insulin-like growth factor 1 (IGF-1). *J. Am. Chem. Soc. Perkin Trans I.*, 1699–1707.
60. Jones, J.K.N., Perry, M.B., Shelton, B., and Walton, D.T. (1961) Carbohydrate-protein linkage in glycoproteins. I. Syntheses of some model substituted amides and a (2-amino-2-deoxy-D-glucosyl)-L-serine. *Can. J. Chem.* **39**, 1005–1016.
61. Vargas-Berenguel, A., Meldal, M., Paulsen, H., and Bock, K. (1994) Convenient synthesis of *O*-(2-acetamido-2-deoxy-β-D-glucopyranosyl)-serine and -threonine building blocks for solid-phase glycopeptide assembly. *J. Am. Chem. Soc. Perkin Trans I.*, 2615–2619.
62. Salvador, L., Elofsson, M., and Kihlberg, J. (1995) Preparation of building blocks for glycopeptide synthesis by glycosylation of Fmoc amino acids having unprotected carboxyl groups. *Tetrahedron Lett.* **51**, 5643–5656.
63. Schultz, M. and Kunz, H. (1992) Enzymatic glycosylation of *O*-glycopeptides. *Tetrahedron Lett.* **33**, 5319–5322.
64. Meinjohanns, E., Meldal, M., and Bock, K. (1995) Efficient synthesis of *O*-(2-acetamido-2-deoxy-β-D-glucopyranosyl)-Ser/Thr building blocks for SPPS of *O*-GlcNAc glycopeptides. *Tetrahedron Lett.* **36**, 9205–9208.
65. Saha, U. and Schmidt, R. (1997) Efficient synthesis of *O*-(2-acetamido-2-deoxy-β-D-glucopyranosyl)-serine and -threonine building blocks for glycopeptide formation. *J. Am. Chem. Soc. Perkin Trans I.*, 1855–1860.
66. Meinjohanns, E., Vargas-Berenguel, A., Meldal, M., Paulsen, H., and Bock, K. (1995) Comparision of *N*-Dts and *N*-Aloc in the solid-phase synthesis of *O*-GlcNAc glycopeptide fragments of RNA-polymerase II and mammalian neurofilaments. *J. Am. Chem. Soc. Perkin Trans I.*, 2165–2175.
67. Jensen, K., Hansen, P., Venugopal, D., and Barany, G. (1996) Synthesis of 2-acetamido-2-deoxy-β-D-glucopyranose *O*-glycopeptides from *N*-dithiasuccinoyl-protected derivatives. *J. Am. Chem. Soc.* **118**, 3148–3155.
68. Norberg, T., Luning, B., and Tejbrant, J. (1994) Solid-phase synthesis of *O*-glycopeptides. *Methods Enymol.* **247**, 87–106.
69. Kuduk, S., Schwarz, J., Chen, X.-T., et al. (1998) Synthetic and immunological studies on clustered modes of mucin-related Tn and TF *O*-linked antigens: The preparation of a glycopeptide-based vaccine for clinical trials against prostate cancer. *J. Am. Chem. Soc.* **120**, 12474–12485.
70. Cudic, M., Ertl, H.C.J., and Otvos, L.J. (2002) Synthesis, conformation, and T-helper cell stimulation of an *O*-linked glycopeptide epitope containing extended carbohydrate side-chains. *Bioorg. Med. Chem.* **10**, 3859–3870.

71. Vuljanic, T., Bergquist, K., Clausen, H., Roy, S., and Kihlberg, J. (1996) Piperidine is preferred to morpholine for Fmoc cleavage in solid phase glycopeptide synthesis as exemplified by preparation of glycopeptides related to HIV gp120 and mucins. *Tetrahedron Lett.* **52**, 7983–8000.

72. Kunz, H. and Birnbach, S. (1986) Synthesis of *O*-glycopeptides of the tumor-associated Tn- and T-antigen type and their binding to bovine serum albumin. *Agnew. Chem Int Ed.* **98(4)**, 360–362.

73. Paulsen, H. and Adermann, K. (1989) Synthesis of *O*-glycopeptides of the *N*-terminus of interleukin-2. *Liebigs Ann. Chem.*, 751–769.

74. Liebe, B. and Kunz, H. (1997) Solid-phase synthesis of a tumor-associated sialyl-TN antigen glycopeptide with a partial sequence of the "tandem repeat" of the MUC-1 mucin. *Angew. Chem. Int. Ed.* **36**, 618–621.

75. Kragol, G. and Otvos, L.J. (2001) Orthogonal solid-phase synthesis of tetramannosylated peptide constructs carrying three independent branched epitopes. *Tetrahedron* **57**, 957–966.

76. Grundler, G. and Schmidt, R. (1984) Glycosyl imidates, 13. Application of the trichloroacetimidate procedure to 2-azidoglucose and 2-azidogalactose derivatives. *Liebigs Ann. Chem.*, **(11)**, 1826–1847.

77. Kinzy, W. and Schmidt, R. (1987) Glycosylimidates. Part 24. Application of the tricholoroacetimidate method to the synthesis of glycopeptides of the mucin type containing a β-D-Galp-(1–3)-D-GalpNAc unit. *Carbohydr. Res.* **164**, 265–276.

78. Kinzy, W. and Schmidt, R. (1989) Glycosylimidates. Part 39. Synthesis of glycopeptides of the mucin type containg a β-D-GlcpNAc-(1–3)-D-GalpNAc unit. *Carbohydr. Res.* **193**, 33–47.

79. Yule, J.E., Wong, T.C., Gandhi, S.S., Qiu, D., Riopel, M.A., and Koganty, R.R. (1995) Steric control of *N*-acetylgalactosamine in glycosidic bond formation. *Tetrahedron Lett.* **36**, 6839–6842.

80. Qiu, D., Gandhi, S.S., and Koganty, R.R. (1996) β-Gal(1–3)GalNAc block donor for the synthesis of TF and α-sialyl(2–6)TF as glycopeptide building blocks. *Tetrahedron Lett.* **37**, 595–598.

81. Schmidt, R. and Kinzy, W. (1994) Anomeric-oxygen activation for glycoside synthesis: the trichloroacetimidate method. *Adv. Carbohydr. Chem. Biochem.* **50**, 21.

82. Norberg, T. (1996) Glycosylation properties and reactivity of thioglycosides, sulfoxides, and other S-glycosides: current scope and future prospects. *Mod. Meth. Carbohydr. Synth.* 82–106.

83. Paulsen, H., Rauwald, W., and Weichert, U. (1988) Building units of oligosaccharides. LXXXVI. Glycosidation of oligosaccharide thioglycosides to *O*-glycoprotein segments. *Liebigs Ann. Chem.* **1**, 75–86.

84. Braun, P., Waldmann, H., and Kunz, H. (1992) Selective enzymatic removal of protecting functions: heptyl esters as carboxy protecting groups in glycopeptide synthesis. *Synlett* **1**, 39–40.

85. Paulsen, H. and Brenken, M. (1988) Synthesis of L-alanyl-3-*O*-(β-D-xylopyranosyl)-L-seryl-glycyl-L-isoleucine. Direct glycosidation of peptides. *Liebigs Ann. Chem.* **7**, 649–654.

86. Kunz, H., Waldman, H., and Marz, J. (1989) Synthesis of partial structures of *N*-glycopeptides representing the linkage regions of the transmembrane neuraminidase of an influenza virus and of factor B of the human complement system. *Liebigs Ann. Chem.* (1), 45–49.

87. Thiem, J. and Wiemann, T. (1990) Combined chemoenzymic structure of *N*-glycoprotein synthons. *Angew. Chem. Int. Ed.* **29**, 80.

88. Tropper, F., Andersson, F., Braun, S., and Roy, R. (1992) Phase transfer catalysis as a general and stereoselective entry into glycosyl azides from glycosyl halides. *Synthesis* 618–620.

89. Marks, G. and Neuberger, A. (1961) Synthetic studies relating to the carbohydrate-protein linkage in egg albumin. *J. Am. Chem. Soc.* **958**, 4872–4879.

90. Nakabayashi, S., Warren, C.D., and Jeanloz, R.W. (1988) The preparation of a partially protected heptasaccharide-asparagine intermediate for glycopeptide synthesis. *Carbohydr. Res.* **174**, 279–289.

91. McDonald, F.E. and Danishefsky, S. (1992) A stereoselective route from glycals to asparagine-linked *N*-protected glycopeptides. *J. Org. Chem.* **57**, 7001–7002.

92. von dem Bruch, K. and Kunz, H. (1994) Synthesis of *N*-glycopeptide clusters with Lewis antigen side chains and their binding of carrier proteins. *J. Org. Chem.* **33**, 101–103.

93. Saha, U.K. and Roy, R. (1995) First synthesis of *N*-glycopeptoid as new glycopeptidomimetics. *Tetrahedron Lett.* **36**, 3635–3638.

94. Unverzagt, C. (1996) Chemoenzymatic synthesis of a sialylated undecasaccharide-asparagine conjugate. *Angew. Chem. Int. Ed.* **35**, 2350–2353.

95. Kunz, H. and Unverzagt, C. (1988) Protecting-group-dependant stability of inter-saccharide bonds-Synthesis of a fucosyl-chitobiose glycopeptide. *Angew. Chem. Int. Ed.* **27**, 1697–1699.

96. Likhosherstov, L.M., Novikova, O.S., Derevitskaja, V.A., and Kochetkov, N.K. (1986) A new simple synthesis of amino sugar β-D-glycosylamines. *Carbohydr. Res.* **146**, c1–c5.

97. Urge, L., Kollat, E., Hollosi, M., et al. (1991) Solid-phase synthesis of glycopeptides: synthesis of *N*-α-fluorenylmethoxycarbonyl L-asparagine *N*-β-glycosides. *Tetrahedron Lett.* **32**, 3445–3448.

98. Clark, R.S., Banerjee, S., and Coward, J.K. (1990) Yeast oligosaccharyltransferase: glycosylation of peptide substrates and chemical characterization of the glycopeptide product. *J. Org. Chem.* **55**, 6275–6285.

99. Anisfeld, S. and Lansbury, P., Jr. (1990) A convergent approach to the chemical synthesis of asparagine-linked glycopeptides. *J. Org. Chem.* **55**, 5560–5562.

100. Schultz, M. and Kunz, H. (1993) Synthetic *O*-glycopeptides as model substrates for glycosyltransferases. *Tetrahedron: Asymmetry* **4**, 1205–1220.

12

Synthesis of *O*-Phosphopeptides on Solid Phase

David Singer and Ralf Hoffmann

Summary

Most mammalian proteins are transiently phosphorylated on seryl, threonyl, or tyrosyl positions by kinases and dephosphorylated by phosphatases in response to specific intra- or extracellular signals. Many diseases, such as cancer, are caused by altered kinase and phosphatase activities or changed expression levels of either of these enzymes. Thus phosphopeptides are important and universal tools to study disease-specific changes, such as protein–protein/DNA interactions or hyperactive kinases, using phosphopeptides or the corresponding mimics. Here we describe two generally applicable techniques to synthesize multiply phosphorylated peptides as basic tools for drug development. The phosphopeptides were purified to homogeneity and characterized by mass spectrometry.

Key Words: Global phosphorylation; phosphopeptide; phosphoramidite; phosphoserine; phosphothreonine; phosphotyrosine.

1. Introduction

Phosphorylation of hydroxyl groups of seryl, threonyl, and tyrosyl residues is a very common reversible posttranslational protein modification catalyzed by kinases in response to external or internal signals. Thereby, the proteins often undergo conformational changes that can influence the interaction with other biomolecules, relocate the protein to a different part of the cell, or regulate the activity of an enzyme. A single mammalian cell expresses hundreds of different kinases and phosphatases that form a complex signaling network to regulate all cell processes. Disturbances of a single pathway can seriously affect the cellular environment, causing different diseases, such as cancer. As phosphoproteins cannot be simply expressed in cells and phosphorylation of recombinant proteins

From: *Methods in Molecular Biology, vol. 494: Peptide-Based Drug Design*
Edited by: L. Otvos, DOI: 10.1007/978-1-59745-419-3_12, © Humana Press, New York, NY

by kinases is not specific for a single position and often not quantitative, it is necessary to synthesize phosphopeptides to study the underlying mechanisms or to develop novel drugs, e.g., specific kinase inhibitors.

Currently, two strategies to synthesize multiply phosphorylated peptides by 9-fluorenylmethoxycarbonyl (Fmoc) chemistry are well established and routinely used in many laboratories (1). The first Fmoc-based phosphopeptide synthesis on solid phase were reported more than 20 years ago using different reagents to phosphorylate partially unprotected peptides (i.e., free hydroxyl groups at serine, threonine, or tyrosine positions to be phosphorylated) after completion of the synthesis and before cleaving all other side-chain–protecting groups (2). Among the many reagents tested, phosphoramidites (3) offered the best yields and highest purities. The building-block strategy to directly incorporate phosphoamino acid derivatives was introduced first for tyrosine (4,5) which is stable against the basic conditions during Fmoc cleavage. On the other hand, it was only 10 years ago when Vorherr and Bannwarth introduced partially protected phosphoserine and phosphothreonine derivatives compatible to the Fmoc chemistry (6).

Whereas both approaches typically yield the targeted sequences in high purities for peptide lengths up to 30 residues (7), they face the same limitations described for some peptide syntheses, i.e., the "difficult sequences." The lower coupling efficiencies obtained at certain positions or even truncated sequences are a result of steric hindrance or the formation of secondary structures on the solid phase that lower the efficiency of either the following coupling or deprotection steps. This is especially true for the bulky phosphoamino acid building blocks, although the global phosphorylation can also be hampered. In spite of these limitations, peptides with up to three phosphorylation sites can be synthesized routinely by the described protocols. In our experience, the analytics is often more challenging for multiphosphorylated sequences than the synthesis (8). In RP-HPLC these peptides often elute in broad peaks not separated from lower phosphorylated versions or deleted sequences. In mass spectrometry, the high content of negatively charged groups reduces the ionization efficiency in positive ion mode typically applied to analyze peptides. As a rule of thumb the signal intensity of phosphopeptides in a mass spectrum decreases about 10 times with each additional phosphate group. Therefore, low signal intensities of peptides carrying several phosphate groups do not necessarily indicate that the targeted sequences are obtained in low yields.

2. Materials

2.1. Peptide Synthesis

1. Synthesize peptides with the base-labile Fmoc group to temporarily protect the α-amino group and acid-labile permanent side-chain protecting groups: tert-butyl

('Bu) for aspartic acid (Asp), glutamic acid (Glu), serine (Ser), threonine (Thr), and tyrosine (Tyr), *tert*-butyloxycarbonyl (Boc) for lysine (Lys) and tryptophan (Trp), triphenylmethyl (trityl, Trt) for cysteine (Cys), histidine (His), asparagine (Asn), and glutamine (Gln), as well as 2,2,4,6,7-pentamethyldihydrobenzofuran-5-sulfonyl (Pbf) for arginine (Arg) (ORPEGEN Pharma, Heidelberg, Germany).

2. As solid support use Rink-amide MBHA resin (0.6 mmol/g, 100–200 mesh, MultiSynTec, Bochum, Germany) to obtain C-terminal peptide amides and Wang-resin (1.2 mmol/g, 100–200 mesh, MultiSynTec) or 2-chlorotrityl chloride resin (1.5 mmol/g, 100–200 mesh, Novabiochem, Merck KGaA, Darmstadt, Germany) to produce C-terminal free acids.

3. Instrument: Multiple peptide synthesizer SYRO (MultiSynTec) equipped with a 48-reaction vessel block (*see* **Note 1**).

4. Carbodiimide activation:

   a) *N,N*′-Diisopropylcarbodiimide (DIC), 2 *M* in DMF (Fluka Neu-Ulm, Germany).

   b) 1-Hydroxybenzotriazole hydrate (HOBt), 0.5 *M* in DMF (Fluka).

5. Uronium-reagents:

   a) *O*-(benzotriazol-1-yl)-*N,N,N*′,*N*′-Tetramethyluronium hexafluorophosphate (HBTU), 0.5 *M* in DMF (MultiSynTec).

   b) 1-Hydroxybenzotriazole hydrate (HOBt), 0.5 *M* in DMF (Fluka).

   c) *N,N*-Diisopropylethylamine (DIPEA), 1.33 *M* in NMP (Fluka).

6. Fmoc cleavage:

   a) Mixture of 20% piperidine in DMF (Fluka).

   b) Mixture of 40% piperidine in DMF (Fluka).

7. Dimethylformamide (DMF, Biosolve, Valkenswaard, the Netherlands; *see* **Note 2**).

8. *N*-Methylpyrrolidone (NMP, Biosolve).

9. Dichloromethane (DCM, Biosolve).

10. Methanol (Biosolve).

11. Ether (Merck KGaA, Darmstadt, Germany).

12. Reagents for peptide cleavage (Fluka):

    a) Trifluoroacetic acid (TFA) 10 mL (*see* **Note 3**).

    b) Scavenger mixture: 500 µL water, 500 µL thioanisole, 500 µL m-cresol, 250 µL 1,2-ethanedithiol.

2.2. Direct Incorporation of Fmoc Phosphoamino Acid Derivatives

1. Partially protected phosphoamino acids (Novabiochem, *see* **Note 4**):

   a) *N*-α-Fmoc-*O*-benzyl-L-phosphoserine (Fmoc-Ser(PO(OBzl)OH)-OH) (*see* **Note 5**).

b) *N*-α-Fmoc-*O*-benzyl-L-phosphothreonine (Fmoc-Thr(PO(OBzl)OH)-OH).

c) *N*-α-Fmoc-*O*-benzyl-L-phosphotyrosine (Fmoc-Tyr(PO(OBzl)OH)-OH).

d) *N*-α-Fmoc-*O*-benzyl-D-phosphoserine (Fmoc-D-Ser(PO(OBzl)OH)-OH).

e) *N*-α-Fmoc-*O*-benzyl-D-phosphothreonine (Fmoc-D-Thr(PO(OBzl)OH)-OH).

f) *N*-α-Fmoc-*O*-benzyl-D-phosphotyrosine (Fmoc-D-Tyr(PO(OBzl)OH)-OH).

g) -N-α-Fmoc-*O*-(bis-dimethylamino-phosphono)-L-tyrosine   (Fmoc-Tyr(PO(NMe$_2$)$_2$-OH).

h) *N*-α-Fmoc-*O*-phospho-L-tyrosine (Fmoc-Tyr(PO$_3$H$_2$)-OH).

2.3. Global Phosphorylation of Partially Protected Peptides

1. Side-chain–unprotected hydroxyamino acids (Novabiochem) (*see* **Note 4**):

    a) Fmoc-Ser-OH / Fmoc-D-Ser-OH (*see* **Note 5**).
    b) Fmoc-Thr-OH / Fmoc-D-Thr-OH.
    c) Fmoc-Tyr-OH.

2. Hydroxyamino acids with selectively cleavable side-chain–protecting groups (Novabiochem) (*see* **Note 4**):

    a) Fmoc-Ser(Trt)-OH / Fmoc-D-Ser(Trt)-OH (*see* **Note 5**).
    b) Fmoc-Thr(Trt)-OH / Fmoc-D-Thr(Trt)-OH.
    c) Fmoc-Tyr(2-ClTrt)-OH / Fmoc-D-Tyr(2-ClTrt)-OH.

3. Dibenzyl-*N,N*-diisopropylphosphoramidite (Fluka).
4. ^1H-Tetrazole solution in acetonitrile (purum, ~0.45 *M*, Fluka; *see* **Note 6**).
5. *tert*-Butylhydroperoxide solution, 5.5 *M* in decan, anhydrous (Fluka).
6. Acetonitrile (Biosolve; *see* **Note 1**).
7. Fifteen- or 50-mL polypropylene tubes, septa, syringes, and cannulae (6 and 12 cm).

2.4. Liquid Chromatography

1. Instrument: ÄKTA purifier 10 HPLC System (GE Healthcare, Munich, Germany) consisting of pump P-900, detector UV-900, autosampler A-905, and fraction collector Frac-950.
2. Analytical columns:

    a) ZORBAX Eclipse XDB-C8 (C$_8$ stationary phase, 5 μm particle size, 80 Å pore size, 4.6 mm internal diameter and 150 mm long; Agilent Technologies GmbH, Böblingen, Germany).
    b) Jupiter C18 (C$_{18}$ stationary phase, 5 μm particle size, 300 Å pore size, 4.6 mm internal diameter and 150 mm long; Phenomenex Inc, Torrance, CA).

3. Preparative column: Jupiter C18 (C$_{18}$ stationary phase, 15 μm particle size, 300 Å pore size, 21.2 mm internal diameter and 250 mm long; Phenomenex Inc).
4. Eluents:

a) Eluent A: Water (*see* **Notes 7** and **8**) containing 0.1% TFA as ion pair reagent.
b) Eluent B: 60% aqueous acetonitrile containing 0.1% TFA (*see* **Note 9**).

5. Analytical separation: Inject 20- to 50-μg sample and keep the starting conditions of 5% eluent B for 2 min. Elute the peptides with a linear increase of 3% eluent B per min to 95% eluent B using a flow rate of 1 mL/min. Hold these conditions for 5 min before equilibrating the column again with 5% eluent B for at least 15 min.
6. Preparative separation: Purify up to 40 mg crude peptide in a single separation using 5% eluent B as starting conditions. After 2 min increase the content of eluent B to 95% using a linear gradient of 1–3% acetonitrile per minute and a flow rate of 10 mL/min (*see* **Note 10**). The slope of the gradient depends on the number of by-products eluting close to the target peptide. Hold the final conditions of 95% eluent B for 5 min before equilibrating the column with the starting conditions of 5% eluent B.
7. Detection: Absorption at 220 nm (*see* **Note 11**).

2.5. Mass Spectrometry

1. MALDI-TOF/TOF-MS (4700 Proteomics Analyzer, Applied Biosystems GmbH, Darmstadt, Germany).
2. α-Cyano 4-hydroxy-cinnamic acid (CHCA) matrix:

a) Weigh 2 mg α-cyano 4-hydroxy-cinnamic acid (Bruker Daltonics GmbH, Bremen, Germany, store at 4°C) in a microcentrifuge tube and dissolve in 250 μL water containing 0.1% TFA.
b) Vortex the solution for 10 s.
c) Add 250 μL acetonitrile (HPLC grade).
d) Vortex well until the matrix is completely dissolved.
e) Store at −20°C. This solution can be used for one month.

3. α-Cyano 4-hydroxy-cinnamic acid with phosphate (pCHCA) matrix:

a) Prepare a fresh matrix solution as described above.
b) Add ammoniumphosphate (ultra ≥99.5%, Fluka) to the matrix solution to a final concentration of 10 mM.
c) Store at −20°C. This solution can be used for one month.

4. 2,5-Dihydroxybenzoic acid (DHB) matrix:

a) Weigh approximately 5 mg 2,5-dihydroxybenzoic acid (Bruker Daltonics, store at 4°C) exactly in a 1.5-mL Eppendorf tube.
b) Add 450 μL water and 450 μL methanol (calculated for 5 mg matrix, adjust according the weighed amount).
c) Vortex well until the matrix is completely dissolved.
d) Store at −20°C. This solution can be used for up to one month.

3. Methods

There are two general strategies to synthesize phosphopeptides by Fmoc chemistry. First, the postsynthetic global phosphorylation approach *(7,9)* incorporates an Fmoc-protected Ser, Thr, or Tyr without side-chain protection. An obvious advantage of this approach is that both the unmodified and phosphorylated peptides can be obtained from a single synthesis by splitting the resin after the peptide synthesis prior to phosphorylation. The second approach incorporates phosphorylated Fmoc amino acid derivatives carrying TFA-labile protecting groups in the side chain, referred to as the building block strategy. Both methods have inherent advantages and disadvantages, but usually work well for singly phosphorylated peptides up to 20 residues in length. For longer or multiply phosphorylated peptides, the appropriate strategy has to be selected depending on the peptide sequence and the number of residues to be phosphorylated. Moreover, it might be necessary in some cases to optimize the here-described standard protocols for "difficult" sequences by using longer reaction times or a higher reagent excess. Alternatively, the purification strategy has to be optimized to separate the desired phosphopeptide from by-products.

For the synthesis of monophosphorylated peptides with "easy-to-synthesize" sequences containing no bulky side chains or protecting groups proximal to the phosphorylation site, the direct incorporation of monobenzyl protected phosphoserine *(10)*, phosphothreonine *(6)*, and phosphotyrosine *(4)* is usually preferred, as standard peptide synthesis reagents are used and no additional postsynthetic handling steps are required (*see* **Note 12**). For multiply phosphorylated peptides and especially for proximal phosphorylation sites, we prefer the global phosphorylation approach. This strategy is also more efficient and cost saving if several phosphopeptides or the corresponding unmodified peptides are synthesized, for example, in a parallel peptide synthesis.

3.1. Multiple Solid Phase Peptide Synthesis

1. Swell the resin in DMF for 30 min (e.g., 40 mg Rink-amide MBHA resin, 25 μmol amino groups).
2. Add 1 mL piperidine (40% by vol. in DMF) and discard after 3 min.
3. Add 1 mL piperidine (20% by vol. in DMF) for 10 min and remove.
4. Wash the resin six times with 1 mL DMF for 1 min each.
5. Prepare a solution of 0.5 *M* HOBt in DMF or in a mixture of NMP and DMF (1:1 by vol.).
6. Dissolve the desired amount of the Fmoc amino acids in the HOBt-solution to a final concentration of 0.5 *M* (*see* **Note 13**).
7. Add the amino acid solution from **step 6** at a 4- to 10-fold molar excess over the amino groups of the resin (e.g., 200 μmol [8 eq] amino acid derivative corresponding to 400 μL Fmoc amino acid solution).

8. Add the required volume of coupling reagents equimolar to the Fmoc amino acid derivative (e.g., 200 μmol DIC, 100 μL).
9. Incubate for 1–1.5 h with frequent mixing.
10. Wash the resin four times with DMF for 1 min each.
11. Add a mixture of 40% piperidine in DMF for 3 min and discard.
12. Add a mixture of 20% piperidine in DMF for 10 min and discard.
13. Wash the resin six times with DMF for 1 min each.
14. Repeat **steps 5–13** for the different amino acid derivatives until the target sequence is completed.
15. Wash the resin five times with DCM for 1 min each.
16. Dry the resin under vacuum.

3.2. Building Block Strategy

The direct coupling of the Fmoc- and monobenzyl-protected phosphoamino acids is accomplished by standard synthesis protocols applied for the Fmoc/^{t}Bu-strategy (*see* **Subheading 3.1.**). The recommended solvent for dissolving phosphoamino acids is NMP or a 1:1 (by volume) mixture of NMP and DMF. Existing protocols for automated peptide synthesis can be used without any further modification, although it might be useful to increase the coupling time and the amino acid excess.

3.3. Global Phosphorylation

1. Synthesize the peptides with side-chain–unprotected hydroxy amino acids (*see* **Note 14**) or amino acids with selectively cleavable side-chain–protecting groups (such as trityl cleaved with 1 to 5% TFA in dichloromethane at room temperature for 1 h *[11]*) in each position to be modified.
2. After completion of the synthesis (including cleavage of the Fmoc group (*see* **Note 15**), wash the resin five times with DMF and again five times with DCM (1 mL each).
3. Place the resin in a 15- or 50-mL polypropylene tube (*see* **Note 16**), dry the resin overnight in a vacuum chamber to remove DCM and any remaining water traces.
4. Close the tube with a septum and flush the tube with nitrogen (*see* **Note 17**).
5. Swell the resin in DMF for 30 min. The volume of DMF should slightly exceed the volume of the swollen resin.
6. Add 40 eq of ^1H-tetrazole (0.45 M in acetonitrile) with a syringe to the resin suspended in DMF.
7. Dilute 20 eq of dibenzyl-N,N-diisopropylphosphoramidite with the ninefold volume of DMF under inert conditions (*see* **Note 18**) and add the solution immediately to the resin (*see* **Note 19**).
8. Incubate for 90 min at room temperature with gentle shaking.

9. Remove as much as possible of the reagent mixture with a syringe without losing any resin, add the same amount of fresh DMF and reagents **(steps 6–8)** and incubate overnight at room temperature.
10. Open the tube, remove the reaction solution, and wash the resin twice with DMF.
11. For oxidation add 100 equivalents of dry *tert* -butylhydroperoxide solution (5.5 *M* in decan; *see* **Note 20**), close the tube and incubate at room temperature for 90 min by gentle shaking.
12. Remove the reaction mixture, add once more the same amount of fresh *tert*-butylhydroperoxide solution, close the tube, and incubate with gentle shaking at room temperature for 90 min more.
13. Wash the resin carefully four times with DMF and four times with DCM. Dry the resin in a vacuum chamber or desiccator.

3.4. Peptide Cleavage

1. Add a freshly prepared TFA-scavenger mixture to the resin (1.5 mL for 40 mg resin) and incubate for 90 min.
2. Remove the TFA-scavenger mixture (store at room temperature) and add fresh TFA-scavenger mixture (0.5 mL for 40 mg resin) and incubate again for 90 min.
3. Remove the TFA-scavenger mixture, combine it with the first cleavage mixture, and place the tube on ice.
4. For precipitation of the peptide add the cold cleavage mixture to 8 mL of ice cold diethyl ether.
5. Centrifuge for 5 min (1600 g) and discard the ether phase.
6. Wash the precipitate three times with ice-cold ether.
7. Dry the precipitate on air.
8. Dissolve the peptide in water or 0.1% aqueous TFA and store at −20°C (*see* **Note 21**).

3.5. Analysis of Synthetic Phosphopeptides

The synthesis of phosphopeptides is typically confirmed mass spectrometrically using either a MALDI (matrix-assisted laser desorption/ionization) or an ESI (electrospray ionization) source, and the peptide purity is determined by reversed-phase chromatography coupled to an UV detector (**Figs. 1** and **2B**). Whenever possible, phosphopeptide analyses should be complimented by mass spectra recorded in negative ion mode. For most biochemical applications it is necessary to purify the peptides by HPLC techniques (**Fig. 2A**).

3.5.1. HPLC

Although the described protocol was optimized for a C_8-Zorbax and a C_{18}-Jupiter column on an Äkta-HPLC system, it can be applied to C_8- and C_{18}-stationary phases from other companies and other HPLC equipment

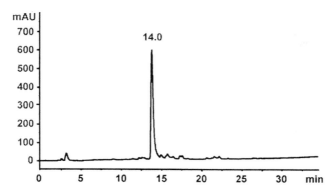

Fig. 1. Reversed-phase chromatogram of the crude unphosphorylated peptide H-CXVAVVRTPPKSPSSAK-NH$_2$ (X = ε-aminohexanoic acid) obtained after TFA cleavage and diethyl ether precipitation using the Jupiter C$_{18}$-column and the gradient described in **Subheading 2.4.** Absorbance was recorded at 220 nm.

without changing the conditions (*see* **Subheading 2.3.**). However, hydrophobic sequences may require a higher acetonitrile content in eluent B (up to 80%; *see* **Note 22**), and difficult sequences yielding truncated or deleted sequences may require shallower gradients (0.5–1% increase of acetonitrile per min).

Due to the high polarity of the phosphate group, phosphopeptides typically elute at a 1–3% lower acetonitrile content than the corresponding unmodified peptides (**Figs. 1** and **2B**). However, this shift depends on the position within the peptide chain. Sometimes phosphopeptides and their non- or lower phosphorylated analogs coelute or the elution order is even reversed. For such difficult separations a more polar stationary phase (C$_4$ or C$_8$), a different ion-pair reagent (formic acid or heptafluorobutyric acid), or a different organic solvent (methanol or isopropanol) might accomplish separation.

3.5.2. Mass Spectrometry

For details on MALDI or ESI mass spectrometry, *see* Chapter 3. For MALDI-MS the phosphopeptides are cocrystallized with pCHCA and measured in positive and negative ion mode (**Fig. 3**). Addition of phosphate to CHCA, which is commonly used for peptides, increases the signal intensities of phosphopeptides relative to unphosphorylated peptides. For multiphosphorylated peptides we typically use DHB to further increase the signal intensity. However, these matrix effects depend on the instrument and cannot be generalized.

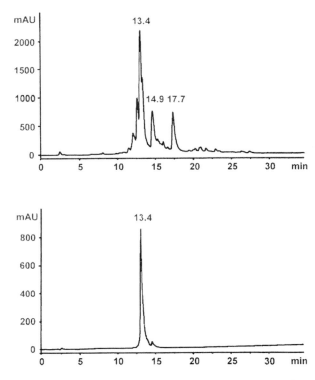

Fig. 2. Chromatograms of the double phosphorylated peptide H-CXVAVVRT (P)PPKS(P)PSSAK-NH₂ (X = ε-aminohexanoic acid, T(P) = phosphothreonine, and S(P) = phosphoserine) synthesized by the global phosphorylation strategy (*see* **Subheading 3.3.**) after precipitation with diethyl ether (upper panel) and after purification on a preparative Jupiter C₁₈-column (lower panel). For chromatographic conditions see **Fig. 1**. The by-products eluting at 14.9 min and 17.7 min are most likely impurities of the amidite reagent and not formed by the peptide.

4. Notes

1. The protocols described here for multiple peptide synthesis can be transferred to any other instrument by correcting the volumes and synthesis scale accordingly without changing the chemistry, reagent excess, etc.
2. Whereas some solvents (MeOH, dichloromethane, etc.) can be used at lower purities without significant loss of the peptide purities, it is obligatory to use only dry and amine-free dimethylformamide (peptide synthesis grade) to dissolve the amino acids and activating reagents as well as at least for the final washing step prior to coupling the next amino acid. DMF-contaminated with amines can be used to dilute piperidine for Fmoc cleavage and for all washing steps after coupling the amino acid and for the first four washing steps after the piperidine cleavage.

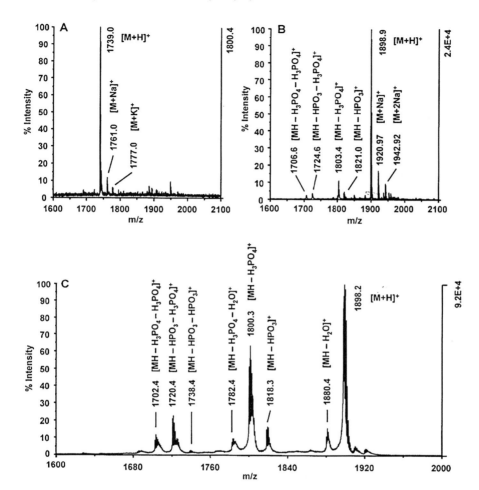

Fig. 3. MALDI- mass spectrum of the crude unphosphorylated peptide H-CXVAVVRTPPKSPSSAK-NH$_2$ (**A**) and the corresponding purified doubly phosphorylated peptide H-CXVAVVRT(P)PPKS(P)PSSAK-NH$_2$ (**B**) using the pCHCA matrix (*see* **Subheading 2.5.**). The expected monoisotopic peptide masses for the unphosphorylated analogue ([M+H]$^+$ = 1739.0) and the phosphopeptide ([M+H]$^+$ = 1898.9) were confirmed. Characteristic for peptides phosphorylated on seryl- and threonyl-residues are the consecutive mass losses of 98 u (-H$_3$PO$_4$) and 80 u (-HPO$_3$) in MALDI-MS by post-source decay (PSD; **C**) using a PSD calibration or the MS/MS mode in TOF/TOF instruments. In reflector TOF mode using the regular TOF calibration (**B**) the [MH-HPO$_3$]$^+$ and [MH- H$_3$PO$_4$]$^+$ signals are not displayed at the correct masses but at slightly higher m/z values, i.e., at m/z 1821.0 and 1803.4 corresponding to mass losses of 77.9 and 95.5, respectively.

3. Whereas the scavenger mixture can be stored for one month, the reagent mixture with TFA should be prepared fresh for each cleavage.

4. The three-letter code used for all amino acids refers always to the corresponding L-amino acids. D-Enantiomers are abbreviated by D- followed by the corresponding three-letter code, such as D-Ser for D-serine.

5. Store all amino acid derivatives and activation reagents dry at 4°C in the darkness in aliquots of 25 g or less. Before opening a container, place the derivatives in the laboratory for at least 2 h to reach room temperature. Thus the derivatives stay dry even when opened several times at higher humidity. Dissolve all solid compounds just before starting the synthesis without prolonged storage. It should be noted that amines are formed slowly in DMF in the presence of water traces. Therefore it is important to store DMF under nitrogen atmosphere or to consume it within 1 wk. We use older DMF only to dilute piperidine.

6. It is important to use amidite reagent, ^{1}H-tetrazole, and all solvents for the phosphitilation step of high purities and to keep all vials and solutions dry. Thereby, many by-products described in the literature for the global amidite approach (i.e., H-phosphonopeptides, dimers, etc.) can be omitted yielding the corresponding peptides in high purities accompanied only by species with lower phosphorylation degrees.

7. All buffers and solutions were prepared in desalted water with a conductivity below 18.2 mS and a total organic content below 2 ppb (parts per billion) using a ELGA PURELAB Ultra system (ELGA LabWater, Celle, Germany).

8. Eluents should be filtered through a 0.2-μm filter (GH Polypro, Hydrophilic Polypropylene membrane filter, Pall Life Sciences, Ann Arbor, MI) and stored in clean glass bottles for less than 1 mo. Solvents should be degassed by vacuum (1 mbar) for 1–2 min if a low-pressure gradient mixture is used (do not use longer times to minimize loss of TFA and acetonitrile). For high-pressure gradient HPLC systems this degassing is not necessary if a backpressure valve is used after the detector to prevent formation of air bubbles in the detector cell.

9. As acetonitrile has a higher absorption than water at 220 nm, the UV trace slightly climbs with increasing acetonitrile content. Whereas this linear climb of the background is a good measure for judging the reproducibility of the formed gradient, it can be advantageous to compensate it by reducing the TFA content in eluent B to 0.09%. If 100% acetonitrile is used as eluent B, the TFA content should be further decreased to 0.085%. The resulting stable base line is especially useful to analyze small peptide amounts with low signal intensities.

10. For most crude peptides a linear increase of 3% per min eluent B is sufficient to separate the target peptide from all by-products. In order to obtain high purities and load larger sample amounts on the preparative column use a linear increase of only 1% per min eluent B at least 10 min before and after the targeted peptide elutes. Calculate the retention time on the preparative column from the retention time of the analytical separation using the known dead volumes of the system and the different flow rates.

11. To distinguish peptides and partially protected peptidic by-products carrying an aromatic protecting group (e.g., Fmoc, benzyl), it is advantageous to record the

absorption additionally in the 280- to 300-nm region. For larger peptide amounts producing UV signals out of range at 220 nm, use higher wave lengths at 226 or 230 nm. Thereby, impurities eluting close to the target peptide can be distinguished.

12. To our experience, it is difficult to couple more than one phosphoamino acid derivative in one peptide. This is due to a reduced reactivity and/or sterical hindrance, especially if two phosphoamino acids are in close proximity.

13. Do not use higher concentrations of the amino acid derivatives to prevent the risk of precipitation of some derivatives, which may block the instrument.

14. If amino acids with unprotected hydroxyl groups are incorporated, it is better to use carbodiimide activation (such as DIC/HOBt-activation) instead of more reactive procedures to avoid acylation of the hydroxyl-groups.

15. The N-terminal position can be incorporated as a Boc-protected amino acid to omit piperidine cleavage. Otherwise, the Fmoc group is cleaved prior to the global phosphorylation procedure, as fully protected phosphoserine and -threonine are base labile yielding dehydroalanine and dehydrobutyric acid, respectively. Although the N-terminal amino group is also modified by the reagent, this phosphoramidite is cleaved quantitatively during the peptide workup. Usually we phosphorylate peptides with free N-termini.

16. The size of the tube depends on the amount of resin to be phosphorylated resulting in certain volumes of the reagents. For appropriate mixing the reagent volume should not exceed half of the total tube volume.

17. All reagents are applied through the septum using syringes. To keep a constant nitrogen pressure a balloon filled with nitrogen is connected to the tube via a needle through the septum.

18. It is obligatory to use clean and dry syringes. Add first the required volume of DMF (dry, free of amines and oxygen) to the syringe and discharge any gas bubbles from the syringe. Before adding the amidite reagent, place the bottle containing the amidite under nitrogen atmosphere (*see* **Note 17**). Add the required volume of amidite reagent and mix the solution by rotating or shaking the syringe.

19. If the tetrazol precipitates (formation of clear white crystals) after addition of the amidite, add more DMF until the precipitate is completely dissolved.

20. Peptides containing no amino acids prone to oxidation (i.e., neither Cys nor Met) the stronger oxidation reagent *meta*-chloroperbenzoic acid (MCPBA, 20 eq) is preferred as a 0.3 M solution in dry acetonitrile.

21. If the peptide is not completely dissolved add some acetonitrile.

22. If polar sequences or short phosphopeptides fail to bind to the stationary phase upon loading, the acetonitrile content of the eluent could be reduced to 3%. A further decrease is not possible for standard C_{18}-columns, as the stationary phase could collapse reducing the loading capacity and the reproducibility. Instead, TFA could be replaced by the more hydrophobic ion-pair reagent HFBA (heptafluorobutyric acid), or a stationary phase with a special coating compatible with pure aqueous eluents could be used (e.g., Aqua-column from Phenomenex).

Acknowledgment

We thank the Deutsche Forschungsgemeinschaft (DFG) and the European Regional Development Fund (EFRE, European Union and Free State Saxonia) for financial support.

References

1. McMurray, J.S., Coleman IV, D.R., Wang, W., and Campbell, M.L. (2001) The synthesis of phosphopeptides. *Biopolymers* **60**, 1, 3–31.
2. Bannwarth, W. and Trzeciak, A. (1987) A simple and effective dhemical phosphorylation procedure for biomolecules. *Helv. Chim. Acta* **70**, 175–186.
3. Perich, J.W. and Johns, R.B. (1988) Di-*tert*-butyl *N,N*-diethylphosphoramidite. A new phosphitylating agent for the efficient phosphorylation of alcohols. *Synthesis* 142–144.
4. Handa, K.B. and Hobbs, C.J. (1998) An efficient and convenient procedure for the synthesis of N_α-Fmoc-*O*-monobenzyl phosphonotyrosine. *J. Pept. Sci*, **4**, 138.
5. Kitas, E.A., Perich, J.W., Wade, J.D., Johns, R.B., and Treagear, G.W. (1989) Fmoc-polyamide solid phase synthesis of an *O*-phosphotyrosine-containing tridecapeptide. *Tetrahedron Lett.* **30**, 45, 6229–6232.
6. Vorherr, T. and Bannwarth, W. (1995) Phospho-serine and phospho-threonine building blocks for the synthesis of phosphorylated peptides by the Fmoc solid phase strategy. *Bioorg. Med. Chem. Lett.* **5**, 2661–2664.
7. Singer, D., Lehmann, J., Hanisch, K., Härtig, W., and Hoffmann, R. (2006) Neighbored phosphorylation sites as PHF-tau specific markers in Alzheimer's disease. *Biochem. Biophys. Res. Comm.* **346**, 819–828.
8. Zeller, M. and König, S. (2004) The impact of chromatography and mass spectrometry on the analysis of protein phosphorylation sites. *Anal. Bioanal. Chem.* **378**, 898–909.
9. Hoffmann, R., Wachs, O.W., Berger, G.R., et al. (1995) Chemical phosphorylation of the peptides GGXA (X = S,T,Y): an evaluation of different chemical approaches. *Int. J. Pept. Prot. Res.* **45**, 26–34.
10. Wakamiya, T., Saruta, K., Yasuoka, J., and Kusumoto, S. (1994) An efficient procedure for solid-phase synthesis of phosphopeptides by the Fmoc strategy. *Chem. Lett.* 1099–1102.

13

Peptidomimetics: Fmoc Solid-Phase Pseudopeptide Synthesis

Predrag Cudic and Maciej Stawikowski

Summary

Peptidomimetic modifications or cyclization of linear peptides are frequently used as attractive methods to provide more conformationally constrained and thus more stable and bioactive peptides. Among numerous peptidomimetic approaches described recently in the literature, particularly attractive are pseudopeptides or peptide bond surrogates in which peptide bonds have been replaced with other chemical groups. In these peptidomimetics the amide bond surrogates possess three-dimensional structures similar to those of natural peptides, yet with significant differences in polarity, hydrogen bonding capability, and acid-base character. The introduction of such modifications to the peptide sequence is expected to completely prevent protease cleavage of amide bond and significantly improve peptides' metabolic stability.

In this chapter we consider Fmoc solid-phase synthesis of peptide analogs containing the amide surrogate that tend to be isosteric with the natural amide. This includes synthesis of peptidosulfonamides, phosphonopeptides, oligoureas, depsides, depsipeptides, and peptoids.

Key Words: Peptidomimetics; pseudopeptides; isosteres; Fmoc solid-phase synthesis; peptidosulfonamides; phosphonopeptides; oligourea; depsides; depsipeptides; peptoids.

1. Introduction

In recent years peptides have gained momentum as therapeutic agents. According to Frost and Sullivan around 720 peptide drugs and drug candidates were reported in 2004, among which 5% are already marketed worldwide, 1% in registration, 38% in clinical trials, and 56% in advanced preclinical phases (*1*). Peptides' potential high efficacy combined with minimal side effects made them

From: *Methods in Molecular Biology, vol. 494: Peptide-Based Drug Design*
Edited by: L. Otvos, DOI: 10.1007/978-1-59745-419-3_13, © Humana Press, New York, NY

widely considered as lead compounds in drug development, and at present peptide-based therapeutics exists for a wide variety of human diseases, including osteoporosis (calcitonin), diabetes (insulin), infertility (gonadorelin), carcinoid tumors and acromegaly (octreotide), hypothyroidism (thyrotropin-releasing hormone [TRH]), and bacterial infections (vancomycin, daptomycin) *(2)*. However, despite the great potential, there are still some limitations for peptides as drugs per se. Major disadvantages are short half-life, rapid metabolism, and poor oral bioavailability. Nevertheless, pharmacokinetic properties of peptides can be improved by different types of modifications *(2,3)*. Peptidomimetic modifications or cyclization of linear peptides are frequently used as attractive methods to provide more conformationally constrained and thus more stable bioactive peptides *(4–11)*. Among numerous peptidomimetic approaches used for the design and synthesis of peptide analogues with improved pharmacological properties, pseudopeptides or peptide bond surrogates, in which peptide bonds have been replaced with other chemical groups, are especially attractive. This is mainly because such approaches create amide bond surrogates with defined three-dimensional structures similar to those of natural peptides, yet with significant differences in polarity, hydrogen bonding capability, and acid-base character. Also importantly, the structural and stereochemical integrities of the adjacent pair of α-carbon atoms in these pseudopeptides are unchanged. The psi-bracket ($\Psi[\;]$) nomenclature is used for this type of modifications *(12)*. The introduction of such modifications to the peptide sequence is expected to completely prevent protease cleavage of amide bond and significantly improve peptides' methabolic stability. But besides these positive effects, such modifications may also have some negative effects on peptides' biophysical and biochemical properties, in particular their conformation, flexibility, and hydrophobicity. Therefore, the choice of an amide bond surrogate is a compromise between positive effects on pharmacokinetics and bioavailability and potential negative effects on activity and specificity. In particular, the conservation of the stereochemistry of the parent peptide should be an important criterion in selection of the surrogate. The ability of the surrogate to mimic the steric, electronic, and solvation properties of the amide bond is certainly the most important characteristic determining the potency of pseudopeptide analogs.

In this chapter we consider the synthesis of peptide analogs containing the amide surrogate that tend to be isosteric with the natural amide. This includes synthesis of peptidosulfonamides, phosphonopeptides, oligoureas, depsides, depsipeptides, peptoids, and peptomers (**Fig. 1**). Given the importance of peptides as lead compounds for drug discovery and development, it is not surprising that the above-mentioned peptidomimetic strategies to enhance peptide stability have attracted a great deal of attention in the last few decades *(2,11)*.

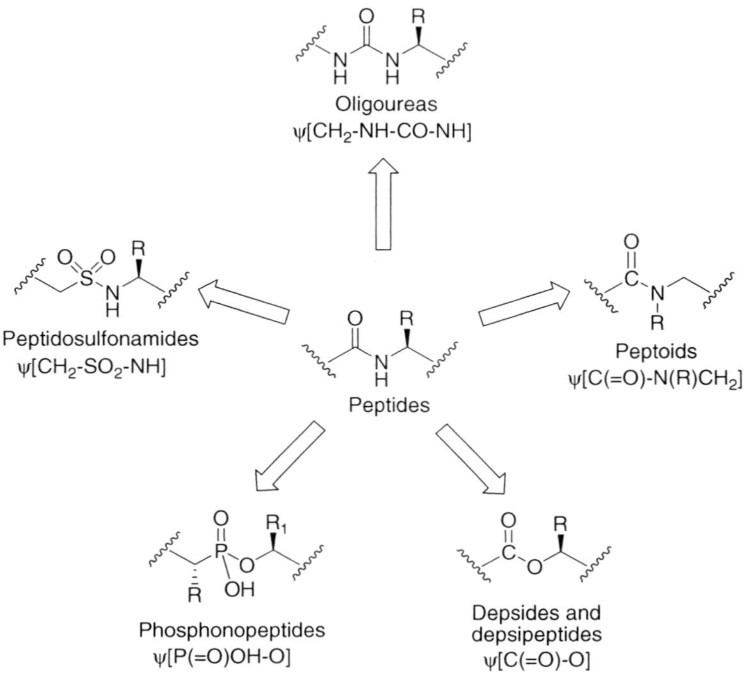

Fig. 1. Schematic representation of peptidomimetics containing the amide surrogates that are isosteric with the natural peptidic amide bonds.

2. Materials

1. All materials and reagents are commercially available and used as received.
2. Synthesis solvents, such as dichloromethane (DCM), dimethoxyethane (DME), acetic acid (AcOH), ethyl acetate (EtOAc), *tert*-butanol (*t*-BuOH), 1-methyl-2-pyrrolidinone (NMP), tetrahydrofurane (THF), methanol (MeOH), toluene, benzene, *N,N'*-dimethylformamide (DMF), methyltertbutylether (MTBE), hexane, dimethylsulfoxide (DMSO), and diethyl ether, were synthesis or peptide synthesis grade (if necessary) and can be obtained from Sigma-Aldrich, Fisher, VWR, or other commercial sources.
3. Peptide synthesis reagents such as diisopropylcarbodiimide (DIC), benzotriazole-1-yl-oxy-tris-pyrrolidino-phosphonium hexafluorophosphate (PyBOP), bromo-tris-pyrrolidino-phosphonium hexafluorophosphate (PyBroP), 2-(1H-benzotriazole-1-yl)-1,1,3,3-tetramethyluronium hexafluorophosphate (HBTU), 1-hydroxybenzotriazole (HOBt), piperidine, *N*-methylmorpholine (NMM), trifluoroacetic acid (TFA), triisopropylsilane (TIS), *N,N*-diisopropylethylamine (DIPEA, DIEA), 1,8-diazabicyclo[5.4.0]undec-7-ene (DBU), 1,2-ethanedithiol (EDT), and 4-dimethylaminopyridine (DMAP) may be obtained from Sigma-Aldrich, ChemImpex, and Novabiochem.

4. Specific reagents including diethyl azodicarboxylate (DEAD) and derivatives, triphenylphosphine (PPh$_3$), 9-fluorenylmethoxycarbonyl chloride (Fmoc-Cl), isobutyl chloroformate (iBu-CO-Cl), methanesulfonyl chloride (MsCl), 4-nitrophenylethyl chloroformate (NPEOC), 4-nitrophenyl chloroformate, hydrogen peroxide, thionyl chloride, thioacetic acid (HSAc), cesium carbonate (Cs$_2$CO$_3$), palladium on activated carbon (10% Pd) (Pd/C), and celite (diatomaceous earth, diatomaceous silica) can be purchased from Sigma-Aldrich, Fisher, VWR, or other commercial sources.

5. Resins for solid-support synthesis can be obtained from Rapp-Polymere (Germany), Novabiochem (USA), Advanced ChemTech (USA), ChemImpex, and other suppliers.

3. Methods

From the synthetic point of view, the methods for assembly of peptidosulfonamides, phosphonopeptides, oligoureas, depsides, depsipeptides, peptoids, and peptomers parallel those for standard solid-phase peptide synthesis, although different reagents and different coupling and protecting strategies need to be used. Since these peptidomimetics can be constructed in a modular way from orthogonally protected monomeric building blocks and they are therefore suitable for potential combinatorial chemistry diversity, the solid-phase methodology is the method of choice for their synthesis. Particularly attractive is Fmoc solid-phase methodology since it is now a standard approach for the routine peptide synthesis. Therefore, in this chapter we describe only modifications of the peptide main chain using the solid-phase methodology that includes or is fully compatible with the Fmoc chemistry.

3.1. Peptidosulfonamide Synthesis, Ψ [CH$_2$-SO$_2$-NH]

The tetrahedral achiral sulfur atom bonded to the two oxygen atoms possesses geometry similar to the high-energy intermediate formed during the amide-bond hydrolysis or amide-bond formation *(13,14)*. Therefore, peptidosulfonamides are at the same time stable to proteolytic hydrolysis and capable of significantly altering polarity and H-bonding patterns of native peptides. Because of the relative acidic N-H in a sulfonamide moiety, it can be expected that H-bonds involving this amide surrogate will be stronger as compared to the amide analogs. These sulfonamides' properties made the peptidosulfonamides attractive building blocks for synthesis of peptidomimetics with enhanced metabolic stability and potentially potent enzyme inhibitory activities. Due to the intrinsic chemical instability of α-peptidosulfonamides, most of the peptidomimetics containing a sulfonamide bond have been limited to more stable β-peptidosulfonamides *(15)*. However, this peptidomimetic approach is not without its disadvantages. The greatest disadvantage of peptidosulfonamide peptidomimetics is the ability of sulfonamide moiety to disrupt any defined secondary structure of the parent peptide in solution, even if present at the *N-terminus (16,17)*. The following structural features of the sulfonamide moiety may contribute to this phenomenon:

1. Sulfonamide N-H is more acidic than amide N-H and therefore a better H-bond donor but a poorer H-bond acceptor.
2. The presence of two sulfonamide oxygens as H-bond acceptors may also impair a H-bonding network which holds together a secondary structure.
3. The sulfonamide oxygens can assume varying positions due to less energy-demanding rotation about the sulfonamide bond. This may also prevent a proper alignment of the H-bonds necessary for a particular secondary structure.

Despite these sulfonamide peptidomimetic disadvantages, the pharmacological properties of peptides may be improved by introducing the sulfonamide residue at one of several possible positions within the peptide sequence while maintaining the desired biological activity *(18,19)*.

3.1.1. Synthesis of Peptidosulfonamide Monomeric Building Blocks

The key step in the synthesis of peptidosulfonamides is the conversion of an amino acid into corresponding activated sulfonic acid derivatives such as sulfinyl- and sulfonyl chloride (**Fig. 2**) *(16,20)*. Typical synthesis of Fmoc-protected β-substituted-β-aminoalkylsulfonyl chlorides include reduction of Fmoc-protected amino acid to corresponding alcohol, conversion of the alcohol to sulfonic acid, and, in a final step, chlorination with thionyl chloride *(21,22)*, phosgene *(17,23–25)*, or triphosgene *(18,26)*. Since traces of triphosgen can significantly lower coupling yields during the peptidosulfonamide solid-phase assembly *(23)* and phosgene is highly toxic, the use of thionyl chloride as a chlorinating agent represents the most practical approach. Another interesting

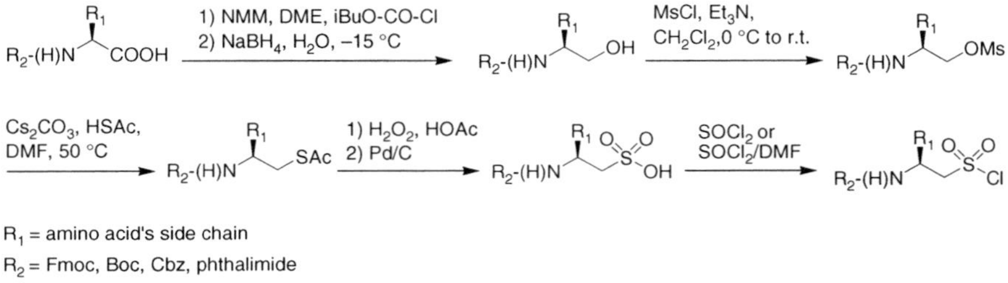

R_1 = amino acid's side chain
R_2 = Fmoc, Boc, Cbz, phthalimide

Fig. 2. Synthesis of protected sulfonyl chloride building blocks.

approach for the synthesis of β-aminosulfonamides requires synthesis of sulfonyl chlorides followed by coupling of an amino acid or peptide via amino group and subsequent oxidation using an OsO$_4$/N-methylmorpholine-N-oxide (NMMO) mixture *(27)*. The advantage of the sulfonyl chloride approach is their relative high reactivity in the coupling reactions compared to the sulfonyl chlorides, but the disadvantage is sulfonyl chloride's limited stability and requirements for the oxidation step after completion of the coupling reaction. Therefore, preparation of sulfonyl chloride monomeric building blocks will not be described in this chapter.

3.1.1.1. SULFONYL CHLORIDE SYNTHESIS

These protocols were adopted from **refs. *(21–23)*** and ***(28)***.

Reduction step:

1. To a cold ($-15°C$) solution of N-protected α-amino acid (10 mmol) in DME (10 mL) add NMM (10 mmol) and isobutyl chloroformate (10 mmol).
2. Remove precipitated N-methylmorpholine hydrochloride by filtration and wash it with DME (5×2 mL).
3. Combine washings in a large flask and cool to $0°C$.
4. Add solution of NaBH$_4$ (15 mmol) in water (5 mL) (*see* **Note 1**).
5. Add additional 250 mL of water.
6. If alcoholic product precipitates, collect precipitate by filtration and wash thoroughly with water and hexane (*see* **Note 2**).

Mesylation step:

1. To a solution of alcohol (50 mmol) in DCM (150 mL) add Et$_3$N (57 mmol).
2. Cool solution to $0°C$ and add dropwise methanesulfonyl chloride (57 mmol).
3. Stir reaction mixture overnight at room temperature.
4. Add 100 mL of DCM and wash the mixture with 5% NaHCO$_3$ (2×100 mL), water (2×100 mL), and saturated NaCl solution (80 mL).
5. Dry organic phase over Na$_2$SO$_4$, concentrate the filtrate, and purify by column chromatography over silica gel (eluent: DCM/CH$_3$OH = 9/1).

Thioacylation step:

1. Add thioacetic acid (51 mmol) to a suspension of Cs$_2$CO$_3$ (47 mmol) in DMF (70 mL).
2. Add in one portion previously prepared mesylate (43 mmol).
3. Stir reaction mixture at $50°C$ for 24 h (*see* **Note 3**).
4. Pour reaction mixture in to water (250 mL) and extract with EtOAc (3×150 mL).
5. Combined organic layers wash with water (150 mL), 5% NaHCO$_3$ (150 mL), and saturated NaCl solution (150 mL).
6. Dry organic phase over Na$_2$SO$_4$, concentrate the filtrate and purify by column chromatography over silica gel (eluent: EtOAc/hexane = 1/1).

Oxidation step:

1. Add a mixture of 30% H_2O_2 (30 mL) and AcOH (60 mL) to a solution of thioacetate (34 mmol) in AcOH (30 mL).
2. Stir reaction mixture for 24 h at room temperature.
3. Add 10% Pd/C (*see* **Note 4**).
4. Filter resulting mixture through celite, concentrate the filtrate, and co-evaporate with toluene and diethyl ether in order to remove traces of AcOH.
5. Resulting product can be used in the next step without further purification.

Chlorination step:

1. To a cooled (ice bath) mixture of protected sulfonic acid (18 mmol), add excess of $SOCl_2$ (10 mL), add dropwise DMF (1 mL).
2. Remove the ice bath.
3. Stir reaction mixture at reflux for 5 h.
4. Remove chlorinating reagents by evaporation, followed by co-evaporating with toluene and diethyl ether.
5. Dissolve the residue in EtOAc (100 mL) and wash with water (60 mL), saturated $NaHCO_3$ (60 mL), and saturated NaCl solution (50 mL).
6. Dry organic phase over Na_2SO_4, concentrate the filtrate, add DCM and purify by filtration through a short pad of silica gel.

3.1.2. Solid-Phase Peptidosulfonamide Synthesis

For the synthesis of peptidosulfonamides, especially long and complex ones, a solid-phase methodology is indispensable (**Fig. 3**). Literature reports on methods for solid-phase synthesis of peptidosulfonamides include both sulfonyl- *(17,18)* and sulfinyl chloride amino acid analog couplings *(20,27)*. However, the sulfinyl chloride method turned out to be inferior mainly because the yield of peptidosulfinamide oxidation to corresponding peptidosulfonamide strongly depends on the peptidomimetic sequence and the type of resin used in the synthesis. Sulfinyl chloride methodology therefore represents an attractive approach for the synthesis of very short peptidosulfonamides rather than a general methodology for solid-phase synthesis of these peptidomimetics.

This protocol was adopted from **refs.** *(17)* and *(18)*:

1. Place resin in dry reaction vessel.
2. Remove the Fmoc protecting group by agitating the resin with 20% piperidine in NMP (3 × 10 min).
3. Wash resin with NMP (25 mL/mmol, 5 × 2 min) and DCM (25 mL/mmol, 5 × 2 min) (*see* **Note 5**).
4. Dissolve Fmoc-aa-Ψ(CH_2SO_2)-Cl (4 eq) and NMM (6 eq) in DCM (15 mL/mmol). Add this solution to the resin.
5. Allow the resin to agitate at room temperature for 3 h.

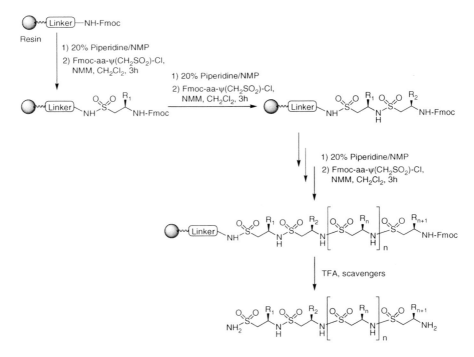

Fig. 3. Solid-phase peptidosulfonamide synthesis using sulfonyl chloride methodology.

6. Transfer small amount of resin to the test tube, wash resin with DCM, and perform the bromophenol blue (BPB) test *(29)* for presence of free amine (*see* **Note 6**).

7. If the resin gives the positive test, prolong the reaction time for an additional 3 h.

8. Cleave peptidosulfonamide product from the resin using a 95% TFA, 2.5% TIS, 2.5% H_2O mixture.

9. To the solution containing cleaved peptidosulfonamide, add cold diethyl ether and collect the resulting precipitate by filtration.

10. Purify the crude product by RP-HPLC.

3.2. Phosphonopeptide Synthesis, Ψ [P(=O)OH-O]

Phosphonopeptides containing a transition state analog of the hydrolysis of the amide bond represent another attractive approach for the preparation of proteolitically stable peptides *(10,30,31)*. In addition to increased stability, incorporation of a phosphonate moiety into the peptide sequence provides access to additional binding interactions within the transition-state conformation of the enzyme/substrate complex *(13)*. This peptidomimetic approach is used to design very effective protease inhibitors *(31–34)*. As in the case of peptidosulfonamides, preparation of phosphorous amino acid analogues suitable for incorporation into peptidic backbone represents a first step toward preparation of phosphonopeptide peptidomimetics. Methods for obtaining α-aminoalkylphosphonic acid analogs of amino acids in stereochemically pure form have been summarized in the literature *(30)*. Moreover, some of these analogs are also commercially available (*see* **Note 7**). Therefore, methods for preparation of phosphorous amino acid analogues will not be addressed in this chapter.

Typical methods for solid-phase incorporation of α-aminoalkylphosphonic acid into the peptide main chain include modified Mitsunobu condensation, which proceeds with an inversion of configuration of alcohol-bearing carbon *(35–37)* (**Fig. 4**). Another interesting approach includes solution synthesis of aminophosphonate dipeptide analogs by BOP- or PyBOP-promoted reaction between appropriate phosphonic acid monoesters and hydroxyl acid followed by their coupling using standard Fmoc chemistry. The latest synthetic methodology results in no epimerization *(38,39)*. In order to efficiently apply the Mitsunobu coupling methodology to phosphonopeptide solid-phase synthesis, the following steps need to be considered: Fmoc protection of hydroxyl group

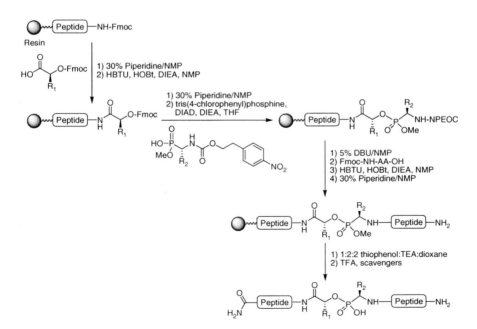

Fig. 4. Solid-phase phosphonopeptide synthesis using modified Mitsunobu coupling procedure.

in α-hydroxy acid (*see* **Note 7**) and activation of aminoalkylphosphonate with a 4-nitrophenethyloxycarbonyl (NPEOC) group.

These protocols were adopted from **refs. *(35–37)*,*(40)*, and *(41)***:

Fmoc protection of hydroxyl group:

1. Dissolve α-hydroxy acid (1 eq) in anhydrous pyridine (20 mL) at 0°C.
2. Add Fmoc-chloroformate (3 eq).
3. Stir reaction mixture for 3 h at 0°C.
4. Concentrate reaction mixture under vacuum and purify by column chromatography over silica gel (eluent: $DCM/CH_3OH/AcOH=9/0.9/0.1$).

Aminoalkylphosphonate activation:

1. Add 1-(*N*-benzyloxycarbonyl-amino)-alkylphosphonate (1 eq) to aqueous KOH (5 eq) at room temperature. Monitor reaction by TLC.
2. Wash reaction mixture with diethyl ether.
3. Acidify water phase with concentrated HCl to pH 2 and then extract with EtOAc (3x).
4. Dry organic phase over $MgSO_4$, filter, and concentrate under vacuum.
5. Dissolve concentrated residue in EtOH and hydrogenate in the presence of 10% Pd/C catalyst.
6. Upon completion of reaction, filter the mixture through celite, wash catalyst with MeOH, and concentrate filtrate under vacuum.
7. Dissolve obtained material in H_2O.
8. Add Na_2CO_3 (1.1 eq) and 4-nitrophenylethyl chloroformate (1.1 eq) dissolved in dioxane.
9. Monitor reaction by TLC.
10. Upon completion of the reaction wash the mixture with diethyl ether, acidify with concentrated HCl to pH 2 and extract with EtOAc (3x).
11. Dry organic phase over $Mg\ SO_4$, filter, and concentrate under vacuum.

This protocol was adopted from **refs. *(35–37)***.

1. Place resin in dry reaction vessel.
2. Synthesize the desired peptide sequence using standard Fmoc chemistry.
3. Couple α-O-Fmoc–protected hydroxyl acid using HBTU/HOBt or PyBroP protocol.
4. Remove the Fmoc-protecting group by agitating the resin with 30% piperidine in NMP (25 mL/mmol, 3 × 10 min).
5. Wash resin with NMP (25 mL/mmol, 5 × 2 min) and diethyl ether (25 mL/mmol, 5 × 2 min) and dry overnight under vacuum in the presence of P_2O_5.
6. Suspend NPEOC-aa-Ψ(HO-P(=O)-OMe) (1 eq) and NMP (1 eq) in anhydrous THF, then add tris(4-chlorophanyl)phosphine (1 eq) and DIAD (1 eq).
7. Leave reagent mixture for 5 min and then add it to the resin (*see* **Note 8**).

8. Allow the resin to agitate at room temperature for 0.5–4 h (*see* **Note 9**).
9. Wash the resin with NMP (5x) and diethyl ether (5x) and dry overnight under vacuum in the presence of P_2O_5.
10. Remove the NPEOC group by treatment with 5% DBU in NMP for 2 h (*see* **Note 10**).
11. Proceed with the rest of the peptide synthesis using standard Fmoc chemistry.
12. Wash resin with NMP (3x), CH_3OH (3x), and DCM (3x).
13. Add thiophenol/triethylamine/dioxane (1/2/2) mixture and agitate resin at room temperature for 1 h (*see* **Note 11**).
14. Wash resin with NMP (5x), diethyl ether (5x), and dry under vacuum.
15. Cleave phosphonopeptide product from the resin using TFA and the appropriate scavengers (*see* **Note 12**).
16. To the solution containing cleaved phosphonopeptide, add cold diethyl ether and collect the resulting precipitate by filtration.
17. Purify crude product by RP-HPLC.

3.3. Oligourea Synthesis, Ψ [CH$_2$-NH-CO-NH]

Considering a planar conformation of urea moiety, urea replacement of the amide linkage in native peptides represents a conformationally more conservative type of peptidomimetics *(43–50)*. N,N´-Linked oligoureas are readily accessible by Fmoc solid-phase synthetic methodology with a variety of appropriate building blocks *(51–53)*. For the reasons of synthetic efficiency and stability of the product, each urea repeating unit in these peptidomimetics is extended by one additional carbon atom in comparison with the amino acid counterpart. This extra carbon atom may increase the lipophylicity and flexibility of the main chain, which could therefore make it easier for these peptidomimetics to cross the cell wall or the blood-brain barrier. On the other hand, the hydrogen-bond forming urea units may also increase their water solubility and provide additional binding sites for interaction with their biological targets *(43)*.

3.3.1. Synthesis of Oligourea Monomeric Building Blocks

The monomeric building blocks for oligourea peptidomimetic solid-phase synthesis were prepared from Fmoc-protected amino acids *(27,53)*. This synthetic strategy includes reduction of Fmoc-protected amino acid into the corresponding alcohol, its conversion into an azide by a Mitsunobu reaction, followed by reduction of the azide to amine using catalytic hydrogenation (**Fig. 5**). In the final step, obtained Fmoc-protected monomeric amino building blocks were activated by conversion to corresponding carbamate with 4-nitrophenyl chloroformate.

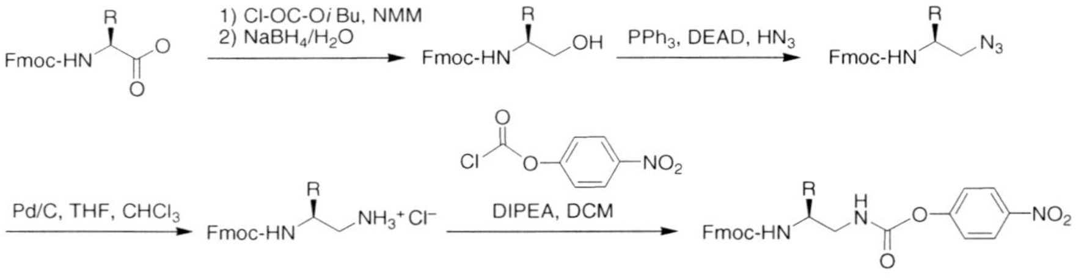

Fig. 5. Synthesis of Fmoc-protected urea building blocks.

These protocols were adopted from **ref. *(53)*.**

Reduction step:

Fmoc-protected amino alcohols can be prepared according to the protocol described in **Subheading *3.1.1.1.***

3.3.1.1. PREPARATION OF FMOC-PROTECTED AZIDES

1. To a cooled solution (0°C) of PPh$_3$ (10 mmol) in THF (30 mL) add dropwise DEAD (1 eq) under an inert atmosphere.
2. Add dropwise 1.5 M solution of HN$_3$ in benzene (1 eq) (*see* **Note 13**).
3. Add in one portion Fmoc-protected amino alcohol (1 eq).
4. Stir reaction mixture overnight at room temperature.
5. Concentrate reaction mixture under vacuum and purify by column chromatography over silica gel (eluent: EtOAc/hexane = 1/6).

3.3.1.2. ACTIVATION STEP

1. Dissolve azide in the mixture of THF (50 mL) and CHCl$_3$ (1 mL), add 10% Pd/C and stir overnight at room temperature under H$_2$ atmosphere (1 bar).
2. After completion of hydrogenation, add water (4 mL) to reaction mixture to redissolve precipitated material.
3. Filter reaction mixture over celite and evaporate the solvents.
4. Dry obtained product under vacuum over P$_2$O$_5$.
5. Add dry product (1 eq) and 4-nitrophenyl chloroformate (1.1 eq) to DCM.
6. Cool resulting suspension to 0°C (ice bath) and add dropwise solution of DIPEA (2 eq) in DCM.
7. Stir reaction mixture for 30 min, then allow warming to room temperature and continuing stirring for another 2 h.
8. Wash reaction mixture with 1 M KHSO$_4$ (3x), water, and saturated NaCl solution.
9. Dry organic phase dry over sodium sulfate, filter, and concentrate under vacuum.
10. Purify crude product by column chromatography over silica gel (eluent: EtOAc/hexane = 1/1).

3.3.2. Solid-Phase Oligourea Synthesis

The use of the Fmoc-protected 4-nitrophenyl carbamate building blocks and resins with acid-labile linkers allows synthesis of the final products with *C*-terminal carboxylic acid or amide groups (**Fig. 6**). Unfortunately, Fmoc solid-phase synthesis of oligourea peptidomimetics with *C*-terminal carboxylic acid also leads to formation of corresponding hydantoin byproducts *(53–56)* (**Fig. 7**). In this case hydantoin formation arises as a result of an acid-catalyzed intramolecular cyclization reaction. It has been reported that the ratio of desired oligourea peptidomimetic acid product and hydantoin byproduct is approximately 2:1 *(53)*. However, these two compounds are in principle separable by preparative HPLC.

This protocol was adopted from **ref.** *(53)* (*see* **Note 14**):

1. Place resin in dry reaction vessel.
2. Synthesize desired peptide sequence using standard Fmoc chemistry.
3. Remove the Fmoc-protecting group by agitating the resin with 20% piperidine in DMF (3 × 5 min).
4. Wash resin with DCM (2x) and DMF (3x).

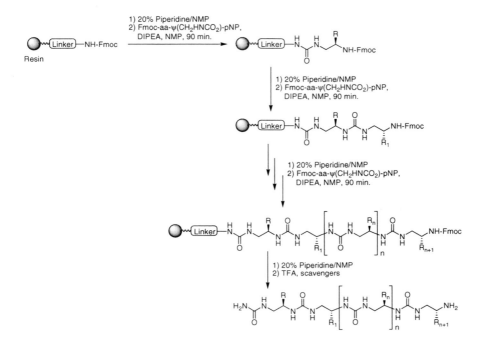

Fig. 6. Solid-phase synthesis of oligourea amide peptidomimetics.

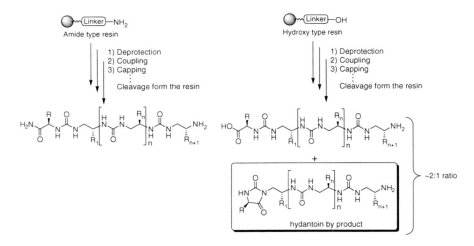

Fig. 7. Solid-phase synthesis of oligourea peptidomimetics on amide and hydroxy-type resins.

5. Dissolve Fmoc-aa-Ψ(CH₂NHCO₂)-pNP (4 eq, relative to resin loading) in small amount of DMF and DIPEA (8 eq).

6. Transfer that mixture to the reaction vessel and agitate the resin for 1.5 h.

7. Filter off the resin and wash it with DMF (3x).

8. Block unreacted amino groups by treatment with capping solution (acetic anhydride, DIPEA, HOBt, and catalytic amount of DMAP in DMF) for 15 min.

9. Wash resin with DCM (2x) and DMF (3x).

10. Repeat **steps 3–9** until desired sequence is obtained.

11. Wash resin with DMF (3x), DCM (2x), and dry it in vacuum over P₂O₅ for 3 h.

12. Cleave the oligourea from the resin by agitating a mixture of TFA/H₂O/TIS/EDT (95.2/2.5/2.5/2.5) with the resin for 3 h.

13. Filter off the resin and rinse it with small amount of cold TFA.

14. Add cold solution of MTBE/hexane (1/1) to a filtrate until precipitate is formed.

15. Collect precipitate by centrifugation, lyophilize it, and purify by RP-HPLC.

3.4. Depsipeptide and Depside Synthesis, Ψ [C(=O)-O]

Replacement of the amide groups that undergo proteolytic hydrolysis with ester groups may also lead to longer acting compounds not so prone to proteolysis *(57–60)*. Naturally occurring depsipeptides, which contain one or more ester bonds in addition to the amide bonds, have been found in many natural organisms such as fungi, bacteria, and marine organisms *(61,62)*. It is very well known that these natural products and their derivatives exhibit a diverse spectrum of biological activities including insecticidal, antiviral, antimicrobial, antitumor, tumor-promotive, anti-inflammatory, and immunosuppressive actions.

On the other hand, depsides are peptide analogs entirely build up by hydroxyl acids mutually connected through ester bonds. A representative example of depsides is the naturally occurring macrotetralide antibiotic nonactin *(63)*. Nonactin has been shown to possess activity against the P170-glycoprotein efflux pump associated with multiple drug-resistant cancer cells.

Among many approaches described in the literature for depsipeptide synthesis, carbodiimide/4-dimethylaminopyridine (DMAP) coupling method is the most efficient and fully compatible with Fmoc solid-phase synthetic methodology *(64–66)*. Carbodiimide reagents have been widely used in peptide synthesis because of their moderate activity and low cost *(67)*. They are used as a coupling reagents and esterification reagents during loading first amino acid on the resin. The most commonly used carbodiimide reagent is 1,3-diisopropylcarbodiimide (DIC, DIPCI). By using a 2:1 molar ratio of amino acid to DIC the symmetrical anhydride is formed, which in turn reacts with free hydroxyl group and the ester bond is formed **(Fig. 8)**. The reaction is catalyzed by the presence of DMAP, which increases the nucleophilicity of the hydroxyl group *(68)*. The best results for this coupling method were obtained using PEG-based resins such as TentaGel S RAM (Advanced Chemtech) or PAL-ChemMatrix (Matrix Innovation) and DCM as a solvent *(69)*. Use of polar *N,N*-dimethylformamide (DMF), a typical solvent for SPPS, leads to no ester bond formation. Also, poor coupling yields were obtained on PS-based Rink-MBHA resins (Novabiochem) regardless of the solvent used. These results could be attributed to a better swelling of PEG-based resins *(70)*, rapid DIC activation of the carboxylic group *(71)*, and significant suppression of *N*-acylurea byproduct formation *(72)* in a nonpolar solvent such as DCM. When carbodiimide is used in 1:1 molar ratio with amino acid, the reaction proceeds via *O*-acylisourea mechanism and the corresponding ester bond is formed **(Fig. 9)**.

The following general solid-phase synthetic protocol can be used for depside and depsipeptide bond formation using DIC/DMAP coupling methodology *(64–66)*.

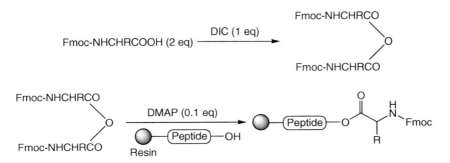

Fig. 8. Symmetrical anhydride method of ester bond formation.

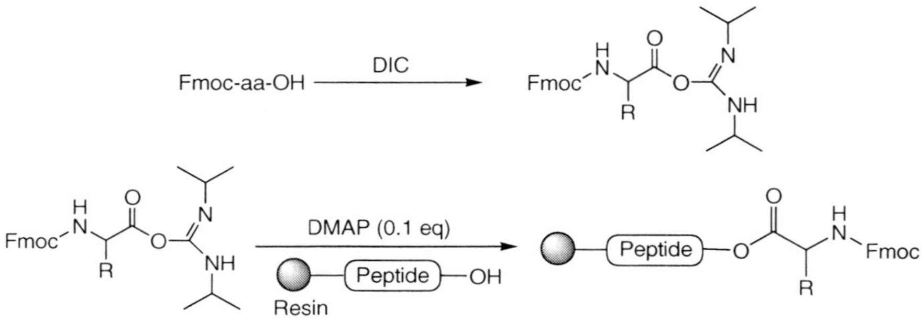

Fig. 9. Mechanism of ester bond formation via *O-acylisourea.*

3.4.1. Solid-Phase Depsipeptide Ester Bond Formation Using the DIC/DMAP Method

This protocol was adopted from **refs. (65)** and **(66)** (*see* **Note 15**).

1. Place PEG-based resins in dry reaction vessel.
2. Swell resins in DCM.
3. Dissolve DIC (4 eq), DMAP (0.1 eq relative to resin loading), and protected amino acid derivative (4 eq) in DCM (1 mL/100 mg of the resin). Add this solution to the resin.
4. Allow the resin to agitate at room temperature for 1 h.
5. Wash the resin (3 min each) with 3×10 mL of DMF and 5×10 mL of DCM.
6. Transfer a small amount of resin to the test tube and perform test for the presence of free hydroxyl groups (*see* **refs. (65)** and **(66)**) or monitor the coupling by MALDI-TOF analysis.
7. If the resin gives the positive test, repeat **steps 2–5** with newly prepared reagents.
8. Cleave synthesized depsipeptide from the resin using TFA and appropriate scavengers (*see* **Note 12**).
9. To the solution containing cleaved depsipeptide, add cold diethyl ether and collect the resulting precipitate by filtration.
10. Purify the crude product by RP-HPLC.

3.5. Peptoids and Peptide-Peptoid Hybrid Synthesis, $\Psi [C(=O)-N(R)CH_2]$

The oligomeric peptidomimetics such as peptoids are particularly interesting compounds since they provide access to an enormous molecular diversity by variation of the building blocks. Peptoids represent a class of polymers that are not found in nature. They differ from the peptides in the manner of side-chain attachment and thus can be considered as peptide mimetics in which the side chain has been shifted from the chiral α-carbon atom in a peptide to the achiral nitrogen *(73)*. Peptoids lack the hydrogen of the peptide secondary amides and are thus incapable of forming the same type of hydrogen bond networks that stabilize peptide helices and β-sheets. In addition, the peptoid polymer backbone is achiral. However, a chiral center can be introduced in the side chains in order to obtain preferred secondary structures. Among many peptoid properties, an improved bioavailability *(74,75)* and protease resistance *(76,77)* are especially attractive for the peptidomimetic design.

Two methods for the synthesis of peptoids are described in the literature: monomeric and submonomeric (**Fig. 10**). The monomeric method is analogous to the standard solid-phase peptide synthesis *(78–80)*. In general, this method includes activation of the N^α-Fmoc protected N-substituted glycine using standard reagents for the Fmoc solid-phase chemistry followed by its coupling to the secondary amino group of the resin-bound peptoid chain. This step is repeated until the desired peptoid sequence is synthesized. However, a major disadvantage of this method is the requirement for the separate synthesis of suitably protected N-substituted glycine monomeric building blocks. On the other hand, the submonomeric method represents a more practical approach since this method does not require use of Fmoc-protected N-substituted glycines. Moreover, any amine (side-chain protected, if necessary) can be used instead. In the submonomeric method each cycle of monomeric building block addition consists of an acylation step using haloacetic acid (bromoacetic acid) and a nucleophilic displacement step. This method is particularly useful for the synthesis of peptoid-based combinatorial libraries *(81)*. Protection of carboxyl, thiol, amino, and other reactive side-chain functionalities is required to minimize undesired side reactions. However, the mild reactivity of some side-chain moieties toward displacement or acylation may allow their use without protection (e.g., indole, imidazole and phenol).

Since the monomeric method for peptoid synthesis is in principle identical to the standard Fmoc solid-phase peptide synthesis and the methods for preparation

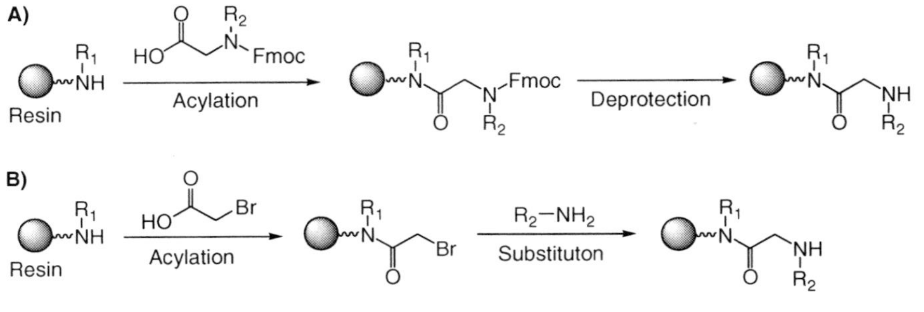

Fig. 10. Two methods of peptoid synthesis: monomeric (**A**) and submonomeric (**B**).

of fully protected monomeric building blocks are described in the literature *(78,79)*, it will not be further described in this chapter. It is noteworthy that coupling of monomeric units using this method is more difficult to perform in comparison to using peptides due to secondary amine's low reactivity if electron-withdrawing groups are attached, and also due to the sterical hindrance around this atom. Therefore, for these difficult couplings, reagents such as PyBroP, PyBOP, or HATU are recommended.

Since the standard solid-phase peptide synthesis starts from the *C-terminus* and finishes at the *N-terminus*, solid-phase peptoid and peptide synthesis could be combined, giving peptide–peptoid hybrid polymers. Ostergaard and Holm named such hybrids peptomers *(82)*. This approach may also be used in the conversion of biologically active peptide ligands, e.g., peptide hormones, protease inhibitors, into an active peptomeric version by ensuring that the essential amino acids comprising the lead motif are included in the synthesis *(77)*. Both peptoids and peptomers can be easily sequenced using modified Edman degradation conditions *(83)*.

This protocol was adopted from **refs.** *(80)* and *(81)*.

1. Place the resin in dry reaction vessel.
2. Remove the Fmoc-protecting group (if necessary) by agitating the resin with 20% piperidine in DMF (3 × 6 min).
3. Wash the resin (3 min each) with 3 × 10 mL of DMF; and 3 × 10 mL of DCM.
4. Dissolve bromoacetic acid (10 eq, relative to resin loading) and DIC (11 eq) in minimal amount of DMF. Add this solution to the resin.
5. Allow the resin to agitate at room temperature for 30 min.
6. Wash the resin (3 min each) with 3 × 10 mL of DMF.
7. Dissolve amine (25 eq) in DMSO, NMP, or DMF.
8. Allow the resin to agitate at room temperature for 2 h (*see* **Note 16**).
9. Wash the resin (3 min each) with 3 × 10 mL of DMF; and 3 × 10 mL of DCM.
10. Repeat these steps for each coupling cycle.
11. Cleave synthesized peptoid from the resin using TFA and appropriate scavengers (*see* **Note 12**).
12. To the solution containing cleaved peptoid product, add cold diethyl ether and collect the resulting precipitate by filtration.
13. Purify crude product by RP-HPLC.

4. Notes

1. Addition of NaBH$_4$ causes strong evolution of H$_2$. Use of a fume hood is strongly recommended.
2. If alcoholic product does not precipitate, purify the product by extraction with the appropriate organic solvent.

3. Cover the reaction flask with aluminum foil. *O*-Mesyl derivatives are light sensitive *(84)*.
4. Addition of 10% Pd/C is required to destroy the excess of peroxide.
5. DCM is the solvent of choice. NMP causes decomposition of the sulfonyl chlorides.
6. In some cases double coupling needs to be carried out to ensure completion of the reaction. Monitoring the coupling efficiency with the Kaiser test may be misleading due to the relatively acidic sulfonamide NH.
7. Various α-aminoalkylphosphonic acids and α-hydroxy acids are commercially available from Sigma-Aldrich, International Laboratory, Ltd., TCI America, Acros Organics, etc.
8. NPEOC-*N*-phosphonic acid should be completely dissolved within a 5-min period.
9. Reaction rate is insensitive to the steric bulk of the side chain. Only steric effects from the hydroxyl component influence the reaction rate. Therefore, coupling time for bulky hydroxyl residues needs to be longer.
10. Coupling efficiency of α-hydroxy acid and phosphonic acid segments can be determined by spectrophotometric analysis of the released 4-nitrostyrene (λ_{max} = 308 nm, $\varepsilon = 13200\ M^{-1}cm^{-1}$) upon treatment of peptidyl-resin with 5% DBU in NMP *(85)*.
11. This step is required for dimethylation of the phosphonate diester.
12. Peptide sequence determines selection of the scavengers.
13. HN_3 is an extremely toxic and explosive substance. Other general protocols for the conversion of hydroxy to azido groups may be more suitable.
14. Boc solid-phase methodology leads exclusively to formation of hydantoins *(54)*.
15. The identical protocol can be used for the synthesis of depsides.
16. In some cases the displacement step has to be prolonged for up to 12 h.

References

1. Marx, V. (2005) Watching peptide drugs grow up. *C&EN*, **83**, 17–24.
2. Adessi, C., and Soto, C. (2002) Converting a peptide into drug: strategies to improve stability and bioavailability. *Curr. Med. Chem.* **9**, 963–978.
3. Sawyer, T. K. (2000) Peptidomimetic and nonpeptide drug discovery: chemical nature and biological targets, in *Peptide and Protein Drug Analysis* (R. E. Reid, ed.), Marcel Dekker, New York, pp. 81–115.
4. Davies, J. S. (2003) The cyclization of peptides and depsipeptides. *J. Pept. Sci.* **9**, 471–501.
5. Lambert, J. N., Mitchell, J. P., and Roberts, K. D. (2001) The synthesis of cyclic peptides. *J. Chem. Soc, Perkin Trans.* **1**, 471–484.
6. Li, P., and Roller, P. P. (2002) Cyclization strategies in peptide derived drug design. *Curr. Top. Med. Chem.* **2**, 325–341.
7. Blackburn, C., and Kates, S. A. (1997) Solid-phase synthesis of cyclic homodetic peptides. *Methods Enzymol* . **289**, 175–198.

8. Hruby, V. J., and Bonner, G. G. (1994) Design of novel synthetic peptides including cyclic conformationally and topographically constrained analogs. *Methods. Mol. Biol.* **35**, 201–240.

9. Kates, S. A., Sole, N. A., Albericio, F., and Barany, G. (1994) Solid-phase synthesis of cyclic peptides, in *Peptides: Design, Synthesis, and Biological Activity*, Birkhauser, Boston, pp. 39–59.

10. Ahn, J. M., Boyle, N. A., MacDonald, M. T., and Janda, K. D. (2002), Peptidomimetics and peptide backbone modification. *Mini Rev. Med. Chem.*, **2**, 463–473.

11. Goodman, M., Felix, A., Moroder, L., and Toniolo, C., eds. (2004) *Synthesis of Peptides and Peptidomimetics*, Thieme, Stuttgart.

12. Spatola, A. F. (1983) Peptide backbone modifications in chemistry and biochemistry of amino acids, in *Peptides and Proteins* (Weinstein, B., ed.), Marcel Dekker, New York, pp. 267–357.

13. Radkiewicz, J., McAllister M. A., Goldstein, E., and Houk, K. N. (1998) A theoretical investigation of phosphonoamidates and sulfonoamides as protease transition state isosteres. *J. Org. Chem.* **63**, 1419–1428.

14. Moree, W. J., Schouten, A., Kroon, J., and Liskamp, R. M. J. (1995) Peptides containing the sulfonamide transition-state isostere: synthesis and structure of *N*-acetyl-taryl-L-proline methylamide. *Int. J. Pept. Prot. Res.* **45**, 501–507.

15. Paik, S., and White, E. H. (1996) α-Aminosulfonopeptides as possible functional analogs of penicillin; evidence for their extreme instability. *Tetrahedron* **52**, 503–5318.

16. Liskamp, R. M. J., and Kruijtzer A. W. (2004) Peptide transformastion leading to peptide-peptidosulfonamide hybrids and oligo peptidosulfonamides. *Mol. Divers.*, **8**, 79–87.

17. De Jong, R., Rijkers, D. T. S., and Liskamp, R. M. J. (2002) Automated solid-phase synthesis and structural investigation of β-peptidosulfonamides and β-peptidosulfonamide/β-peptide hybrids: β-peptidosulfonamide and β-peptide foldamers are two of a different kind. *Helv. Chim. Acta* **85**, 4230–4243.

18. De Bont, D. B. A., Dijkstra, G. D. H., Den Hartog, J. A. J., and Liskamp, R. M. J. (1996) Solid-phase synthesis of peptidosulfonamide containing peptides derived from Leu-enkephalin. *Bioorg. Med. Chem. Lett.* **6**, 3035–3040.

19. De Bont, D. B. A., Sliedregt, K. M., Hofmeyer, L. J. F., and Liskamp, R. M. J. (1999) Increased stability of peptidosulfonamide peptidomimetics towards protease catalyzed degradation. *Bioorg. Med. Chem.* **7**, 1043–1047.

20. Moree, W. J., Vand der Marel, G. A., and Liskamp R. M. J. (1995) Synthesis of peptidosulfinamides and peptidosulfonamides: peptidomimetics containing the sulfinamide or sulfonamide transition-state isostere. *J. Org. Chem.* **60**, 5157–5169.

21. Humljan, J., Kotnik, M., Boniface, A., et al. (2006) A new approach towards peptidosulfonamides: synthesis of potential inhibitors of bacterial peptidoglycan biosynthesis enzymed MurD and MurE. *Tetrahedron* **62**, 10980–10899.

22. Humljan, J., and Gobec, S. (2005) Synthesis of *N*-phthalimido β-aminoethanesulfonyl chlorides: the use of thionyl chloride for a simple and

efficient synthesis of new peptidosulfonamide building blocks. *Tetrahedron Lett.* **46**, 4069–4072.

23. Brouwer, A. J., Monnee, M. C. F., and Liskamp R. M. J. (2000) An efficient synthesis of *N*-protected β-aminoethanesulfonyl chlorides: versatile building blocks for the synthesis of oligopeptidosulfonamides. *Synthesis*, 1579–1584.

24. Monnee, M. C. F., Marijne, M. F., Brouwer, A. J., and Liskamp R. M. J. (2000) A practical solid phase synthesis of oligopeptidosulfonamide foldamers. *Tetrahedron Lett.* **41**, 7991–7995.

25. Van Ameijde, J., and Liskamp R. M. J. (2000) Peptidomimetic building blocks for the synthesis of sulfonamide peptoids. *Tetrahedron Lett.* **41**, 1103–1106.

26. Gennari, C., Gude, M., Potenza, D., and Piarulli, U. (1998) Hydrogen-bonding donor/acceptor scales in β-sulfonamidopeptides. *Chem. Eur. J.* **4**, 1924–1931.

27. De Bont, D. B. A., Moree, W. J., and Liskamp, R. M. J. (1996) Molecular diversity of peptidomimetics: approaches to the solid-phase synthesis of peptidosulfon-amides. *Bioorg. Med. Chem.* **4**, 667–672.

28. Rodriguez, M., Llinares, M., Doulut, S., Heitz, A., and Martinez, J. (1991) A facile synthesis of chiral *N*-protected β-amino alcohols. *Tetrahedron Lett.* **32**, 923–926.

29. Krchnak, V., Vagner, J., and Lebl, M. (1988) Noninvasive continuous monitoring of solid-phase peptide synthesis by acid-base indicator. *Int. J. Pept. Protein Res.* **32**, 415–416.

30. Bartlett, P. A., Hanson, J. E., Morgan, B. P., and Ellsworth, B. A., (2004) Synthesis of peptides with phosphorus-containing amide bond replacements, in *Synthesis of Peptides and Peptidomimetics*, 4th ed. (Goodman, M., Felix, A., Moroder, L., and Toniolo, C., eds.), Thieme, Stuttgart, pp. 492–528.

31. Palacios, F., Alonso, C., and de los Santos, J. (2004) β-Phosphono- and phosphinopeptides derived from β-amino-phosphonic and phosphinic acids. *Curr. Org. Chem.* **8**, 1481–1496.

32. Bird, J., De Mello, R. C., Harper, G. P., et al. (1994) Synthesis of novel *N*-phosphonoalkyl dipeptide inhibitors of human collagenase. *J. Med. Chem.* **37**, 158–169.

33. De Lombeart, S., Erion, M. D., Tan, J., et al. (1994) *N*-Phosphonomethyl dipeptides and their phosphonate prodrugs, a new generation of neutral endopeptidase (NEP, EC 3.4.24.11) inhibitors. *J. Med. Chem.* **37**, 1498–1511.

34. Barlett, P. A., Hanson, J. E., and Giannousis, P. P. (1990) Potent inhibition of pepsin and penicillopepsin by phosphorus-containing peptide analogs. *J. Org. Chem.* **55**, 6268–6274.

35. Campbell, D. A., and Bermak, J. C. (1994) Solid-phase synthesis of peptidylphos-phonates. *J. Am. Chem. Soc.*, **116**, 6039–6040.

36. Campbell, D. A., and Bermak, J. C. (1994) Phosphonate ester synthesis using a modified Mitsunobu condensation. *J. Org. Chem.* **59**, 658–660.

37. Campbell, D. A. (1992) The synthesis of phosphonate esters, an extension of the Mitsunobu reaction. *J. Org. Chem.* **57**, 6331–6335.

38. Champagne, J.-M., Coste, J., and Jouin, P. (1995) Solid phase synthesis of phospho-nopeptides. *Tetrahedron Lett.* **36**, 2079–2082.

39. Champagne, J.-M., Coste, J., and Jouin, P. (1993) Synthesis of mixed phosphonate diester analogues of dipeptides using BOP or PyBOP reagents. *Tetrahedron Lett.* **34**, 6743–6744.

40. Seebach, D., Charczuk, R., Berber, C., Renaud, P., Bener, H., and Schneider, H. (1989) Electrochemical decarboxylation of L-threonine and oligopeptide derivatives with formation of N-acyl-N,O-acetals: preparation of oligopeptides with amide or phosphonate C-terminus. *Helv. Chim. Acta* **72**, 401–425.

41. Nicolau, K. C., Winssinger, N., Pastor, J., and DeRoose, F. (1997) A general and highly efficient solid phase synthesis of oligosaccharides. Total synthesis of a heptasaccharide phytoalexin elicitor (HPE). *J. Am. Chem. Soc.* **119**, 449–450.

42. Gademann K., and Seebach D. (2001) Synthesis of cyclo-β-tripeptides and their biological in vitro evaluation as antiproliferatives against the growth of human cancer cell lines. *Helv. Chim. Acta* **84**, 2924–2937.

43. Tamilarasu, N., Hoq, I., and Rana, T. M. (2001) Targeting RNA with peptidomimetic oligomers in human cells. *Bioorg. Med. Chem. Lett.* **11**, 505–507.

44. Appella, D.H., Christianson, L.A., Karle, I.L., Powell, D.R., and Gellman, S. H. (1996) β-Peptide foldamers: robust helix formation in a new family of β-amino acid oligomers. *J. Am. Chem. Soc.* **118**, 13071–13072.

45. Wang, X., Espinosa, J. F., and Gellman, S. H. (2000) 12-Helix formation in aqueous solution with short β-peptides containing pyrrolidine-based residues. *J. Am. Chem. Soc.* **122**, 4821–4822.

46. Hanessian, S., Luo, R., Schaum, S., and Michcnik, S. (1998) Design of decondary dtructures in unnatural peptides: stable helical γ-tetra-, hexa-, and octapeptides and consequences of α-substitution. *J. Am. Chem. Soc.* **120**, 8569–8570.

47. Hanessian, S., Luo, X., and Schaum, R. (1999) Synthesis and folding preferences of γ-amino acid oligopeptides: stereochemical control in the formation of a reverse turn and a helix. *Tetrahedron Lett.* **40**, 4925–4929.

48. Seebach, D., Ciceri, P., Overhand, M., et al. (1996) Probing the helical secondary structure of short-chain β-peptides. *Helv. Chim. Acta* **79**, 2043–2066.

49. Seebach, D. Sifferlen, T., Mathieu, P. A., et al. (2000) CD spectra in methanol of β-oligopeptides consisting of β-amino acids with functionalized side chains, with alternating configuration, and with geminal backbone substituents—fingerprints of new secondary structures. *Helv. Chim. Acta* **83**, 2849–2864.

50. Skurski, P., and Simons. J. (2001) An excess electron bound to urea. I. Canonical and zwitterionic tautomers. *J. Chem. Phys.* **115**, 8373–8380.

51. Burgess, K., Ibarzo, J., Linthicum, D. S., et al. (1997) Solid phase synthesis of oligoureas, *J. Am. Chem. Soc.* **119**, 1556–1564.

52. Kim, J.-M., Bi, Y., Paikoff, S. J., and Scultz, P. G. (1996) The solid phase synthesis of oligoureas. *Tetrahedron Lett.* **37**, 5305–5308.

53. Boeijen, A., van Ameijde, J., and Liskamp, R. M. (2001) Solid-phase synthesis of oligourea peptidomimetics employing the Fmoc protection strategy. *J. Org. Chem.* **66**, 8454–8462.

54. Boeijen, A., and Liskamp, R. M. (1999) Solid-phase synthesis of oligourea peptidomimetics. *Eur. J. Org. Chem.* 2127–2135.

55. Alsina, J., Scott, W. L., and O'Donell, M. J., (2005) Solid-phase synthesis of α-substituted proline hydantoins and analogs. *Tetrahedron Lett.* 46, 3131–3135.
56. Nefzi, A., Giulianotti, M., Truong, L., Rattan, S., Ostresh, J. M., and Houghten, R. A., (2002) Solid-phase synthesis of linear ureas tethered to hydantoins and thiohydantoins. *J. Comb. Chem.* **4**, 175–178.
57. Shemyakin, M. M., Shchukina, L. A., Vinogradova, E. I., Ravidel, G. A., and Ovchinnikov, Y. A. (1966) Mutual replaceability of amide and ester groups in biologically active peptide and depsipeptides. *Experimentia* **22**, 535–536.
58. Bramson, H. N., Thomas, N. E., and Kaiser, E. T. (1985) The use of *N*-methylated peptides and depsipeptides to probe the binding of heptapeptide substrates to cAMP-dependent protein kinase. *J. Biol. Chem.* **260**, 15452–15457.
59. Arad, O., and Goodman, M., (1990) Depsipeptide analogues of elastin repeating sequences: synthesis. *Biopolymers* **29**, 1633–1649.
60. Coombs, G. S., Rao, M. S., Olson, A. J., Dawson, P. E., and Madison, E. L. (1999) Revisiting catalysis by chymotrypsin family serine proteases using peptide substrates and inhibitors with unnatural main chains. *J. Biol. Chem.* **274**, 24074–24079.
61. Davidson, B. S. (1993) Ascidians: producers of amino acid-derived metabolites. *Chem. Rev.* **93**, 1771–1791.
62. Fusetani, N., and Matsunaga, S. (1993) Bioactive sponge peptides. *Chem. Rev.* **93**, 1793–1806.
63. Woo, A. J., Strohl, W. R., and Priestley, N. D. (1999) Nonactin biosynthesis: the product of *nonS* catalyzes the formation of the furan ring of nonactic acid. *Antimicrob. Agents Chemother.* **43**, 1662–1668.
64. Stawikowski, M., and Cudic, P. (2006) Depsipeptide synthesis, in *Peptide Characterization and Application Protocols* (Fields, G. B., ed.), Humana Press, Totowa, NJ, pp. 321–339.
65. Kuisle, O., Lolo, M., Quinoa, E., and Riguera, R. (1999) Solid phase synthesis of depsides and depsipeptides. *Tetrahedron* **55**, 14807–14812.
66. Kuisle, O., Quinoa, E., and Riguera, R., (1999) A general methodology for automated solid-phase synthesis of depsides and depsipeptides. Preparation of a valinomycin analogue. *J. Org Chem.* **64**, 8063–8075.
67. Marder, O., and Albericio, F. (2003) Industrial application of coupling reagents in peptides. *Chim. Oggi* **6**, 35–40.
68. Berry J. D., Digiovanna V. C., Metrick S. S., and Murugan R. (2001) Catalysis by 4-dialkylaminopyridines. *Arkivoc* **i**, 201–226.
69. Stawikowski, M., and Cudic, P. (2006) A novel strategy for the solid-phase synthesis of cyclic lipodepsipeptides. *Tetrahedron Lett.* **47**, 8587–8590.
70. Park, B.-D., and Lee, Y.-S. (2000) The effect of PEG groups on swelling properties of PEG-grafted-polystyrene resins in various solvents. *React. Funct. Polymers* **44**, 41–46.
71. Hudson, D. (1988) Methodological implications of simultaneous solid-phase peptide synthesis. 1. Comparison of different coupling procedures. *J. Org. Chem.* **53**, 617–624.

72. Podlech J. (2001) Carbodiimides, in *Synthesis of Peptides and Peptidomimetics*, 4th ed. (Goodman, M., Felix, A., Moroder, L., and Toniolo, C., eds.), Thieme, Stuttgart, pp. 517–533.

73. Patch, J. A., Kirshenbaum, K., Seurynck, S., Zuckermann, R., and Barron, A. E. (2004) Versatile oligo (*N*-substituted) glycines: the many roles of peptoids in drug discovery, in *Pseudo-Peptides in Drug Development* (Neilsen, P. E., ed.), Wiley-VCH, Weinheim, pp. 1–31.

74. Kwon, Y. U., and Kodadek, T. (2007) Quantitative evaluation of the relative cell permeability of peptoids and peptides. *J. Am. Chem. Soc.* **129**, 1508–1509.

75. Schröder, T., Schmitz, K., Niemeier, N., et al. (2007) Solid-phase synthesis, bioconjugation, and toxicology of novel cationic oligopeptoids for cellular drug delivery. *Bioconjugate Chem.* ASAP Article; DOI: 10.1021/bc0602073.

76. Miller, S. M., Simmon, R. J., Ng, S., Zuckermann, R. N., Kerr, J. M., and Moos, W. H. (1995) Comparison of the proteolytic susceptibilities of homologous L-amino acid, D-amino acid, and *N*-substituted glycine peptide and peptoid oligomers. *Drug Dev. Res.* **35**, 20–32.

77. Stawikowski, M., Stawikowska, R., Jaskiewicz, A., Zablotna, E., and Rolka, K. (2005) Examples of peptide-peptoid hybrid serine protease inhibitors based on the trypsin inhibitor SFTI-1 with complete protease resistance at the P1-P1' reactive site. *Chembiochem.* **6**, 1057–1061.

78. Simon, R. J., Kania, R. S., Zuckermann, R. N., et al. (1992) Peptoids: a modular approach to drug discovery. *Proc. Natl. Acad. Sci. USA* **89**, 9367–9371.

79. Kruijtzer, J. A. W., Hofmeyer, L. J. F., Heerma, W., Versluis, C., and Liskamp, R. M., J. (1998) Solid-phase syntheses of peptoids using Fmoc-protected *N*-substituted glycines: the synthesis of (retro)peptoids of Leu-enkephalin and substance P. *Chem. Eur. J.* **4**, 1570–1580.

80. Uno, T., Beausoleil, E., Goldsmith, R. A., Levine, B.H., and Zuckermann, R. N. (1999) New submonomers for poly *N*-substituted glycines (peptoids). *Tetrahedron Lett.* **40**, 1475–1478.

81. Zuckermann, R. N., Kerr, J.M., Kent, S. B. H., and Moos, W. H. (1992) Efficient method for the preparation of peptoids oligo(*N*-substituted glycines) by submonomer solid-phase synthesis. *J. Am. Chem. Soc.* **114**, 10646–10647.

82. Ostergaard, S., and Holm, A. (1997) Peptomers: a versatile approach for the preparation of diverse combinatorial peptidomimetic bead libraries. *Mol. Diver.* **3**, 17–27.

83. Boeijen, A., and Liskamp, R. M. J. (1998) Sequencing of peptoid peptidomimetics by Edman degradation. *Tetrahedron Lett.* **39**, 3589–3592.

84. Greene, W. T., and Wuts, G. M. P. (1999) *Protective Groups in Organic Synthesis*, 3rd ed., John Wiley & Sons, Inc., New York, p. 198.

85. Bernatowicz, M. S., Daniels, S. B., and Koster, H. (1989) A comparison of acid labile linkage agents for the synthesis of peptide *C*-terminal amides. *Tetrahedron Lett.* **30**, 4645–4648.

14

Synthesis of Toll-Like Receptor-2 Targeting Lipopeptides as Self-Adjuvanting Vaccines

Brendon Y. Chua, Weiguang Zeng, and David C. Jackson

Summary

Effective Th1- and Th2-type immune responses that result in protective immunity against pathogens can be induced by self-adjuvanting lipopeptides containing the lipid moiety dipalmitoyl-S-glyceryl cysteine (Pam$_2$Cys). The potent immunogenicity of these lipopeptides is due to their ability to activate dendritic cells by targeting and signaling through Toll-like receptor-2 (TLR-2). In addition, the simplicity and flexibility in their design as well as their ease of chemical definition and characterisation makes them highly attractive vaccine candidates for humans and animals. We describe in this chapter the techniques involved in the synthesis of an immunocontraceptive lipopeptide vaccine as well as the experimental assays carried out to evaluate its efficiency.

Key Words: Self-adjuvanting lipopeptide; peptide synthesis; ELISA; antibody; immunocontraception.

1. Introduction

Immunization with peptide epitopes in adjuvant can elicit both antibody- and cellular-mediated immune responses that are able to provide protective immunity against a repertoire of targeted pathogens that have included viruses *(1–4)*, bacteria *(4–7)*, parasites *(4,8–10)*, tumors *(11–14)*, and self-hormones *(4,15–18)*. However, a major problem associated with the use of subunit

From: *Methods in Molecular Biology, vol. 494: Peptide-Based Drug Design*
Edited by: L. Otvos, DOI: 10.1007/978-1-59745-419-3_14, © Humana Press, New York, NY

vaccines, including peptide epitope-based vaccines, is the weak immuno-genicity that they display in the absence of adjuvant. Although many exper-imental adjuvants are available, only oil-in-water emulsions, aluminum-based salt adjuvants, and virosomes are currently licensed for use in humans, while surfactant-based and microbial-derived formulations are presently still at the stage of preclinical and clinical trial testing *(19,20)*.

An effective way to enhance the immunogenicity of peptide epitopes is to incorporate lipid groups that have self-adjuvanting properties into the peptide structure. Of particular interest to many groups are palmitic acid–based adjuvants ranging from palmitic acid itself *(21,22)* to the somewhat more sophis-ticated derivatives dipalmitoyl-*S*-glyceryl cysteine (Pam$_2$Cys) *(1,4,23–25)* and tripalmitoyl-*S*-glyceryl cysteine (Pam$_3$Cys) *(26–28)*. These simple lipid moieties have been shown to be effective in adjuvanting epitope-based vaccines and, importantly, do not exhibit the harmful side effects commonly associated with many other adjuvant formulations *(29)*.

Lipopeptides bearing Pam$_2$Cys in particular have been shown to be able to induce effective Th1- and Th2-type immune responses as demonstrated by the generation of both short- and long-term protective cytotoxic T lympho-cytes and induction of antibody. Pam$_2$Cys is a synthetic analogue of MALP-2 (macrophage-activating lipopeptide-2) derived from the cytoplasmic membrane of *Mycoplasma fermentans* *(24)*. The potent immunogenicity of lipopeptides bearing Pam$_2$Cys can be explained by their ability to activate dendritic cells *(4,23,25)*, key cells that are responsible for initiating immune responses. This event occurs through a mechanism consistent with signal transduction via Toll-like receptor-2 (TLR-2) *(4)*, one of many receptors involved in the first line of host defense against pathogenic threats.

This chapter describes in detail the techniques involved in the synthesis of a Pam$_2$Cys-based lipopeptide vaccine that can induce the production of specific antibodies in vaccinated mice. As an example, we have used the sequence for luteinizing hormone-releasing hormone (LHRH) as the target antigen, but this can be substituted with other B-cell epitopes depending on the requirements of the investigator. LHRH is a 10-amino-acid peptide hormone that is secreted by the hypothalamus to initiate a cascade of events that regulates gametogenesis and reproductive function. It does this by stimulating the synthesis and secretion of follicle-stimulating hormone and luteinizing hormone from the anterior pituitary *(30)*. The LHRH sequence is an ideal immunocontraceptive target because its sequence, HWSYGLRPG, is conserved in all mammals. Such a vaccine could be of benefit as an alternative to surgical castration in companion animals and particularly in the livestock industry *(31)*. Although its use in humans is limited because of the behavioral problems associated with the suppression of sex hormone production, it may nonetheless still prove useful as a supplementary treatment for the control of hormone-dependent malignancies *(32,33)*.

2. Materials

2.1. Synthesis of Lipopeptides

1. Solid phase supports are from Rapp Polymere GmbH (Germany). The substitution levels of supports used in this laboratory are in the range 0.15–0.4 mmol/g.
2. Fluorenylmethoxycarbonyl (Fmoc) derivatives of amino acids (obtained from Auspep, Australia or Merck, Germany).
3. Solvents and reagents used should be of analytical grade or equivalent:

    a. *N,N´*-Dimethylformamide (DMF; Auspep, Australia or Merck, Germany).
    b. Dichloromethane (DCM; Merck, Australia).
    c. Diethylether (Merck, Germany).
    d. Diisopropylethylamine (DIPEA; Sigma-Aldrich, Australia).
    e. 1,8-Diazabicyclo-[5.4.0]undec-7-ene (DBU; Sigma-Aldrich, Germany).
    f. O-benzotriazole-*N,N,N´,N´*-tetramethyl-uronium-hexafluoro-phosphate (HBTU; Merck, Germany).
    g. 1-Hydroxybenzotriazole (HOBT; Auspep, Australia).
    h. 2,4,6-Trinitrobenzenesulfonic acid (TNBSA; Fluka Biochemika, Switzerland).

4. Pam_2Cys attachment to peptides:

    a. Diisopropylcarbodiimide (DICI; Auspep, Australia).
    b. Palmitic acid (Merck, Germany) .
    c. 4-[Dimethylamino]-pyridine (DMAP; Fluka Biochemika, Switzerland).
    d. 3-Bromo-1,2-propanediol (Fluka Biochemika, Switzerland).
    e. Fluorenylmethoxylcarbonyl-*N*-hydroxysuccinimide (Novabiochem, Germany).
    f. Triethylamine (Merck, Germany).
    g. L-Cysteine hydrochloride monohydrate (Sigma-Aldrich, USA).

5. Cleavage of lipopeptide from resin:

    a. Phenol (Sigma-Aldrich, USA).
    b. Triisopropylsilane (TIPS; Sigma-Aldrich, Australia).
    c. Trifluoroacetic acid (TFA; Auspep, Australia).

6. Purity analysis and characterization of synthesized lipopeptide:

    a. Acetonitrile (Merck, Germany).
    b. High-performance liquid chromatography (HPLC) system (Waters, Australia).
    c. C4 column (4.6 × 300 mm; Vydac, USA).
    d. Ion-trap mass spectrometer (Agilent 1100 Series LC/MSD) (Agilent Technologies, Australia).

2.2. Induction of Antibody Response in Mice

2.2.1. Inoculation of Mice

1. 6- to 8-Week-old female mice (BALB/c or C57BL6), 5 mice per group.
2. Saline.

3. Complete Freund's adjuvant (CFA; Sigma-Aldrich, USA).
4. Incomplete Freund's adjuvant (IFA; Sigma-Aldrich, USA).

2.2.2. Enzyme-Linked Immunosorbent Assay

1. Flat-bottom well polyvinyl plates (Thermo, USA).
2. $PBSN_3$, phosphate-buffered saline (PBS) containing v/v 0.1% sodium azide.
3. LHRH peptide (HWYSGLRPG) dissolved at 5 μg/mL in $PBSN_3$.
4. PBST, PBS containing v/v 0.05% Tween-20 (Sigma-Aldrich, USA).
5. $BSA_{10}PBS$, PBS containing 10 mg/mL bovine serum albumin (Sigma-Aldrich, USA).
6. BSA_5PBST, PBST containing 5 mg/mL bovine serum albumin (Sigma-Aldrich, USA).
7. Horseradish peroxidase-conjugated rabbit anti-mouse IgG antibodies (Dako, Denmark).
8. Enzyme substrate: 0.2 mM 2,2′-azino-bis 3-ethylbenzthiazoline-sulfonic acid (ABTS; Sigma-Aldrich, USA) in 0.05 M citric acid (M&B, England) containing 0.004% hydrogen peroxide(Ajax Chemicals, Australia).
9. 0.05 M Sodium fluoride (NaF; BDH Chemicals, Australia).
10. ELISA plate reader (Labsystems Multiskan, USA).

3. Methods

3.1. Synthesis of Lipopeptide

The lipopeptide vaccine described in this study consists of a CD4$^+$ helper T-cell epitope ([T]) and a B-cell epitope ([B]). These two epitopes are separated by a lysine residue (K) to which is attached the lipid moiety via two serine residues **(Fig. 1B)**. The CD4$^+$ T-helper epitope KLIPNASLIENCTKAEL used is derived from the fusion protein of the morbillivirus canine distemper virus *(34)* and is recognized by T cells from BALB/c and C57BL6 mouse strains *(4)*. The B-cell epitope is LHRH and has the sequence HWYSGLRPG. The presence of anti-LHRH antibodies can render vaccinated animals sterile.

We have found that although the use of an automatic synthesizer can save time and can be relatively unproblematic for simple sequences, the synthesis of peptides manually allows for more flexibility and control over the assembly process. This is particularly important for the synthesis of difficult sequences as it permits quick and easy intervention at any point.

The apparatus routinely used in this laboratory for the manual synthesis of peptides consists of a flask attached to a glass manifold that can support up to four sintered funnels, thereby permitting the simultaneous synthesis of up to four peptides. The side arm of the flask is attached to a vacuum pump to allow for solvents to be aspirated from each funnel. The manifold also contains valves

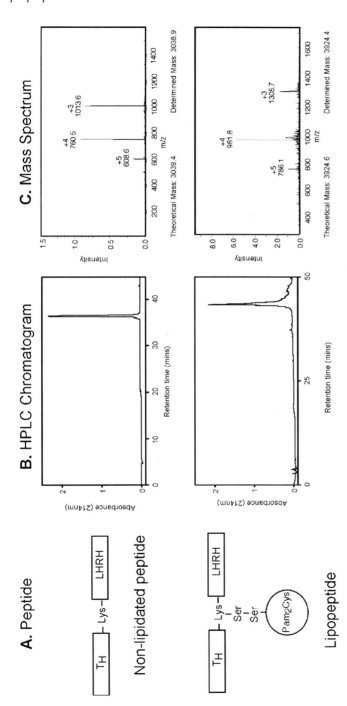

Fig. 1. Structural representations and fidelity of synthesis of lipopeptide and nonlipidated peptides. (**A**) Structure of the lipopeptide vaccines: palmitic acid, Pam. (**B**) HPLC chromatograms of purified lipopeptides. Analysis by HPLC was performed on a Waters HPLC System using a Vydac C4 column with 0.1% TFA in water and 0.1% TFA in acetonitrile as a gradient mobile phase. A flow rate of 1 mL/min was used at a gradient of 2%/min. Chromatograms were obtained by the detection of absorbance at a wavelength of 214 nm. (**C**) Mass spectral analysis of purified lipopeptides. Lipopeptides were analyzed by mass spectrometry using an Agilent 1100 Series LCM/MSD ion trap.

that are arranged so that a vacuum can be applied either universally to all four funnels or restricted to individual funnels.

There are a very large number of choices of solid phase supports available for peptide synthesis, and the prospective peptide chemist should spend a little time familiarizing themselves with the possibilities. For the purposes of the B-cell epitope sequence assembled here, S RAM resin (Tentagel, Rapp Polymere) was used so that the synthesized peptide will contain a carboxamide, $CONH_2$ group at the C-terminus of the peptide. In the case of cytotoxic T-cell epitopes designed to fit in the groove of MHC Class I molecules, however, resins should be used to assemble peptides containing a free carboxyl, COOH group, at the C-terminus (i.e., Tentagel S PHB resin, Rapp Polymere).

For synthesis of 0.23 mmol of peptide (e.g., 698.7 mg of peptide KLIPNASLIENCTKAEL-K-HWSYGLRPG with a molecular weight of 3038), weigh 1 g Tentagel S RAM resin (substitution factor = 0.23 mmol/g) into a sintered funnel and allow to swell in DMF at room temperature for at least 30 min.

1. To expose the Fmoc-protected NH_2 group on the resin, treat with either piperidine or 2.5% DBU in DMF for 2 × 5 min, followed by four washes with DMF (*see* **Note 1**).
2. Weigh 0.92 mmol of Fmoc-amino acid (i.e., a fourfold excess of amino acid relative to the substitution level of the support) into clean and dry plastic tubes (Sarstedt, Germany); tubes with a 10-mL volume capacity are ideal. Add an equimolar amount of HOBT and HBTU relative to the amount of amino acid to 2 mL of DMF and a sixfold excess of DIPEA over the substitution level of the solid phase support. Dissolve fully by vortex and sonication.
3. Remove DMF from the swollen resin in the glass sinter filter funnel by aspiration using the vacuum pump and add the activated amino acid solution. Stir with a spatula and incubate at room temperature for 30–45 min, stirring occasionally.
4. After 30–45 min, aspirate the amino acid solution and followed by two washes of the resin with DMF.
5. Transfer a few beads of resin with a Pasteur pipet into an Eppendorf tube and add two drops of DIPEA followed by five drops of TNBSA solution. Inspect the beads by eye or under a microscope if possible. If the beads are colorless after 1 min, then acylation is complete and the next step can be carried out. Any trace of orange color in the beads indicates the presence of free amino groups and incomplete coupling. In these cases **steps 2–5** should be repeated until a negative TNBSA test is returned.
6. Remove the N-α Fmoc group of the coupled amino acid by carrying out **step 1**. Confirmation of the removal of the Fmoc group can be determined by performing a TNBSA test that should result in a positive orange color change.
7. Repeat **steps 2–6** with the next amino acid until completion of the HWYSGLRPG sequence.

8. To enable lipid attachment, repeat **steps 3–6** using (Fmoc)-K(Mtt)-OH to enable lipid attachment between the two epitopes.
9. Repeat **steps 2–6** with the amino acids corresponding to the sequence KLIPNASLIENCTKAEL.
10. Repeat **steps 3–6** using (Boc)-Gly-OH to temporarily block the N-terminus of the peptide. The Boc-protective group is resistant to removal by the conditions used for lipid attachment until cleavage of the assembled product from the resin and concomitant removal of the side-chain–protecting groups.
11. At this point, the completed peptide on resin should be washed sequentially in DMF, DCM, and methanol, dried under vacuum, and stored in a dessicated atmosphere at room temperature until ready to be cleaved for use as a nonlipidated peptide control (*see* **Subheading 3.3.**).
12. Continuing from **step 11**, the resin is treated with 1% TFA in DCM 5 × 12 min to remove the Mtt group from the side chain of the lysine residue situated between the two epitopes.
13. Repeat **steps 2–6** in order to couple two serines to the exposed ε-amino group of the intervening lysine residue and remove the Fmoc group from the second serine residue. The peptide is now ready for lipid attachment (*see* **Subheading 3.2.**).

3.2. Attachment of Pam₂Cys to Peptide

3.2.1. Synthesis of S-(2,3-dihydroxypropyl) Cysteine

1. Triethylamine (6 g, 8.2 mL, 58 mmol) is added to L-cysteine hydrochloride (3 g, 19 mmol) and 3-bromo-propan-1,2-diol (4.2 g, 2.36 mL, 27 mmol) in water. This homogeneous solution is kept at room temperature for 3 d.
2. The solution is reduced in vacuo at 40°C to a white residue which is washed three times with acetone and dried to give S-(2,3-dihydroxypropyl) cysteine as a white amorphous powder (2.4 g, 12.3 mmol, 64.7%). This product can be used for the next step without further purification.

3.2.2. Synthesis of N-Fluorenylmethoxycarbonyl-S-(2,3-dihydroxypropyl) Cysteine (Fmoc-Dhc-OH)

1. Dissolve S-(2,3-dihydroxypropyl)cysteine (2.45 g, 12.6 mmol) in 9% sodium carbonate (20 mL).
2. Add a solution of fluorenylmethoxycarbonyl-N-hydroxysuccinimide (3.45 g, 10.5 mmol) in acetonitrile (20 mL) and stir the mixture for 2 hours. Dilute with water (240 mL), and extract with diethyl ether (25 mL × 3).
3. Acidify the aqueous phase to pH 2 with concentrated hydrochloric acid and then extract with ethyl acetate (70 mL × 3).
4. Wash the extract with water (50 mL × 2) and saturated sodium chloride solution (50 mL × 2). Dry over sodium sulfate and evaporate to dryness. Recrystallize from ether and ethyl acetate at -20°C to yield a colorless powder (2.8 g, 6.7 mmol, 63.8%).

3.2.3. Coupling of Fmoc-Dhc-OH to Resin-Bound Peptide

1. Activate Fmoc-Dhc-OH (100 mg, 0.24 mmol) in DMF (3 mL) with HOBT (36 mg, 0.24 mmol) and DICI (37 µal, 0.24 mmol) at 0°C for 5 min.
2. Add this mixture to a vessel containing the resin-bound peptide (0.06 mmol, 0.25 g amino-peptide resin). After shaking for 2 h remove the solution by filtration and wash the resin with DCM and DMF (3 × 30 mL). Completeness of reaction is monitored using the TNBSA test. If necessary double coupling can be performed.

3.2.4. Palmitoylation of the Two Hydroxy Groups of the Fmoc-Dhc-Peptide Resin

1. Dissolve palmitic acid (307 mg, 1.2 mmol), DICI (230 µL, 1.5 mmol) and DMAP (14.6 mg, 0.12 mmol) in 3 mL of DCM.
2. Suspend the resin-bound Fmoc-Dhc-peptide resin (0.06 mmol, 0.25 g) in this solution and shake for 16 h at room temperature. Remove the supernatent by filtration and thoroughly wash with DCM and DMF to remove any residue of urea. The removal of the Fmoc group is accomplished with 2.5% DBU (2 × 5 min).

3.3. Cleavage of Lipopeptide from Solid Phase Support

This procedure simultaneously cleaves the lipopeptide or peptide from the solid phase support and removes side-chain–protecting groups from those amino acids that have them.

1. Transfer the vacuum-dried resin into a clean dry McCartney glass bottle and add 3 mL cleavage reagent (88% TFA, 5% phenol, 5% water, and 2% TIPS) (*see* **Note 2**).
2. Gently flush with nitrogen and leave for at least 3 h with occasional mixing. The use of an automatic mixing rack is ideal for this step (*see* **Note 3**).
3. Transfer the mixture into the barrel of a 5-mL syringe plugged with nonadsorbent cotton wool and use the plunger to drive the peptide-containing supernatant into a clean dry 10-mL centrifuge tube.
4. Evaporate the solution to a volume of approximately 500 µL under a gentle stream of nitrogen.
5. Add 10 mL cold diethyl ether to the peptide solution and vortex vigorously to precipitate the peptide.
6. Centrifuge to sediment the peptide material, wash by aspirating the diethyl ether, and resuspend the precipitate in cold diethyl ether. Wash twice in cold diethyl ether.
7. After the final wash, aspirate the remaining diethyl ether and allow the pellet to dry in a fume hood for approximately 1 h.
8. Dissolve the precipitate in 0.1% aqueous TFA and lyophilize (*see* **Note 4**).
9. Assess product purity using reversed-phase chromatography and fidelity of the target sequence by mass spectrometry.

Representative results of HPLC chromatograms and mass spectra that we have obtained are shown in **Fig. 1**. Also shown are schematic representations of the peptide and lipopeptide structures.

3.4. Evaluation of Vaccine Efficacy

There are many techniques that can be employed to assess the immunogenicity of a vaccine candidate, and laboratories interested in pursuing this objective should also explore other assays suited to their purposes. In this model, vaccine immunogenicity is measured by determining levels of anti-LHRH antibodies induced in mice immunized with lipopeptide or a nonlipidated peptide control. It is important to correlate vaccine immunogenicity with biological function, and we measure the reproductively capability of vaccinated female mice as an indication of vaccine efficacy. Techniques that determine testosterone and oestrogen levels are also useful.

3.4.1. Immunization of Mice

1. Dissolve peptide and lipopeptide vaccines in saline to a concentration of 200 nmol/mL. If necessary, encourage dissolution by warming in a water bath and/or by sonication. Do not mix by pipet if the inoculant is not soluble; this can result in insoluble peptide or lipopeptide being trapped in the pipet tip.
2. Immunise 6- to 8-week-old female BALB/c or C57BL6 mice either intranasally or subcutaneously in the base of the tail with 20 nmol (100 μL) of lipopeptide or nonlipidated peptide in saline per dose.
3. Perform booster inoculations using the same amount of peptide and lipopeptide 4 wk later (*see* **Note 5**).
4. Obtain blood 2 wk following the second dose of vaccine (*see* **Note 6**).
5. Prepare sera by leaving blood overnight at 4°C to coagulate.
6. Centrifuge tubes containing coagulated blood to separate sera.
7. Collect sera and store at −20°C or use immediately.

3.4.2. Enzyme-Linked Immunosorbent Assay (ELISA)

The working principle of the ELISA described here involves the use of LHRH peptide-coated wells to capture any anti-LHRH antibodies present in the sera of vaccinated mice. Captured antibodies are detected through the use of a horseradish peroxidase-conjugated antibody that is specific for mouse immunoglobulins and a substrate is then added to induce a color change. In this case, the substrate utilized is 2,2′-azino-bis 3-ethylbenzthiazoline-sulfonic acid (ABTS; Sigma-Aldrich, USA), which is converted to a green soluble end product by horseradish peroxidase (*see* **Note 7**). The level of substrate-induced colour change is then measured at an appropriate wavelength where the amount of specific antibodies present is proportional to the intensity of the reading.

1. Prepare LHRH peptide antigen at 5μg/mL in PBSN$_3$ (allow for a minimum of 5.2 mL per 96-well flat-bottomed plate, i.e., 50 μL/well).
2. Add 50 μL of antigen to each well and incubate for 16–24 h at room temperature in a humidified environment (*see* **Note 8**).
3. Aspirate the antigen solution and add 100 μL of BSA$_{10}$PBS to each well and incubate for 1 h at room temperature.
4. Discard BSA$_{10}$PBS and wash wells by pipetting 200 μL PBST into each well followed by expelling the contents. Wash four times.
5. Add 50 μL of BSA$_5$PBST to each well on the plate except for the first row of wells.
6. Add 73 μL of a $^1/_{100}$ dilution of sera in BSA$_5$PBST to the first row of wells.
7. Remove 23 μL of this initial dilution and transfer it into the next row of wells containing 50 μL of BSA$_5$PBST. Perform this consecutively across the remaining wells, ensuring that each dilution is mixed thoroughly in between transfers. Each transfer of sera in diluent represents an increase in the dilution of 3.17-fold ($\log_{10} = 0.5$) of sera.
8. Incubate for 3 h at room temperature in a humidified environment.
9. Discard antibody dilutions. Wash wells with PBST six times as in **step 5**.
10. Dilute horseradish peroxidase-conjugated rabbit anti-mouse IgG antibodies in BSA$_5$PBST according to the manufacturer's recommendation. We typically use a $^1/_{400}$ dilution with this particular brand of antibody conjugate (*see* **Subheading 2.2.2.**).
11. Add 50 μL of antibody conjugate to each well and incubate for 1 h at room temperature.
12. Discard the antibody conjugate. Wash wells with PBST six times as in **step 5**.
13. Add 100 μL of the enzyme substrate (*see* **Subheading 2.2.2.**) to each well and allow the color to develop for 15 min.
14. Add 50 μL of 0.05 M NaF and mix well to stop the color reaction.
15. Determine the amount of color development using an ELISA plate reader or equivalent at a wavelength of 450 nm.
16. Titers of antibody are expressed as the highest dilution of serum (in log) required to achieve an optical density of 0.2.

The results of an experiment that we have obtained are shown in **Fig. 2**. It can be seen that little or no antibody is elicited when nonlipidated peptide is inoculated in saline, whereas nonlipidated peptide administered in CFA or lipidated peptide administered in saline elicit similar and very high titers of anti-LHRH antibodies.

3.4.3. Fertility Trial

Mouse fertility trials are simple to establish in any animal housing facility because they do not require any special requirements or conditions beyond those

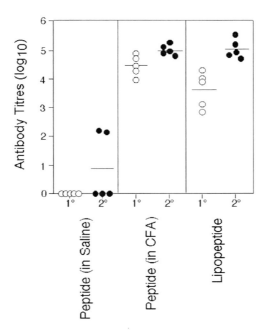

Fig. 2. Immunogenicity of synthesized lipopeptide and nonlipidated peptide. Groups of BALB/c mice (6–8 wk old) were inoculated subcutaneously with 20 nmol of peptide immunogens in both primary and secondary inoculations. All lipopeptides were administered in saline, and the nonlipidated peptide was administered in either CFA or saline. Sera were obtained from blood taken 4 wk following the primary (open circles) inoculation and 2 wk following the secondary (closed circles) inoculation. Antibody levels were determined by ELISA. Individual antibody titers are presented with the mean value represented by the horizontal bar.

already available for the breeding and housing of experimental animals. We have focused on evaluating the reproductive capabilities of vaccinated female mice rather than males because hormone-mediated changes in their behavior may inhibit copulation.

1. Two weeks following the last inoculation, house mice at a 1:2 or 1:3 ratio of males to vaccinated females.
2. Rotate male mice weekly between sample groups so that each group of female mice are exposed to each and every male.
3. Remove male mice after 3 wk and observe female mice over the next 3 wk for pregnancies. The gestation period for female mice is 19–21 d. Some pregnancies may be difficult to detect due to the variation in litter sizes, and it is advisable to keep females housed separately or ear clip them as a means of identification so

Table 1
Incidence of Pregnancy Following Inoculation with Peptide Immunogens

Peptide immunogen	Administered in	Incidence of pregnancy
T_H-Lys-LHRH (in saline)	Saline	$^4/_5$
T_H-Lys (Ser$_2$Pam$_2$Cys)- LHRH	Saline	$^0/_5$
T_H-Lys-LHRH (in CFA)	CFA	$^0/_5$

Inoculated female mice were housed with male mice at 1:2 or 1:3 ratios of males to females for a period of 3 wk. The male mice were rotated weekly between each cage so that each group of female mice was exposed to every male. Following removal of the males, female mice were observed for litter production over a period of 3 wk.

that litter production by a particular female can be confirmed. If female mice are housed together, be sure to remove those that have dropped litters along with their offspring.

Table 1 summarizes the results of a fertility trial using lipidated and nonlipidated peptides. Fertility is totally inhibited in those groups that received lipidated peptide or nonlipidated peptide administered in CFA.

4. Notes

1. Piperidine has a strong odor and many prefer the use of the odorless DBU reagent (*35*).
2. The cleavage reagent should be prepared fresh.
3. Longer time periods for cleavage may be necessary for some sequences, particularly those containing multiple arginine residues.
4. In those cases where peptides do not dissolve easily in 0.1% aqueous TFA, addition of acetonitrile may be necessary.
5. As a positive control, a separate group of mice can be inoculated with the same amount of the nonlipidated peptide emulsified in CFA and boosted with nonlipidated peptide emulsified in IFA
6. If required, mice can be bled before the second inoculation to obtain an early indication of vaccine immunogenicity
7. There are many enzyme–substrate combinations that can be utilized in an ELISA, and investigators should explore which systems work best for their purposes. Ideally, they should be stable, safe, inexpensive, and easy to optimize. Commonly used combinations include p-nitrophenyl phosphate (pNPP), which is converted to the yellow p-nitrophenol by alkaline phosphatase and o-phenylenediamine and 3,3´5,5´- tetramethylbenzidine base, which is converted by peroxidase to orange and blue end products.

8. A large plastic box with an accompanying lid and lined at the bottom with wet dish cleaning sponges provides a humidified environment for incubating ELISA plates.

References

1. Deliyannis, G., Jackson, D.C., Ede, N.J., et al. (2002) Induction of long-term memory CD8(+) T cells for recall of viral clearing responses against influenza virus. *J. Virol.* **76**(9), 4212–4221.
2. Egan, M.A., Chong, S.Y., Hagen, M., et al. (2004) A comparative evaluation of nasal and parenteral vaccine adjuvants to elicit systemic and mucosal HIV-1 peptide-specific humoral immune responses in cynomolgus macaques. *Vaccine* **22**(27–28), 3774–3788.
3. Halassy, B., Mateljak, S., Bouche, F.B., et al. (2006) Immunogenicity of peptides of measles virus origin and influence of adjuvants. *Vaccine* **24**(2), 185–194.
4. Jackson, D.C., Lau, Y.F., Le, T., et al. (2004) A totally synthetic vaccine of generic structure that targets Toll-like receptor 2 on dendritic cells and promotes antibody or cytotoxic T cell responses. *Proc. Natl. Acad. Sci. USA* **101**(43), 15440–15445.
5. Olive, C., Clair, T., Yarwood, P., and Good, M.F. (2002) Protection of mice from group A streptococcal infection by intranasal immunisation with a peptide vaccine that contains a conserved M protein B cell epitope and lacks a T cell autoepitope. *Vaccine* **20**(21–22), 2816–2825.
6. Taouji, S., Nomura, I., Giguere, S., et al. (2004) Immunogenecity of synthetic peptides representing linear B-cell epitopes of VapA of Rhodococcus equi. *Vaccine* **22**(9–10), 1114–1123.
7. Cassataro, J., Estein, S.M., Pasquevich, K.A., et al. (2005) Vaccination with the recombinant Brucella outer membrane protein 31 or a derived 27-amino-acid synthetic peptide elicits a CD4+ T helper 1 response that protects against Brucella melitensis infection. *Infect. Immun.* **73**(12), 8079–8088.
8. Ben-Yedidia, T., Tarrab-Hazdai, R., Schechtman, D., and Arnon, R. (1999) Intranasal administration of synthetic recombinant peptide-based vaccine protects mice from infection by *Schistosoma mansoni*. *Infect. Immun.* **67**(9), 4360–4366.
9. Lougovskoi, A.A., Okoyeh, N.J., and Chauhan, V.S. (1999) Mice immunised with synthetic peptide from N-terminal conserved region of merozoite surface antigen-2 of human malaria parasite *Plasmodium falciparum* can control infection induced by *Plasmodium yoelii yoelii* 265BY strain. *Vaccine* **18**(9–10), 920–930.
10. Audran, R., Cachat, M., Lurati, F., et al. (2005) Phase I malaria vaccine trial with a long synthetic peptide derived from the merozoite surface protein 3 antigen. *Infect. Immun.* **73**(12), 8017–8026.
11. Smith, J.W., 2nd, Walker, E.B., Fox, B.A., et al. (2003) Adjuvant immunization of HLA-A2-positive melanoma patients with a modified gp100 peptide induces peptide-specific CD8+ T-cell responses. *J. Clin. Oncol.* **21**(8), 1562–1573.

12. Valmori, D., Dutoit, V., Ayyoub, M., et al. (2003) Simultaneous CD8+ T cell responses to multiple tumor antigen epitopes in a multipeptide melanoma vaccine. *Cancer Immun.* **3**, 15.

13. Noguchi, M., Itoh, K., Suekane, S., et al. (2004) Phase I trial of patient-oriented vaccination in HLA-A2-positive patients with metastatic hormone-refractory prostate cancer. *Cancer Sci.* **95**(1), 77–84.

14. Chianese-Bullock, K.A., Pressley, J., Garbee, C., et al. (2005) MAGE-A1-, MAGE-A10-, and gp100-derived peptides are immunogenic when combined with granulocyte-macrophage colony-stimulating factor and montanide ISA-51 adjuvant and administered as part of a multipeptide vaccine for melanoma. *J. Immunol.* **174**(5), 3080–3086.

15. Ghosh, S. and Jackson, D.C. (1999) Antigenic and immunogenic properties of totally synthetic peptide-based anti-fertility vaccines. *Int. Immunol.* **11**(7), 1103–1110.

16. Ferro, V.A., Khan, M.A., McAdam, D., et al. (2004) Efficacy of an anti-fertility vaccine based on mammalian gonadotrophin releasing hormone (GnRH-I)—a histological comparison in male animals. *Vet. Immunol. Immunopathol.* **101**(1–2), 73–86.

17. Zeng, W., Ghosh, S., Macris, M., Pagnon, J., and Jackson, D.C. (2001) Assembly of synthetic peptide vaccines by chemoselective ligation of epitopes: influence of different chemical linkages and epitope orientations on biological activity. *Vaccine* **19**(28–29), 3843–3852.

18. Jinshu, X., Jingjing, L., Duan, P., et al. (2005) A synthetic gonadotropin-releasing hormone (GnRH) vaccine for control of fertility and hormone dependent diseases without any adjuvant. *Vaccine* **23**(40), 4834–4843.

19. Mesa, C. and Fernandez, L.E. (2004) Challenges facing adjuvants for cancer immunotherapy. *Immunol. Cell Biol.* **82**(6), 644–650.

20. Pashine, A., Valiante, N.M., and Ulmer, J.B. (2005) Targeting the innate immune response with improved vaccine adjuvants. *Nat. Med.* **11**(4 Suppl), S63–68.

21. BenMohamed, L., Belkaid, Y., Loing, E., Brahimi, K., Gras-Masse, H., and Druilhe, P. (2002) Systemic immune responses induced by mucosal administration of lipopeptides without adjuvant. *Eur. J. Immunol.* **32**(8), 2274–2281.

22. Le Gal, F.A., Prevost-Blondel, A., Lengagne, R., et al. (2002) Lipopeptide-based melanoma cancer vaccine induced a strong MART-27–35-cytotoxic T lymphocyte response in a preclinal study. *Int. J. Cancer* **98**(2), 221–227.

23. Chua, B.Y., Healy, A., Cameron, P.U., et al. (2003) Maturation of dendritic cells with lipopeptides that represent vaccine candidates for hepatitis C virus. *Immunol. Cell Biol.* **81**(1), 67–72.

24. Muhlradt, P.F., Kiess, M., Meyer, H., Sussmuth, R., and Jung, G. (1997) Isolation, structure elucidation, and synthesis of a macrophage stimulatory lipopeptide from *Mycoplasma fermentans* acting at picomolar concentration. *J. Exp. Med.* **185**(11), 1951–1958.

25. Zeng, W., Ghosh, S., Lau, Y.F., Brown, L.E., and Jackson, D.C. (2002) Highly immunogenic and totally synthetic lipopeptides as self-adjuvanting immunocontraceptive vaccines. *J. Immunol.* **169**(9), 4905–4912.

26. Nardin, E.H., Calvo-Calle, J.M., Oliveira, G.A., et al. (1998) *Plasmodium falciparum* polyoximes: highly immunogenic synthetic vaccines constructed by chemoselective ligation of repeat B-cell epitopes and a universal T-cell epitope of CS protein. *Vaccine* **16**(6), 590–600.

27. Obert, M., Pleuger, H., Hanagarth, H.G., et al. (1998) Protection of mice against SV40 tumours by Pam3Cys, MTP-PE and Pam3Cys conjugated with the SV40 T antigen-derived peptide, K(698)-T(708). *Vaccine* **16**(2–3), 161–169.

28. Mora, A.L. and Tam, J.P. (1998) Controlled lipidation and encapsulation of peptides as a useful approach to mucosal immunizations. *J. Immunol.* **161**(7), 3616–3623.

29. Shimizu, T., Ohtsuka, Y., Yanagihara, Y., Kurimura, M., Takemoto, M., and Achiwa, K. (1991) Comparison of biologic activities of synthetic lipopentapeptide analogs of bacterial lipoprotein in mice. *Mol. Biother.* **3**(1), 46–50.

30. Delves, P.J., Lund, T., and Roitt, I.M. (2002) Antifertility vaccines. *Trends Immunol.* **23**(4), 213–219.

31. Oonk, H.B., Turkstra, J.A., Schaaper, W.M., et al. (1998) New GnRH-like peptide construct to optimize efficient immunocastration of male pigs by immunoneutralization of GnRH. *Vaccine* **16**(11–12), 1074–1082.

32. Ferro, V.A. and Stimson, W.H. (1997) Immunoneutralisation of gonadotrophin releasing hormone: a potential treatment for oestrogen-dependent breast cancer. *Eur. J. Cancer* **33**(9), 1468–1478.

33. Talwar, G.P. (1999) Vaccines and passive immunological approaches for the control of fertility and hormone-dependent cancers. *Immunol. Rev.* **171**, 173–192.

34. Ghosh, S., Walker, J., and Jackson, D.C. (2001) Identification of canine helper T-cell epitopes from the fusion protein of canine distemper virus. *Immunology* **104**(1), 58–66.

35. Wade, J.D., Bedford, J., Sheppard, R.C., and Tregear, G.W. (1991) DBU as an N alpha-deprotecting reagent for the fluorenylmethoxycarbonyl group in continuous flow solid-phase peptide synthesis. *Pept. Res.* **4**(3), 194–199.

15

Synthesis of a Multivalent, Multiepitope Vaccine Construct

Laszlo Otvos, Jr.

Summary

A major challenge to contemporary peptide chemistry is to reproduce highly complex active sites or complexes of native proteins. In addition to the need for the combination of different peptide fragments, true protein mimics designed for therapeutic use frequently require the incorporation of multiple copies of given active domains in a biologically relevant spatial arrangement. Perhaps the best examples for this line of drug design are subunit vaccine candidates against extracellular domains of ion-channel proteins. One of our earlier constructs containing four copies of the ectodomain of the M2 protein of influenza virus together with two independent T-helper cell epitopes induced protective antibody production in mice. Here I describe an improved synthesis of the M2-based multiepitope and multivalent peptide construct. In general, the synthetic strategy outlined here can serve as a model for the orthogonal N-terminal and side-chain–protecting scheme during the preparation of large, complex peptides or small, engineered proteins.

Key Words: Automated peptide synthesis; influenza virus; oligolysine scaffold; orthogonal side-chain protection; protective antibody; T-helper cell determinant.

1. Introduction

Peptide fragments represent the biologically active domains of native proteins. However, many recognition processes require the presence of multiple independent protein fragments, co-activators, or chaperones *(1)*. In addition, it was long considered that peptides without carrier modules are unable to enter cells *(2)*, although since the turn of the millennium this assumption has been increasingly challenged *(3)*. Thus, to elicit the desired and full biological

From: *Methods in Molecular Biology, vol. 494: Peptide-Based Drug Design*
Edited by: L. Otvos, DOI: 10.1007/978-1-59745-419-3_15, © Humana Press, New York, NY

response, synthetic peptide constructs designed for therapeutic use need to contain determinants for the target biological activity, bystander activators, and potentially delivery modules.

The requirement of multifunctional peptide complexes is perhaps most obvious for the development of subunit peptide vaccines. Successful immunizations with peptide antigens cannot be achieved without the inclusion of a bystander T-helper cell determinant in the chemical entity *(4)* or in the immunizing cocktail *(5)*. For outbred animals and humans, multiple peptide epitopes, representing determinants of more than one major histocompatibility complex (MHC) proteins, are used to overcome subunit vaccine unresponsiveness, and this also improves antigen presentation in inbred animals *(6)*.

Various possibilities exist for the assembly of multivalent peptide constructs. Earlier strategies preferred a co-linear assembly of chimeric T-B-cell determinant peptide fragments *(7)*, a strategy that was later replaced by a branched-type arrangement of the peptide constituents *(8,9)*. In his pioneering work for homopolymers, Tam coined the phrase "Multiple Antigenic Peptide," or MAP *(10)*, a term that is currently used for both homo- and heteropolymeric structures. In our particular case, we have to incorporate four copies of the B-cell antigen, the ectodomain of the influenza virus M2 protein because the native protein forms a tetrameric transmembrane channel, and the relative orientation of the four chains *(11)* is supposed to play a role in all biochemical and immunological properties, including the specific immunogenicity. The two murine T-helper cell epitopes, S1 and S2, are dominant T-cell antigens from the influenza virus hemagglutinin *(12)*. In our first construct **(Fig. 1)**, the tetrameric form of M2 was mimicked by four chains of the M2 ectodomain (24 amino acid residues, M2e) built on two linear alternating glycine-lysine scaffolds which were connected with a cysteine bridge *(13)*. Each scaffold was additionally decorated with single copies of the two T-helper cell hemagglutinin epitopes. While the synthesis of the almost 18 kDa construct was successful as

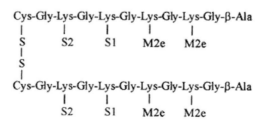

Fig. 1. The original M2e MAP construct *(14)*. The disulfide bridge was formed after assembly and purification of the two half fragments.

Fig. 2. SDS-PAGE of the Fmoc-protected original M2e-S1-S2 dimeric construct used for immunization after deprotection *(14)*. The left lane shows the molecular weight standards (the bands correspond to 42, 30, and 22 kDa, from the top to the bottom, respectively); the right lane documents the purity of the final product (calculated molecular weight: 18,434 Da). While the general quality of the peptide for its size is excellent, insufficient dimer formation and the presence of contaminating peptides can be observed.

documented by reversed-phase–high performance chromatography (RP-HPLC), matrix-assisted laser desorption/ionization mass spectroscopy (MALDI-MS), and sodium dodecyl sulfate–polyacrylamide gel electrophoresis (SDS-PAGE) (**Fig. 2**), disulfide-related significant synthetic difficulties, potentially insurmountable during large-scale preparation, were met *(13)*. Having said that, after purification, mice vaccinated twice by the intranasal route with adjuvanted M2e-MAPs exhibited significant resistance to virus replication in all sites of the respiratory tract *(14)*. Compared to mice primed by two consecutive heterosubtypic infections, resistance was of similar strength in nasal and tracheal tissue but lower in pulmonary tissue.

In ensuing studies, the fine specificities of anti-M2e antibodies were studied *(15)*, along with M2 escape mutants after immunization *(16)*. For these investigations, a new generation of M2e constructs were designed and synthesized. The main feature of the new design is the single chain assembly of the four M2e fragments, attached to two lysines at the carboxy-terminus and two lysines at the amino-terminus on the glycine-lysine alternating scaffold (**Fig. 3**). The two mid-chain lysines carry single copies of the S1 and S2 T-helper cell determinants. Three different M2e constructs were made, according to natural M2 protein mutations, featuring either proline or leucine or histidine in position 9. The increased size of the single chain made the assembly process increasingly difficult, but this was compensated for by the lack of side reactions originating from the presence and bridge formation of the N-terminal cysteine residues

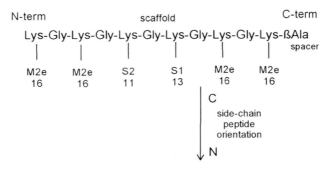

Fig. 3. Schematic presentation of the design of the second-generation (4)M2e-S1-S2 construct. The numbers indicate the number of amino acid residues in each peptide side chain.

during the necessary dimerization reaction. Since an increasing amount of evidence suggests that the immunogenic part of the M2e fragment is embedded at the N-terminal half, the second-generation vaccine candidate is further simplified by eliminating eight C-terminal M2e residues. The linear lysine-based construct appears to be suitable for mimicking the M2 four-helix bundle. An analogous four-helix bundle of the bee venom melittin was successfully assembled on a similar scaffold, where alanines were placed in between melittin side chains 1 and 2 as well as 3 and 4, and a turn-forming Gly-Pro dipeptide between the melittin-carrier lysines 2 and 3 *(17)*. In this chapter the synthetic strategy for the construction of the second-generation (4)M2e-S1-S2 construct is detailed.

2. Materials

2.1. Equipment

1. PS3 automated peptide synthesizer (Protein Technologies) (*see* **Note 1**).
2. Home-assembled gradient high-performance liquid chromatograph (HPLC) consisting of two Beckman-Coulter 110B pumps operated by a 421 gradient controller, interchangeable Phenomenex Jupiter analytical/preparative columns, a Rainin Dynamax ultraviolet detector and a Shimadzu C-R6A integrator.
3. BT53 tabletop lyophilizer (Millrock Technologies).
4. Voyager DE MALDI time of flight (TOF) mass spectrometer (Applied Biosystems).
5. Bio-Rad Protean II ready SDS-PAGE system.
6. Shaker (Fisher Scientific).

2.2. Reagents

2.2.1. Solvents

1. *N,N′*-Dimethyl formamide (DMF) (Fisher Scientific).
2. Piperidine (Sigma) (*see* **Note 2**).
3. *N*-Methyl morpholine (NMM) (Acros).
4. Chloroform (Aldrich).
5. Acetic acid (Aldrich).
6. Acetonitile (Sigma).

2.2.2. Resin

1. TentaGel S RAM resin (Rapp Polymere).

2.2.3. Coupling Reagents

1. *O*-(7-Azabenzotriazole-1-yl)-*N,N,N ′N′*-tetramethyluronium hexafluorophosphate (HATU) (Applied Biosystems) (*see* **Note 3**).
2. 1-Hydroxy benzotriazole (HOBt) (Novabiochem).
3. *O*-(Benzotriazol-1-yl)-*N,N,N',N'*-tetramethyluronium tetrafluoroborate (TBTU) (Novabiochem).
4. *O*-Benzotriazole-*N,N,N ′,N ′*-tetramethyl-uronium-hexafluoro-phosphate (HBTU) (Novabiochem).

2.2.4. Amino Acids

Standard Fmoc-protected amino acids are purchased from Advanced Chemtech (*see* **Note 4**). Unusual amino acids (all from Bachem):

1. Dde-Lys(Fmoc)-OH.
2. Fmoc-Lys(Aloc)-OH.
3. Boc-Lys(Fmoc)-OH.
4. Fmoc-βAla-OH.
5. Boc-His(Boc)-OH.
6. Boc-Ser('Bu)-OH.

2.2.5. Other Reagents

1. Tetrakis triphenylphosphine palladium, (PPh$_3$)$_4$Pd (Chem-Impex).
2. Diethyl dithio-carbamate (Mallinkrodt Baker).
3. Hydrazine (Sigma).
4. Allyl alcohol (Acros).
5. Trifluoroacetic acid (TFA) (Sigma).
6. α-Cyano-4-hydroxycinnamic acid (Fluka).

3. Methods

The main goal of this chapter is to describe the synthesis details of complex, orthogonally protected peptide constructs. Thus, major emphasis is placed on the peptide chain assembly design and practice and the alterations from the solid-phase synthesis of simple, nonmodified peptides. The technology for peptide purification and quality control is not significantly different from that of other peptides, and these methods will be just briefly described. Many chapters of this book focus on the optimization of HPLC and MALDI-MS procedures for peptide separation and analysis and illustrate the expected and/or acceptable quality control parameters.

3.1. Construct Assembly

The basic idea is to build the construct on an oligolysine–oligoglycine scaffold where the active peptides are attached to the side chains of the scaffold lysine amino groups **(Fig. 3)**. The two T-helper cell epitopes are built individually during the assembly of the scaffold, while the four identical M2e fragments are synthesized simultaneously after the scaffold assembly is completed. For this strategy we need to use four different amino protecting groups.

While the amino acid coupling and deprotection steps used here are standard for Fmoc-N-terminal protection-based peptide synthesis, the sheer size of the construct and the fact that for the M2e fragment four peptide side chains are built concomitantly require special considerations. To complete the coupling cycles, the resin has to be loaded with relatively low amounts of reactive amino groups, the reaction times have to be extended along with occasional double coupling and deprotection cycles. The large number of acid-labile protecting groups to be removed at the end of the synthesis requires overnight cleavage from the resin even if a β-alanine moiety is incorporated between the resin and the peptide backbone to spatially separate the growing peptide chain from the solid support.

The three orthogonally removable lysine protecting groups we use here are Fmoc (9-fluorenyl methoxy carbonyl), cleavable with 20% base, preferably with 20% piperidine or 3% DBU (1,8-diazabicyclo[5.4.0]undec-7-ene) *(18)*, Dde (1-(4,4-dimethyl-2,6-dioxocyclohexylidene)-ethyl), cleavable with 2% hydrazine, and Aloc (allyloxy carbonyl), cleavable with palladium. Hydrazine also removes Fmoc, and thus it can be applied only if no Fmoc groups are present on the growing peptide chain. Acid-sensitive amino protecting groups available are Boc (tert-butyloxy carbonyl), cleavable with 90% TFA, and Mtt (4-methyl-trityl), cleavable with 1% TFA. We use Boc for the protection of the N-terminal moieties of N-terminal amino acids in each peptide chain as well as at the N-terminus of the scaffold.

3.2. Synthetic Steps

3.2.1. Basic Instructions

Resin: use low load (0.15 mmol/g) and do not use more than 0.1 g.

Couple each residue for 2 h and remove the Fmoc amino-terminal protection for 2 × 7 min treatment with 20% piperidine in DMF. Double couple isoleucine, histidine, threonine, valine.

Lysines with normal letters to be coupled as Fmoc-Lys(Boc)-OH; **lysines with bold to be coupled as Dde-Lys(Fmoc)-OH;** lysines underlined to be coupled as Fmoc-Lys(Aloc)-OH; *lysines underlined to be coupled as Fmoc-(Aloc)-OH.*

Cycle times: DMF wash 3 × 30 s; piperidine deprotection (20% in DMF) 2 × 7 min; DMF wash, 6 × 30 s; delivery of *N*-methyl morpholine 30 s: Fmoc-amino acid coupling times 2 h, DMF wash, 3 × 30 s.

3.2.2. Construct Assembly

1. Assemble   Boc-Ser-Phe-Glu-Arg-Phe-Glu-Ile-Phe-Pro-Lys-Glu-**Lys**-Gly-Lys-Gly-Lys-β-Ala-resin. Special amino acids needed: Fmoc-βAla-OH; Fmoc-Lys(Aloc)-OH; Dde-Lys(Fmoc)-OH; Boc-Ser('Bu)-OH.
2. Cleave the Dde group from the resin with 2% hydrazine and 1% allyl alcohol in DMF for 3 × 3 min (*see* **Note 5**). Wash the resin with DMF three times. Reagent needed: hydrazine.
3. Take the resin from **step 2** and continue the synthesis: Boc-His-Asn-Thr-Asn-Gly-Val-Thr-Ala-Ala-Ser-Ser-His-Glu-**Lys**-Gly-resin from **step 2**. Special amino acid needed: Boc-His(Boc)-OH.
4. Cleave the Dde group from the resin with 2% hydrazine and 1% allyl alcohol in DMF for 3 × 3 min (*see* **Note 5**). Wash the resin with DMF three times. Reagent needed: hydrazine.
5. Take the resin from **step 4** and continue the synthesis: Boc-*Lys*-Gly-Lys-Gly-resin from **step 4**. Special amino acid needed: Boc-Lys(Fmoc)-OH.
6. Cleave the Aloc groups (*see* **Notes 6, 7**): Add the resin to 1 g (PPh₃)₄Pd in 10 mL chloroform:acetic acid:*N*-methyl morpholine (37:2:1, v/v/v) and shake for 5 h. Wash the resin 2x with 25 mL 0.5% diisopropyl ethyl amine in DMF, 2x with 25 mL of 0.5% sodium diethyl dithiocarbamate in DMF, and 2x with 25 mL DMF alone. Special reagents needed: tetrakis triphenylphosphine palladium; diethyl dithiocarbamate.
7. Take the resin from **step 6**, divide into two parts and continue the synthesis: Boc-Ser-Leu-Leu-Thr-Glu-Val-Glu-Thr-Pro-Ile-Arg-Asn-Glu-Trp-Gly-Ser-resin from **step 6**. This is where we build in the four M2 fragments simultaneously. Special reagent needed: Boc-Ser('Bu)-OH.
8. Cleave the construct overnight with a mixture (2.5 mL per 0.1 g of resin) of TFA:water:thioanisole (90:5:5, v/v/v). Precipitate the peptide in cold ether, centrifuge it, and filter the product.

3.2.3. Purification and Quality Control

Peptides are generally purified on reversed-phase HPLC using C18 derivatized columns and acetonitrile in water gradients in the presence of 0.1% TFA as ion-pairing reagent. This approach is applicable to the multiepitope and multimeric constructs as well, although we observed low resolution of our target first-generation M2e-S1-S2 construct and its serine-deleted analog *(19)*. Slow gradient development (5–65% acetonitrile over a period of 2–2.5 h) improves the separation between structurally closely related peptides. The calculated molecular weight of the (4)M2e-S1-S2 construct described here is 11,110 Da. In our experience, size exclusion chromatography prior to the reversed-phase purification step is unable to satisfactorily separate low molecular weight peptides from the large ones. However, dialysis in 3500 Da even in 10,000 Da cutoff tubing removes all small molecule contaminants and some of the deleted peptide sequences. Although the overall yield is decreased after dialysis, the loss is compensated for by the opportunity to load a higher amount of crude peptide to the reversed-phase HPLC column.

After purification, lyophilize the peptide constructs and check their purity on analytical HPLC and MALDI-TOF mass spectroscopy. The conventional matrix for peptide mass spectroscopy, α-cyano-4-hydroxycinnamic acid, appears to work properly for the vaccine constructs. While it is less frequently used in standard peptide chemistry (and thus not detailed here), the size of these multiple peptide antigens allows quality control by SDS-PAGE as shown in **Fig. 2**. For this purpose we recommend the use of 16.5% precast peptide gels. These gels tend to break easily, so be careful during gel handling.

4. Notes

1. Currently a large number of automated peptide synthesizers are available commercially. We use a batch synthesizer, which mixes the resin with the solvents by bubbling nitrogen through the reaction vessels. More intense mixing methods (vortexing or continuous flow) may offer higher coupling yields. Most recently microwave irradiation was introduced to accelerate the chemical reactions between the resin-bound peptides and the surrounding solvents *(20)*. The automated peptide synthesizer operating on this principle (Liberty, CEM Technologies) has an on-line resin cleavage capability.

2. Piperidine is listed as a controlled substance, and suppliers are required to obtain written information about the end-user and indented application for regulatory agencies.

3. HATU is the most powerful yet most expensive peptide coupling reagent on the market. It is preferentially used for the coupling of sterically hindered amino acids or during the synthesis of complex peptides *(21)* without the addition of HOBt. However, due to its high price, amino acids for which high incorporation

efficacy is expected, HATU can be replaced with TBTU/HObt or HBTU/HOBt combinations.

4. For economical reasons, we purchase bulk amino acids and mix them with the coupling reagents prior starting the synthesis. For our specific peptide synthesizer, Protein Technologies sells prepackaged Fmoc-protected amino acids mixed with coupling reagents, but these are limited to those with standard side chain protection (notably Boc for lysine).

5. Cleavage of Dde protecting group: Transfer the peptide resin from the synthesizer to a small flask fitted with a stopper. Add a solution of 2% hydrazine plus 1% allyl alcohol in DMF. The allyl alcohol is needed to avoid partial hydrogenation of the Aloc groups during Dde removal *(22)*. Place a stopper to the flask and after 3 min decant the hydrazine solution. Repeat this treatment twice, filter the resin, and wash with DMF. Hydrazine is a suspected carcinogen, so do the operations under the hood, and exercise caution.

6. The tetrakis triphenylphosphine palladium fails to cleave the Aloc groups after extended exposure to air. We recommend the use of the reagent from a freshly opened bottle.

7. The generally observed difficulties observed with palladium compounds prompted the development of novel deprotection strategies and selectively removable amino protecting groups *(23)*. Among the alternative allyl deprotection strategies, diisobutyl aluminum hydride (DIBAL) in the presence of dichlorobis (diphenylphosphino) propane nickel [*(dpppNiCl$_2$)*] in an aprotic solvent removes *N*-allyl groups in good yields *(24)*.

References

1. Jiang, S., and Debnath, A.K. (2000) Development of HIV entry inhibitors targeted to the coiled-coil regions of gp41. *Biochem. Biophys. Res. Commun.* **269**, 641–646.

2. Rojanasakul, Y., Luo, Q., Ye, J., Antonini, J., and Toledo, D. (2000) Cellular delivery of functional peptides to block cytokine gene expression. *J. Control. Rel.* **65**, 13–17.

3. Otvos, L. Jr. (2005) Antibacterial peptides and proteins with multiple cellular targets. *J. Pept. Sci.* **11**, 697–706.

4. Dietzschold, B., Gore, M., Marchandiar, D., et al. (1990) Structural and immuno-logical characterization of a linear virus-neutralizing epitope of the rabies virus glycoprotein and its possible use in a synthetic vaccine. *J. Virol.* **64**, 3804–3809.

5. Hsu, S.C., Chargelegue, D., Obeid, O.E., and Steward, M.W. (1999) Synergistic effect of immunization with a peptide cocktail inducing antibody, helper and cytotoxic T-cell responses on protection against respiratory syncytial virus. *J. Gen. Virol.* **80**, 1401–1405.

6. Meyer, D., and Torres, J.V. (1999) Hypervariable epitope construct: a synthetic immunogen that overcomes MHC restriction of antigen presentation. *Mol. Immunol.* **36**, 631–637.

7. Triozzi, P.L., Stoner, G.D., and Kaumaya, P.T. (1995) Subunit peptide cancer vaccines targeting activating mutations of the p21 ras proto-oncogene. *Biomed. Pept. Proteins Nucleic Acids* **1**, 185–192.

8. Fitzmaurice, C.J., Brown, L.E., Kronin, K., and Jackson, D.C. (2000) The geometry of synthetic peptide-based immunogens affects the efficiency of T cell stimulation by professional antigen-presenting cells. *Int. Immunol.* **12**, 527–535.

9. Cruz, L.J., Quintana, D., Iglesias, E., et al. (2000) Immunogenicity comparison of a multi-antigenic peptide bearing V3 sequences of the human immunodeficiency virus type 1 with TAB9 protein in mice. *J. Pept. Sci.* **6**, 217–224.

10. Tam, J. (1988) Synthetic peptide vaccine design: synthesis and properties of a high-density multiple antigenic peptide system. *Proc. Natl. Acad. Sci. USA* **85**, 5409–5413.

11. Wang, J., Kim, S., Kovacs, F., and Cross, T.A. (2001) Structure of the transmembrane region of the M2 protein H(+) channel. *Protein Sci.* **10**, 2241–2250.

12. Shih, F.F., Cerasoli, D.M., and Caton, A.J. (1997) A major T cell determinant from the influenza virus hemagglutinin (HA) can be a cryptic self peptide in HA transgenic mice. *Int. Immunol.* **9**, 249–261.

13. Kragol, G., Otvos, L., Jr., Feng, J., Gerhard, W., and Wade, J.D. (2001) Synthesis of a disulfide-linked octameric peptide construct carrying three different antigenic determinants. *Bioorg. Med. Chem. Lett.* **11**, 1417–1420.

14. Mozdzanowska, K., Feng, J., Eid, M., et al. (2003) Induction of influenza type A virus-specific resistance by immunization of mice with a synthetic multiple antigenic peptide vaccine that contains ectodomains of matrix protein 2. *Vaccine* **21**, 2616–2626.

15. Zhang, M., Zharikova, D., Mozdzanowska, K., Otvos, L., Jr., and Gerhard, W. (2006) Fine specificity and sequence of antibodies directed against the ectodomain of matrix protein 2 of influenza A virus. *Mol. Immunol.* **43**, 2195–2206.

16. Zharikova, D., Mozdzanowska, K., Feng, J., Zhang, M., and Gerhard, W. (2005) Influenza type A virus escape mutants emerge in vivo in the presence of antibodies to the ectodomain of matrix protein 2. *J. Virol.* **79**, 6644–6654.

17. Pawlak, M., Meseth, U., Dhanapal, B., Mutter, M., and Vogel, H. (1994) Template-assembled melittin: Structural and functional characterization of a designed, synthetic channel-forming protein. *Protein Sci.* **3**, 1788–1805.

18. Wade, J.D., Beford, J., Sheppard, R.C., and Tregear, G.W. (1991) DBU as an N-alpha-deprotecting reagent for the fluorenylmethoxycarbonyl group in continuous flow solid-phase peptide synthesis. *Pept. Res.* **4**, 194–199.

19. Kragol, G., and Otvos, L., Jr. (2001) Orthogonal solid-phase synthesis of tetramannosylated peptide constructs carrying three independent branched epitopes. *Tetrahedron* **57**, 957–966.

20. Palasek, S.A., Cox, Z.J., and Collins, J.M. (2007) Limiting racemization and aspartimide formation in microwave-enhanced Fmoc solid phase peptide synthesis. *J. Pept. Sci.* **13**, 143–148.

21. Angell, Y.M., Garcia-Echeverria, C., and Rich, D.H. (1994) Comparative studies of the coupling of N-methylated sterically hindered amino acids during solid-phase peptide synthesis. *Tetrahedron Lett.* **35**, 5891–5894.
22. Rohwedder, B., Mutti, Y., Dumy, P., and Mutter, M. (1998) Hydrazinolysis of Dde: Complete orthogonality with Aloc protecting groups. *Tetrahedron Lett.* **39**, 1175–1178.
23. Spivey, A.C., and Maddaford, A. (1999) Synthetic methods part (v): Protecting groups. *Annu Rep. Prog. Chem. Sect. B.* **95**, 83–95.
24. Taniguchi, T., and Ogasawara, K. Facile and specific nickel-catalyzed D-N-allylation. *Tetrahedron Lett.* **39**, 4679–4682.

16

Cysteine-Containing Fusion Tag for Site-Specific Conjugation of Therapeutic and Imaging Agents to Targeting Proteins

Marina V. Backer, Zoia Levashova, Richard Levenson, Francis G. Blankenberg, and Joseph M. Backer

Summary

Targeted delivery of therapeutic and imaging agents requires conjugation of a corresponding payload to a targeting peptide or protein. The ideal procedure should yield a uniform preparation of functionally active conjugates and be translatable for development of clinical products. We describe here our experience with site-specific protein modification via a novel cysteine-containing fusion tag (Cys-tag), which is a 15-amino-acid (aa) N-terminal fragment of human ribonuclease I with the R4C substitution. Several Cys-tagged proteins and peptides with different numbers of native cysteines were expressed and refolded into functionally active conformation, indicating that the tag does not interfere with the formation of internal disulfide bonds. We also describe standardized procedures for site-specific conjugation of very different payloads, such as functionalized lipids and liposomes, radionuclide chelators and radionuclides, fluorescent dyes, drug-derivatized dendrimers, scaffold proteins, biotin, and polyethyleneglycol to Cys-tagged peptides and proteins, as well as present examples of functional activity of targeted conjugates in vitro and in vivo. We expect that Cys-tag would provide new opportunities for development of targeted therapeutic and imaging agents for research and clinical use.

Key Words: Cysteine-containing tag; site-specific conjugation; targeted imaging; targeted drug delivery; liposomes; dendrimers.

From: *Methods in Molecular Biology, vol. 494: Peptide-Based Drug Design*
Edited by: L. Otvos, DOI: 10.1007/978-1-59745-419-3_16, © Humana Press, New York, NY

1. Introduction

1.1. Fusion Cys-tag for Site-Specific Modification of Targeting Proteins

Coupling therapeutic or imaging agents to a protein or a peptide for targeted "smart" delivery of a payload to a given site within the body is one of the fundamental goals of molecular medicine. There are a variety of tumor-specific antigens and receptors that can be assayed in vitro with recombinant proteins or peptides. However, the number of targeted drugs and contrast agents used for in vivo therapy or imaging is very small. This is due primarily to a lack of efficient technologies for coupling drugs and contrast agents to targeting proteins. Any technology based on random conjugation of "payloads" to carriers, usually to ε-amino groups of lysine residues, is bound to generate highly heterogeneous products with unknown distribution of functional characteristics (**Fig. 1A**). This is particularly true for growth factors or cytokines, whose lysine residues are not readily dispensable. There are several site-specific labeling approaches that are being developed to avoid these problems, including *(1)* insertion of a reactive cysteine in a rationally selected region, where this cysteine will not interfere with refolding or activity, *(2)* insertion of terminal reactive cysteine, where such interference is expected to be less significant, and *(3)* development of fusion tags for site-specific enzymatic modification (e.g., Avi-tag from Rosche Applied Sciences, SNAP-tag from Covalys Biosciences AG). However, the success of the cysteine insertion approach depends on the structure of a protein or peptide of interest, while fusion tags for enzymatic modification are not readily translatable to making clinical products.

We have recently developed a cysteine-containing fusion tag for site-specific conjugation (**Fig. 1B**, also in **ref. *1***). This tag, named Cys-tag, is an α-helical

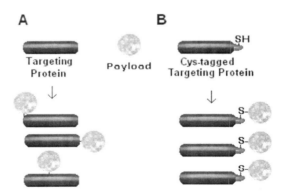

Fig. 1. Conjugation of the payload to a protein for targeted delivery. (A) Random conjugation results in a heterogeneous mixture of products. (B) Site-specific conjugation to Cys-tagged proteins leads to functionally active uniform products.

15-aa-long N-terminal fragment of human ribonuclease I, in which arginine in position 4 is substituted for cysteine. In our experience, Cys-tag is compatible with polypeptides varying in length from 53 to 358 aa (**Table 1**), and can be used for site-specific conjugation of very different payloads (**Table 2**) without affecting protein or peptide functionality. We therefore expect that Cys-tag would provide a standardized approach to coupling therapeutic and contrast agents to various bioactive polyamides and will be translatable into the development of clinical products.

We will briefly review accumulated experience in making and using Cys-tagged peptides and proteins, focusing on Cys-tagged human vascular endothelial growth factor (VEGF). We will discuss technical problems that might arise in using our platform for development of new conjugates for targeted delivery of therapeutic and diagnostic agents.

1.2. Cys-Tagged Proteins

Plasmids for bacterial expression of proteins with N- or C-terminal Cys-tag have been described previously (*1*). They were made by inserting the coding sequence for Cys-tag (Lys-Glu-Ser-Cys-Ala-Lys-Lys-Phe-Gln-Arg-Gln-His-Met-Asp-Ser) and a Ser-Gly linker in the appropriate sites in the commercially available pET29 plasmid for bacterial expression of proteins under control of T7 promoter (Novagen, WI). So far, expression of Cys-tagged proteins in insect, yeast, or mammalian hosts has not been tested.

Table 1
Peptide and Proteins Expressed with Cys-tag

Proteins and peptide	Total number of amino acids	Number of native cysteines	Ref.
VEGF[a]	110–222	16–18	(*1–7,15*)
EGF[b]	53	6	(*6*)
EGF-SLT[c]	358	8	Unpublished
Soluble FLT3 ligand (FLex)	154	6	Unpublished
Annexin V	330	1	Unpublished
Catalytically inactive fragment of anthrax lethal factor (LFn)	254	0	(*5,7*)

[a]Several human VEGF$_{121}$-based Cys-tagged proteins were constructed and tested, including VEGF$_{121}$ dimer, truncated VEGF$_{110}$ dimer, and single-chain (sc) VEGF comprising two 3-110 fragments of VEGF$_{121}$ fused head-to-tail.

[b]Cys-tag was fused to human EGF via (G$_4$S)$_3$ linker.

[c]EGF-SLT comprises Cys-tagged EGF fused to catalytic subunit of *E. coli* Shiga-like toxin.

Table 2
Payloads Conjugated to Cys-Tagged Proteins

Payloads	Applications	Ref.
Lipid/Liposome	Drug delivery	(3,5,7)
Dendrimers	Drug delivery	(1)
DOTA	Targeted PET imaging	(7)
PEG-DOTA	Targeted PET imaging	(7,15)
	Targeted SPECT imaging	Unpublished
	Targeted radiotherapeutics	Unpublished
HYNIC	Targeted SPECT imaging	(2,7)
"Self-chelation" of ^{99m}Tc	Targeted SPECT imaging	(15)
PEG	Changing protein's pharmacokinetics and immunogenicity	(7)
Fluorescent dyes	Targeted imaging in whole animals	(1,5,7)
	Tagging cells with active and accessible receptors in vivo	(7)
	Receptor-mediated endocytosis in vitro	(1,6,7)
Biotin	Targeted imaging with quantum dots and ultrasound microbubbles	Unpublished
Adapter protein	Flexibility in constructing conjugates	(5)
Fibronectin	Tissue culture scaffolds	(6)

To date, all Cys-tagged proteins and peptides were expressed in *Escherichia coli* BL21(DE3) strain. As usual, growth temperature, bacterial density for protein induction, concentration of inducer (IPTG), and duration of induction should be optimized for every protein. Almost all our Cys-tagged proteins were found in inclusion bodies. In our experience, recovery of proteins from inclusion bodies might be greatly improved by a combination of sequential washings and mild sonication prior to solubilization of inclusion bodies.

Expression of proteins with Cys-tag adds an additional cysteine to the set of native cysteines, and therefore it might complicate the formation of native disulfide bonds. In general, protein refolding from a denatured state is achieved by slow removal of denaturing components, while maintaining Red-Ox environment using a mixture of reduced and oxidized forms of glutathione. While peptide is denatured, each cysteine can form a reversible disulfide bond either with other cysteine or with glutathione. Over time, as denaturing components are slowly removed, this process selects stable cysteine–cysteine disulfide bonds that correspond to thermodynamically favored properly folded peptide conformations. Importantly, since cysteine in Cys-tag does not have a "natural" partner, successful refolding leaves the C4 thiol group in a mixed disulfide with glutathione. The key point in refolding is to maintain a reasonable pace of disulfide bond shuffling, so that the native set of disulfide bonds is established before denaturing components are completely removed and protein conformational flexibility is greatly decreased. It should be also noted that recently EMD Biosciences/Novagen (La Jolla, CA) introduced iFOLD™ Protein Refolding System that allows rapid optimization of refolding conditions.

Examples of optimized protocols for recovery from inclusion bodies and refolding of Cys-tagged proteins are presented in **Subheadings 3.1** and **3.2**, respectively. The protocols were worked out for recovery and refolding of single-chain (sc) VEGF that combines two 3- to 112-aa fragments of human $VEGF_{121}$ cloned "head-to-tail" and is fused to a single N-terminal Cys-tag *(4,6)*.

1.3. Site-Specific Conjugation to Cys-Tagged Proteins

After refolding of Cys-tagged proteins in Red-Ox buffer, the thiol group in Cys-tag is not available for conjugation because it is "protected" by glutathione in a form of a mixed disulfide. Deprotection involves a mild dithiothreitol (DTT) treatment (*see* **Subheading 3.3.**) that can be followed by thiol-directed modifications without removing residual DTT from the solution. The critical point in DTT treatment is to obtain protein with a free thiol group in Cys-tag without affecting native disulfide bonds.

Thiol-reactive derivatives of all major payloads, such as maleimide derivatives of fluorescent dyes, PEGs, lipids, biotin, and chelators, are available from commercial vendors. Maleimide derivatives form stable thioester bond with thiols. Alternatively, the thiol group in Cys-tagged protein can be activated with thiol-disulfide exchange reagent DPDS (2,2'-dipyridyl disulfide) and then reacted with thiol-containing payloads. Finally, bifunctional cross-linking agents with one functionality for reaction with a thiol group and another for reaction with another group (e.g., amino, carboxy) can be used for designing more complex constructs, such as dendrimers, nanoparticles, etc. Reaction conditions should be optimized for every specific conjugate with a general caveat that in order to decrease the risk of nonspecific conjugation it is better to use low ratios of payload to protein and the shortest possible incubation time at room temperature. Purification schemes for conjugates are as different as conjugates themselves and should be optimized for a specific purpose.

Site-specific conjugation to Cys-tag can be validated using either enzymatic digestion with Asp-N endopeptidase or CNBr cleavage of the conjugate. Both methods result in the cleavage of Cys-tag between Met-13 and Asp-14 and the release of a labeled N-terminal fragment. Protein digest is separated by RP-HPLC, and all labeled fragments are isolated and subjected to N-terminal sequencing or to mass-spec analysis *(6)*. If conjugation is site-specific, only N-terminal fragments contain payload.

Functional validation of conjugates requires appropriate tissue culture or biochemical assay, where activity of the conjugate can be measured against parental unmodified protein.

1.4. The Use of Cys-Tagged Proteins—Specific Examples

1.4.1. Making Targeted Liposomes

Encapsulation of drugs into liposomes protects drugs from degradation in bodily fluids, delays their clearance, increases the amount of drugs delivered into cells via a single act of endocytosis, and decreases systemic toxicity (8). For example, encapsulation of doxorubicin, one of the most effective chemotherapeutic drugs, within sterically stabilized liposomes improves its stability while decreasing systemic toxicity (9). To enhance selectivity of drug-loaded liposomes, several groups decorated liposomes with peptides that recognize unique cell surface markers on targeted cells. Both in vitro and in vivo studies demonstrated that significant increase in efficacy of liposome-encapsulated drugs is achieved by molecular targeting to specific tumor biomarkers (10,11).

In order to decorate liposomes with proteins, a targeting peptide or protein is derivatized with a synthetic phospholipid that carries a chemically active group, usually N-hydroxysuccinimide, for conjugation to readily available ε-amino groups of lysine residues. At the next step, lipidated protein is either mixed with lipids at the stage of liposome preparation or inserted into preformed liposomes. The functional activity of proteins associated with liposomal membrane depends on their ability to tolerate lipidation and liposome coupling. In turn, efficacy of targeted liposomes depends on the proportion of functionally active proteins on their surface. As we show here, site-specifically lipidated scVEGF inserted into commercially available doxorubicin-loaded liposomes (Doxil®) retains the ability to interact with its receptor VEGFR-2 and efficiently delivers doxorubicin in cells overexpressing this receptor (**Fig. 2**). To date, we have lipidated several Cys-tagged proteins and used them to construct targeted liposomes (3,7). In this work we used a protocol that was optimized for lipidation of scVEGF and construction of scVEGF-driven liposomes (see **Subheadings 3.3.** and **3.4.**).

1.4.2. Making Targeted Radiolabeled Tracers

Radionuclide chelators, such as DOTA (1,4,7,10-tetraazacyclododecane-1,4,7,10-tetraacetic acid) or HYNIC (6-hydrazinopyridine-3-carboxylic acid), can be conjugated to Cys-tag and then "loaded" with radionuclides at the point of use (2,6). Furthermore, chelators can be conjugated via a PEGylated linker, which might improve the tracer's biodistribution (6). Interestingly, Cys-tag apparently has "self-chelating" activity that was described recently for a

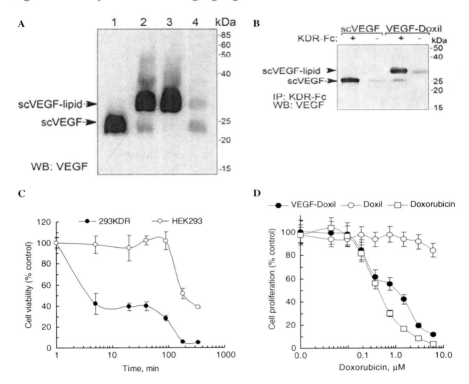

Fig. 2. Using Cys-tagged scVEGF for targeting Doxil®to cells expressing VEGF receptors. (**A**) Western blot analysis of scVEGF-decorated Doxil® (VEGF-Doxil). Lanes: 1, Unmodified scVEGF; 2, lipidation reaction mixture; 3, VEGF-Doxil collected from Sepharose CL-4B column immediately after column void volume; 4, free scVEGF and scVEGF-lipid conjugate eluted from column (pooled fractions). Samples were loaded on a 17.5% gel, separated by SDS-PAGE and analyzed by Western blotting with anti-VEGF antibody. (**B**) VEGF-Doxil or free scVEGF, at equal concentrations of 1 nM, were collected from solution with Protein A Sepharose beads with or without preadsorbed soluble VEGF receptor KDR-Fc and analyzed by Western blotting with anti-VEGF antibody. (**C**) 293/KDR cells expressing ~2.5 × 10⁶ VEGF receptor VEGFR-2 per cell, and HEK293 control cells without VEGF receptors were plated on 96-well plates, 1000 cells/well, 20 h before the experiment. Cells in triplicate wells were exposed to VEGF-Doxil at a final doxorubicin concentration of 1 µM for times indicated, then shifted to fresh culture medium and quantitated 96 h postexposure by an MTT-based assay (Promega). (**D**) VEGFR-2 expressing 293/KDR cells were plated on 96-well plates, 1000 cells/well. Twenty h later cells in triplicate wells were exposed to liposomal or free doxorubicin for 5 min, then shifted to fresh culture medium and quantitated 96 h postexposure as above.

mutant annexin V *(14)*. Thus, Cys-tagged peptide can be labeled with radionuclides in several different ways in order to optimize targeting, clearance, and
biodistribution of a specific tracer. As an example, we have recently compared
three different protocols for radiolabeling Cys-tagged scVEGF with ^{99m}Tc *(15)*
and found that biodistribution of radiolabeled tracer is affected by the choice of a
chelator (**Fig. 3**). For this protein the most advantageous biodistribution with the
lowest nonspecific liver and kidney uptake was achieved with DOTA chelator
linked to Cys-tag via a 3.4 kDa PEGylated linker.

Optimized protocols for radiolabeling scVEGF with ^{99m}Tc for SPECT
imaging and ^{64}Cu for PET imaging are described in **Subheadings 3.5.–3.7.** and
3.8., respectively.

1.4.3. Site-Specifically Biotinylated Cys-Tagged Proteins

Biotin-maleimide (available from Sigma) has been conjugated to deprotected
Cys-tagged proteins using a combination of methods described in **Subheading
3.3.** (C4 deprotection and modification) and **Subheading 3.5.** (separation

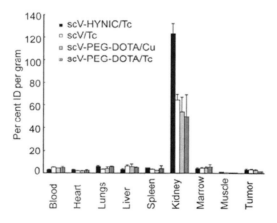

Fig. 3. Biodistribution of scVEGF-based radiotracer in 4T1 tumor-bearing mice
is affected by the nature of chelating functionality. Cys-tag in scVEGF was radiolabeled either directly or via Cys-tag conjugated chelators HYNIC or PEG-DOTA, using
methods described in **Subheadings 3.5–3.8.** The resulting radiotracers, named scV/Tc,
scV-HYNIC/Tc, scV-PEG-DOTA/Tc, and scV/PEG-DOTA/Cu, were injected via the tail
vein into 4T1-tumor bearing mice (*n* = 5) at doses of 80 μCi of Tc^{99m} or 5 μCi of Cu^{64}
per mouse. Mice were sacrificed 1 h after injection, blood and organs were harvested,
weighed and counted in a gamma counter. Data are presented as percent of injected dose
(ID) per gram, an average per group, ± SD.

of unreacted biotin by gel-filtration on PD-10). Site-specifically biotinylated bioactive polyamides can be coupled to streptavidin-coated nanoparticles, such as CdSe/ZnS core/shell quantum dots of different size and emission wavelength (available from Quantum Dot, Hayward, CA) for optical imaging or to biotin-coated microbubbles (available from Targeson, Charlottesville, VA) for ultrasound imaging.

1.4.4. Making Targeted Fluorescent Tracers

Whole animal imaging with near-infrared fluorescent tracers targeted to specific receptors is a new and exciting area of basic and translational research *(12,13)*. Such tracers can be also used for in vivo tagging cells via receptor-mediated endocytosis followed by fluorescent microscopy on histological sections prepared from harvested tissues *(6)*. Importantly, such tagging provides information on cells with accessible and active receptors as opposed to immuno-histochemical staining that visualizes the whole pool of receptors, regardless of their activity or accessibility. There are also translational efforts to develop targeted near-infrared fluorescent tracers and the corresponding hardware for various diagnostic applications.

Fluorescent dyes are relatively large molecules, whose conjugation to targeting proteins can affect interactions with cognate receptors, especially for small peptide ligands. For example, random conjugation of even a single fluorescent dye Cy5.5 to VEGF dramatically inhibits its activity *(5)*. Thus, site-specific conjugation of fluorescent dyes provides an efficient way to generate functionally active fluorescent tracers.

Currently, maleimide derivatives of many dyes are commercially available. In our experience, incubation of deprotected (as described in **Subheading 3.3.**) scVEGF with 1.5- to 2.5-fold molar excess of dye-maleimide for 30 min in 0.1 M Tris-HCl pH 8.0 at room temperature yields 30–60% of conjugates carrying one dye molecule per one protein. Free dye is then readily removed by gel filtration on PD-10 as described in **Subheading 3.5.** However, removal of unmodified and excessively modified protein (two or more molecules of dye per protein) requires additional purification, which is conjugate-specific. For scVEGF-based conjugates with highly charged Cy5.5 and Cy7 dyes, we have successfully used anion-exchange choromatography on Q-Sepharose columns to separate unmodified VEGF, 1:1 conjugate, and overmodified protein. Example of in vivo imaging of mouse tumor vasculature with a combination of functionally active scVEGF site-specifically labeled with AlexaFluor-594-maleimide (Invitrogen) and inactivated scVEGF/Cy (SibTech, Inc.) to evaluate nonspecific tracer uptake is shown in **Fig. 4A.**

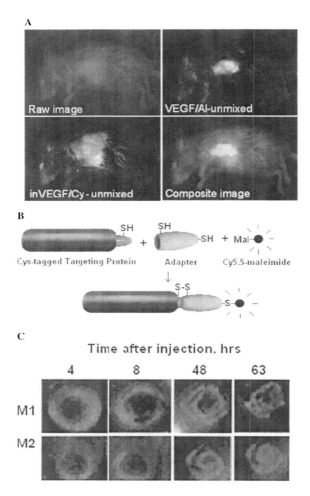

Fig. 4. In vivo fluorescent imaging of angiogenic tumor vasculature with scVEGF-based fluorescent tracers. (**A**) Balb/c mice bearing orthotopic 4T1 tumors were injected via the tail vein with an equimolar mixture of functionally active scVEGF/AlexaFluor-594 (scVEGF/Al) and inactivated inVEGF/Cy5.5 (inVEGF/Cy) fluorescent conjugates, total of 20 µg per mouse. VEGF inactivated by excessive biotinylation does not bind to VEGF receptors (*6*). Mice were anesthetized 5 min postinjection with ketamine (120 mg/kg) and xylazine (8 mg/kg) mixture. Tumor images were obtained using Maestro Imaging Station (CRI, Boston, MA) and processed with CRI software to obtain unmixed and composite images. (**B**) Conjugation of adapter protein to Cys-tag. Adapter can be modified with Cy5.5-maleimide before or after conjugation to Cys-tagged targeting protein. (**C**) Balb/c mice bearing subcutaneous 4T1 tumors were injected via the tail vein with VEGF$_{121}$/Adapter/Cy fluorescent conjugate, 10 µg per mouse. Mice were anesthetized at indicated times postinjection with ketamine (120 mg/kg) and xylazine (8 mg/kg) mixture. Tumor images were obtained on Kodak Image Station 2000 at indicated time after tracer injection. M1, mouse 1; M2, mouse 2.

1.4.5. Protein–Protein Conjugation

A special case is site-specific conjugation of bioactive Cys-tagged polyamides to other peptides or proteins. In some cases, conjugating either bulky or highly charged payloads directly to Cys-tag might interfere with the functional activity of the resulting conjugates. To minimize this interference, payloads can be linked to a recently developed adapter protein, which binds to and forms a disulfide bond with Cys-tag, a so-called "dock-and-lock" system *(5)*. The adapter is a V118C, N88C double mutant C-terminal fragment of human ribonuclease I that naturally forms complexes with Cys-tag (an R4C mutant of the N-terminal fragment of human ribonuclease I), whereby complimentary C4 and C118 spontaneously form a disulfide bond **(Fig. 4B)**. We have recently derivatized this adapter with Cy5.5-maleimide, conjugated it to Cys-tagged VEGF$_{121}$, and then tested the resulting "dock-and-lock" conjugate for tumor imaging in mice with subcutaneous tumors. The conjugates selectively accumulated in the tumor area, and fluorescent dye remained internalized for at least several days, allowing noninvasive monitoring of tumor vasculature remodeling **(Fig. 4C)**.

A different class is conjugation to polymers that can be used as scaffolds for cell growth and/or tissue engineering. It might be beneficial to use functionally active growth factors or a combination of different growth factors for uniform derivatization of collagen, fibronectin, or other supporting proteins or polymers. Recent experiments, in which scVEGF was conjugated to fibronectin and then VEGF-fibronectin-coated tissue culture plates were used for cell growth, established the feasibility of such strategy *(4)*.

2. Materials

2.1. Protein Expression and Purification

1. Concentrated PBS (10X), 0.5 M ethylendiamine tetraacetic acid (EDTA), 1 M Tris-HCl pH 8.0, 5 M NaCl, 3 M NaAc pH 5.2, kanamycin and 1 M isopropyl-β-D-thiogalactopyranoside (IPTG) from Gibco/BRL (Bethesda, MD).
2. LB Broth base bacterial growth medium (Invitrogen, Carlsbad, CA).
3. Na_2SO_3, $Na_2S_4O_6$, Triton X-100, and DTT from Sigma (St. Louis, MO).
4. DTT is dissolved in sterile Milli-Q water at 1 M, stored in small aliquots at −20°C and added to solubilized inclusion bodies as required.
5. Solution of Tris(2-carboxyethyl) phosphine hydrochloride (TCEP) (Pierce, Rockford, IL) in Milli-Q water at 1 M is made and stored as described above for DTT.
6. Urea (Fisher Scientific) is dissolved in DI water at 8 M, sterilized by filtration through 0.45-μm filter and stored at room temperature.
7. High-salt wash buffer: 0.1 M Tris-HCl pH 8.0, 0.5 M NaCl.

8. TES buffer: 0.1 *M* Tris-HCl pH 8.0, 1 m*M* EDTA, 50 m*M* NaCl.
9. TES-T buffer: 0.1 *M* Tris-HCl pH 8.0, 1 m*M* EDTA, 50 m*M* NaCl, 2.5% (v/v) Triton X-100.
10. Solubilization buffer: 7.6 *M* urea, 20 m*M* Tris-HCl pH 8.0, 0.15 *M* NaCl .
11. Sulfonating buffer: Solubilization buffer containing Na_2SO_3 at 17 mg/mL and $Na_2S_4O_6$ at 5.3 mg/mL .
12. Sterilize buffers 7–11 by filtration through 0.45-μm filter.

2.1.1. Refolding of Cys-Tagged Proteins

1. Arginine (Sigma), reduced glutathione (GSH), and oxidized glutathione (GSSG) (Novagen, La Jolla, CA).
2. Refolding buffer: 20 m*M* Tris-HCl pH 8.0, 2 *M* urea, 0.5 *M* arginine, 1 m*M* GSH, 0.4 m*M* GSSG.
3. No salt basic dialysis buffer: 20 m*M* This-HCl pH 8.0.
4. No-salt acidic dialysis buffer: 20 m*M* NaAc pH 5.2. Prior to use, incubate refolding and both dialysis buffers at 4°C for 2–3 h with constant stirring.

2.2. Site-Specific Modifications of Cys-Tagged Proteins

2.2.1. Lipidation for Insertion of Targeting Proteins into Liposomes

1. *N*-(1-Pyrene)-maleimide (Molecular Probes, Eugene, OR). Dissolve *N*-(1-pyrene)-maleimide in dimethylsulfoxide (DMSO) at 10 mg/mL and use immediately for reaction.
2. HPLC column C4 MACROSPHERE 300 (150 × 4.6 mm) from Alltech (Deerfield, IL).
3. Poly(ethylenglycol)-α-distearoyl phosphatidylethanolamine,-ω-maleimide FW 3,400 (mPEG-DSPE-maleimide, from Shearwater Polymers, Huntsville, AL). Dissolve mPEG-DSPE-maleimide in DMSO at 10 mg/mL immediately before reaction.
4. Sepharose CL-4B (Sigma).
5. Doxil® (doxorubicin HCl liposome injection, 2 mg/mL) from Ortho Biotech.
6. HEPES buffer solution (1 *M*) of pH 7.2 (Invitrogen).
7. Running buffer: 10 m*M* HEPES pH 7.2; 0.15 *M* NaCl, 0.1 m*M* EDTA. Sterilize running buffer by filtration through 45 m*M* filter.
8. HPLC column Vydac Diphenyl 219TP5415 (250 × 4.6 mm) (Vydac, Hesperia, CA).

2.2.2. Radiolabeling for PET and SPEC Imaging

1. scVEGF (MW 28 kDa) protein, scVEGF-HYNIC and scVEGF-PEG-DOTA conjugates from SibTech (Newington, CT).
2. Deoxygenate 0.9% NaCl (plastic vials, from Abbott Laboratories, Abbott Park, IL) by purging nitrogen by cannula for 60 min before use.

3. PBS pH 7.4: dilute 10X PBS concentrate (Gibco/BRL) with Milli-Q deoxygenated water (*see* **Note 1**).
4. Tricine buffer (114 m*M*): dissolve 2 g *N*-[tris(hydroxymethyl)methyl]glycine (Tricine, from Sigma) in 100 mL Milli-Q deoxygenated water (*see* **Note 2**).
5. SnCl$_2$ solution: dissolve SnCl$_2$·2H$_2$O (Sigma) in 0.1 *N* HCl at 50 mg/mL.
6. Tin-Tricine reagent: dissolve 2.0 g Tricine in 97 mL of deoxygenated Milli-Q water, adjust pH to 7.1 using approximately 1.3 mL of 1 *N* NaOH. Seal the flask with a cannulated airtight septum and purge with nitrogen for 60 min. Add 1.2 mL of SnCl$_2$ solution, mix and transfer to nitrogen-filled septum-capped vials, 1 mL in each, with a final composition of 20 mg tricine and 0.6 mg of SnCl$_2$ at pH 7.1 (*see* **Note 3**).
7. Radionuclides: ^{99m}TcO4 (5—10 mCi in a volume of approx 50 mL, GE Healthcare, Sunnyvale, CA) and ^{64}Cu (5–10 mCi in 2–10 μL from Washington University Medical School, St. Louis, MO).
8. PD-10 columns (GE Healthcare).
9. Indicator strips pH 4.0–7.0 (Merck, West Point, PA).
10. NaOOCCH$_3$ buffer (1 *M*) pH 6.0: Dilute 3 M NaAc, pH 5.5 (Sigma) with Milli-Q water and adjust pH to 6.0 with 0.1 *N* NaOH.
11. NaOOCCH$_3$ buffer (0.1 *M*) pH 5.5: Dilute 3 *M* NaAc, pH 5.5 with Milli-Q water.

3. Methods

3.1. Recovery of scVEGF from Inclusion Bodies

1. To obtain 1 L of induced bacterial culture, grow *E. coli* BL21(DE3) bacteria transfected with the pET/C4(G4S)/scVEGF plasmid in four 1-L flasks, each flask containing 250 mL LB medium supplemented with 30 mg/mL of kanamycin. Maintain the temperature at 37°C, shaking rate may vary from 220 to 300 rpm.
2. Once bacteria have reached an optical density of 0.4–0.6 optical units at 600 nm, induce scVEGF expression by adding 0.25 mL of 1 *M* IPTG to each flask. Continue incubating at 37°C with 300 rpm shaking for 2 h.
3. Harvest 1 L of induced bacterial culture by centrifugation at 4400 g for 30 min at 4°C (Beckman J2-21 centrifuge, Beckman Coulter, Fullerton, CA). Carefully remove and discard the supernatant. Resuspend bacterial pellet in 1X PBS using a 10-mL disposable plastic pipet. Pipet the solution up and down until no visible pellet is left in the suspension. Adjust the final volume to 25 mL with 1X PBS.
4. Pass bacterial suspension twice through high-pressure homogenizer (EmulsiFlex-C5, Avestin, Canada) at 10,000–15,000 psi.
5. Split homogenized bacteria in two 50-mL sterile centrifuge tubes and centrifuge at 12,000 g for 15 min at 4°C to separate the soluble part of bacterial lysate and inclusion body fraction. Discard the supernatants, trying not to disturb the pellet (inclusion bodies).
6. Resuspend each pellet in 2 mL of ice-cold high-salt wash buffer by pipetting up and down several times. Once the pellet is resuspended, adjust solution volume to 30 mL in each tube, using the same ice-cold buffer.

7. Sonicate briefly (20–30 s) at 40–50% of output power (Virsonic-475 sonicator VirTis, NY). This step is essential for complete homogenization of the pellet, which increases the recovery of recombinant protein (*see* **Note 4**).

8. Centrifuge at 23,500 g for 15 min at 4°C. Discard the supernatants.

9. Repeat resuspension, sonication, and centrifugation (**steps 6–8**) using ice-cold TES-T buffer. At this step, inclusion bodies are purified from detergent-soluble material.

10. Repeat resuspension and centrifugation (**steps 6–8**) using ice-cold TES buffer to wash out TritonX-100.

11. Add 5 mL of solubilization buffer to each pellet and incubate at room temperature for 10–15 min to soften inclusion bodies. Try to break the pellet apart with sterile plastic pestle or by passing it through a glass homogenizer. Some insoluble material can still remain in the solution after this step.

12. Incubate partially solubilized inclusion bodies at 4°C with constant agitation for 60 min; then combine the solutions in one 175-mL disposable centrifuge tube. Adjust the volume to 30 mL with sonication buffer and sonicate for 60 s at maximum output power (*see* **Notes 4, 5**).

13. Add 45 mL of sulfonation buffer. Final volume is now 75 mL; final concentrations of sulfonating agents are: 10 mg/mL for Na_2SO_3 and 3.2 mg/mL for $Na_2S_4O_6$.

14. Add 0.75 mL of 1 *M* DTT to a final concentration of 10 m*M*. Place tube with solution at 4°C on a rocking platform and incubate for 18–22 h with constant rocking. Usually, all residual pieces of inclusion body pellet are completely dissolved during 1–2 h of incubation.

15. Add 0.375 mL of 1 *M* TCEP to a final concentration of 2.5 m*M*. Continue incubation at 4°C for 16–18 h. After this incubation, scVEGF is ready for refolding.

3.2. Refolding of scVEGF

1. Transfer solubilized and completely reduced inclusion bodies into a dialysis bag with molecular weight cut-off pore size of 3500 and dialyze it against 10 volumes of refolding buffer for 24–36 h at 4°C. Once dialysis bag with protein is placed in dialysis beaker, stirring should be stopped to slow down the dialysis rate at this stage.

2. Transfer dialysis bag in a beaker containing 50 volumes of no-salt basic dialysis buffer and continue dialysis for 24–48 h at 4°C with constant slow stirring.

3. Transfer dialysis bag in a beaker containing 50–100 volumes of no-salt acidic dialysis buffer and continue dialysis for 12–16 h at 4°C with constant slow stirring (*see* **Note 6**).

4. After dialysis is complete, carefully transfer protein solution from dialysis bag into three sterile 50-mL centrifuge tubes. Centrifuge at 34,700 g for 30 min at 4°C.

5. Pool the supernatants together and pass the combined solution through 45-mm sterile filter. This protein solution is ready for column purification. scVEGF is

further purified by ion-exchange chromatography on SP-Sepharose Fast-Flow (1-mL prepacked columns from GE Healthcare), according to the manufacturer's instructions.

3.3. Deprotection and Site-Specific Lipidation of scVEGF

1. To "deprotect" C4 thiol group and make it available for SH-directed modification, incubate 0.1 mM scVEGF with 0.1 mM DTT in a buffer containing 0.1 M Tris-HCl pH 8.0 for 30 min at room temperature.
2. Optional: availability of free cysteines can be assayed by reacting with an SH-directed N-(1-pyrene)-maleimide. Add N-(1-pyrene)-maleimide to deprotection reaction mixture to make a molar protein-to-pyrene ratio of 1:2 and incubate for 40 min at room temperature. Load on RP HPLC C4 Alltech MACROSPHERE 300 5-mm column and elute at 0.75 mL/min with 0.1% TFA (v/v) and a linear gradient of acetonitrile (5–50% over 15 min) with detection at 216 nm for protein and 340 nm for pyrene. Calculate the extent of pyrene modification using a ratio of integral peak intensities at 216 nm and 340 nm.
3. To make scVEGF-lipid conjugate, add mPEG-DSPE-maleimide directly to deprotection reaction to a final molar protein-to-lipid ratio of 1:2. Residual DTT will not affect modification reaction. Incubate lipidation reaction mixture at room temperature for 1 h.
4. After incubation, this mixture can be used immediately for insertion into liposomes, or stored in liquid nitrogen or at -70°C in working aliquots for several weeks.

3.4. Insertion of scVEGF–Lipid Conjugate into Preformed Liposomes

1. For insertion into premade doxorubicin-loaded liposomes (Doxil), mix equal volumes of liposomes and lipidation reaction mixture and incubate at 37°C for 12–16 h (*see* **Note 7**). Purification of unreacted lipid and protein is not necessary at this step, because they do not interfere with insertion process and will be removed eventually by gel-filtration of decorated liposomes on Sepharose CL-4B.
2. Equilibrate Sepharose CL-4B column with 5 column volumes of running buffer.
3. After insertion of lipidated protein is complete (*see* **Note 7**), load liposomes on an equilibrated Sepharose CL-4B column. Loaded sample will be separated at the very beginning of gel filtration into two doxorubicin-containing (red-color) peaks. Collect fast-migrating red-colored peak immediately after collecting void column volume. This peak contains liposomes. Place purified liposomes on ice or to 2–8°C for storage, do not freeze. Continue collecting elution fractions until second red-colored peak (free doxorubicin leaked out from Doxil) is eluted from the column. Monitor optical density in eluting material at 280 nm to identify and collect free protein and excess lipid for further analysis (**Fig. 3A**).
4. Analyze the efficiency of protein insertion by RP HPLC on Vydac Diphenyl column with elution at 0.75 mL/min with 50 mM thriethylamine phosphate pH 2.8 and 20% tetrahydrofuran with a linear gradient of acetonitrile (0–70% v/v over 25 min).

5. Alternatively, use Western blot analysis with a specific antibody to analyze and quantify protein inserted into liposomes.
6. This insertion procedure usually results in concentration of liposome-associated Cys-tagged protein in a micromolar range (2–10 μM).

3.5. Deprotection and Direct Radiolabeling of scVEGF with ^{99m}Tc via Cys-tag

1. To deprotect C4 thiol groups and make them available for modification, incubate scVEGF with equimolar DTT as follows: mix 0.1 mL PBS, 50 μg of scVEGF, and 1.8 μL of 1 mM DTT in a 1.5-mL Eppendorf tube and incubate it for 20 min at 25°C.
2. During this incubation, mix 0.25 mL PBS, add 10–15 mCi ^{99m}TcO$_4$ in another glass tube, cover it with parafilm, and purge the mixture with nitrogen for 15 min.
3. Add the entire scVEGF deprotection reaction mixture to the tube with deoxygenated ^{99m}TcO$_4$. There is no need to remove residual DTT, because it will not interfere with direct radiolabeling of scVEGF with ^{99m}Tc.
4. Purge the mixture with nitrogen shortly, add 20 μL of Tin-Tricine reagent, seal the tube with parafilm and incubate it at 37°C for 60 min.
5. Equilibrate PD-10 desalting column with 5 column volumes of PBS.
6. When incubation of scVEGF with ^{99m}TcO$_4$ is over, remove the unreacted technetium by gel filtration on PD-10 column. Load the reaction mixture on equilibrated PD-10 column and elute with PBS. After passing a bed volume, which is 2.4 mL for PD-10, collect 0.5-mL fractions in 1.5-mL microcentrifuge tubes. scVEGF is eluted in fractions 2 through 4–5. Combine fractions with highest activities together. This procedure usually results in incorporation of 100–200 μCi of ^{99m}Tc per μg of protein.
7. ^{99m}Tc-labeled scVEGF should be used for further analysis, tissue culture, or animal experiments immediately. For example, for bio-distribution studies in a mouse model, inject 40–100 μCi of ^{99m}Tc –labeled scVEGF via the tail vein.

3.6. Radiolabeling of scVEGF-HYNIC Conjugate with ^{99m}Tc

1. Add 5 mCi ^{99m}TcO$_4$ to a glass 10-mL tube with 0.3 mL Tricine buffer, cover with parafilm, and purge with nitrogen for 15 min.
2. Add 50 μg of scVEGF-HYNIC conjugate and 10 μL of Tin-Tricine reagent. Purge with nitrogen shortly and seal with parafilm.
3. For loading scVEGF-HYNIC with ^{99m}Tc, incubate the reaction mixture for 60 min at 25°C.
4. Remove free technetium by gel-filtration on PD-10 column as described in **Subheading 3.5.**
5. This procedure usually results in incorporation of 50–100 μCi of ^{99m}Tc per μg of protein.

3.7. Radiolabeling of scVEGF-PEG-DOTA with ^{99m}Tc

1. Add 10 mCi $^{99m}TcO_4$ to a glass tube with 1 mL PBS, cover with parafilm, and purge with nitrogen gas for 15 min.
2. Add 20 μg scVEGF-PEG-DOTA conjugate and 25 μL of Tin-Tricine reagent. Purge the mixture with nitrogen shortly and seal with parafilm.
3. Incubate for 60 min at 37°C
4. Remove free technetium by passing through PD-10 column as described in **Subheading 3.5.**
5. This procedure usually results in incorporation of 100–200 μCi of ^{99m}Tc per μg of protein.

3.8. Radiolabeling of scVEGF/PEG-DOTA with ^{64}Cu

1. Transfer 0.5–1 mCi ^{64}Cu to 1.5-ml Eppendorf tube and adjust pH to ~5.5 with 10–15 μL 0.1 M NaOOCCH$_3$, pH 5.8–6.0. Check pH by spotting ~0.3 μL on indicator paper. Important note: do not use alkaline solution for pH adjustment as Cu ions will form insoluble hydroxides.
2. Add 50 μg scVEGF-PEG-DOTA conjugate, check its pH again, it must be within 5.0–5.5.
3. Incubate for 60 min at 55°C. Add EDTA to a 1 mM final concentration to chelate free ^{64}Cu.
4. Equilibrate PD-10 column with 5 column volumes of 0.1 M NaOOCCH$_3$ pH 5.5.
5. For removal of unreacted ^{64}Cu, load the reaction mixture on equilibrated PD-10 column and elute with 0.1 M NaOOCCH$_3$. Collect protein-containing fractions as described in **Subheading 3.5.**
6. This procedure usually results in incorporation of 0.5–5 μCi of ^{64}Cu per μg of protein.
7. As with ^{99m}Tc-labeled scVEGF, ^{64}Cu -loaded scVEGF should be used for further analysis, tissue culture, or animal experiments immediately. For example, for biodistribution studies in a mouse model, inject 3–5 μCi of ^{64}Cu-labeled scVEGF via the tail vein.

4. Notes

1. Prepare deoxygenated Milli-Q water by boiling fresh Milli-Q water for at least 5 min on a hot plate, followed by flushing it with argon (or nitrogen) for 10 min. Cool the flask on ice to room temperature and bubble with nitrogen for 75 min. Store deoxygenated Milli-Q water in a tightly closed flask at room temperature. If water has been stored for several days between uses, bubble more nitrogen through before use. Filter all solutions prepared with deoxygenated Milli-Q water through 0.2-μm filter and bubble nitrogen through for 15–20 min before use.
2. Check out pH of Tricine buffer, it should be 6.0 ± 2. If necessary, adjust pH with small amounts of 1 M NaOH or 1 M HCl until pH reaches 6.0 ± 2. For making this solution,

use acid-washed, thoroughly rinsed, and dried Erlenmeyer flask. Freshly made Tin-Tricine reagent can be stored frozen at -20°C in 1-mL aliquots for at least several months. Alternatively, freeze 1-mL aliquots at -80°C, then lyophilize. Reconstitute one vial with 1 mL of deoxygenated 1xPBS just before use.

To get a higher specific activity of ^{99m}Tc-labeled scVEGF-HYNIC (for imaging purposes), the amount of ^{99m}Tc in the reaction might be increased up to 50–60 mCi per 50 μg protein, with the respective increase of Tin-Tricine reagent in reaction mixture. This can result in ~1000 μCi/μg labeling efficiency. The increase of ^{99m}Tc in other reactions is much less effective. For ^{64}Cu-labeled scVEGF-PEG-DOTA, a higher specific activity can be achieved by a 2-3 fold decrease of the protein in the reaction mixture.

3. ^{99m}Tc and ^{64}Cu are radionuclides emitting γ-radiation at 140 keV and 511 keV, respectively. Standard shielding and radionuclide-handling procedures must be used. Typically, labeling is done in a lead brick–surrounded area, in a lead-shielded container. Individuals working with the material should monitor their radiation exposure with appropriate devices. As ^{99m}Tc and ^{64}Cu are short-lived isotopes ($t_{1/2}$ = 6.03 h and 12.70 h, respectively), injected animals and their waste products at the doses needed for biodistribution or imaging experiments do not represent any significant radiation hazard after 3–5 d (10 half-lives) of decay.

4. Continuous sonication of bacterial suspension or solubilized inclusion bodies will result in the significant increase of the temperature of the solution, which might be detrimental for the protein activity. To avoid overheating, place tubes with solutions for sonication on ice and start sonication only when the solutions are ice-cold. Importantly, keep every tube completely immersed in ice during the entire sonication step and do not apply ultrasound for longer than 60 s at a time. If additional sonication is needed, let the solution to cool down on ice, and then repeat sonication.

5. It may happen that inclusion bodies are not completely dissolved even after several rounds of sonication. In this case, proceed directly to the reducing step (**Subheading 2.2.1., steps 13–15**). In our experience, all traces of insoluble material disappear after 1–2 h of incubation in the presence of DTT and sulfonating agents.

6. Acidic dialysis provides tremendous purification from bacterial proteins, most of which precipitate at a pH lower than 6. However, not every recombinant protein remains soluble at acidic pH either. To test the solubility of your protein under these conditions, after at least 18–20 h of basic dialysis, transfer a small aliquot of the protein into a separate dialysis bag and put it in a precooled acidic dialysis buffer for 4–6 h, just long enough for bacterial proteins to form visible precipitation. Once the precipitate is formed, separate it by centrifugation in a table-top microcentrifuge for 5–10 min at 23,500 g. Analyze the presence of your protein in the supernatant and in the pellet by SDS-PAGE or Western blotting. If your protein is found in the supernatant, you can transfer the dialysis bag with the bulk of protein from basic dialysis conditions to acidic dialysis.

7. In our experience, as little as 2–4 h of incubation might be enough for insertion of 85–95% of lipid conjugate into liposome. Prolonged incubation (more than 16 h) does not improve the insertion; in fact, it might result in a decreased amount of liposome-associated protein.

Acknowledgments

This work was supported in part by NIH grants R43 CA113080 and R21 EB001946 to J.M.B. and by NIH-R01 EB000898 to F.G.B.

References

1. Backer, M. V., Gaynutdinov, T. I., Patel V., et al. (2005) Vascular endothelial growth factor selectively targets boronated dendrimers to tumor vasculature. *Mol. Cancer. Ther.* **4**, 1423–1429.
2. Blankenberg, F. G., Backer, M. V., Patel, V., and Backer, J. M. (2006) In vivo tumor angiogenesis imaging with site-specific labeled ^{99m}Tc-HYNIC-VEGF. *Eur. J. Nucl. Med. Mol. Imag.* **33**, 841–848.
3. Thirumamagal, B. T. S., Zhao, X. B., Bandyopadhyaya, A. K., et al. (2006) Receptor-targeted liposomal delivery of boron-containing cholesterol mimics for boron neutron capture therapy (BNCT). *Bioconjugate Chem.* **17**, 1141–1150.
4. Backer, M. V., Patel,V., Jehning, B. T., Claffey, K., and Backer, J. M. (2006) Surface immobilization of active vascular endothelial growth factor via a cysteine-containing tag. *Biomaterials* **27**, 5452–5458.
5. Backer, M.V. Patel, V., Jehning, B., and Backer, J. M. (2006) Self-assembled "dock and lock" system for linking payloads to targeting proteins. *Bioconjugate Chem.* **17**, 912–919.
6. Backer, M. V., Levashova, Z., Patel,V., et al. (2007) Molecular imaging of VEGF receptors in angiogenic vasculature with single-chain VEGF driven probes. *Nature Med.* **13**, 504–509.
7. Backer, M. V., Patel,V., Jehning, B. T., Claffey, K., Karginov, V. A., and Backer, J. M. (2007) Inhibition of anthrax protective antigen outside and inside the cell. *Antimicrob. Agent. Chemother.* **51**, 245–251.
8. Torchilin, V. P. (2005) Recent advances with liposomes as pharmaceutical carriers. *Nature Rev. Drug Discov.* **4**, 145–160.
9. Allen, T. M., and Martin, F. J. (2004) Advantages of liposomal delivery systems for anthracyclines. *Semin. Oncol.* **31**, 5–15.
10. Allen, T. M., Sapra, P., Moase, E., Moreira, J., and Iden, D. (2002) Adventures in targeting. *J. Liposome Res.* **12**, 5–12.
11. Sapra, P., Tyagi, P., and Allen, T. M. (2005) Ligand-targeted liposomes for cancer treatment. *Curr. Drug Deliv.* **2**, 369–381.
12. Ntziachristos, V. (2006) Fluorescence molecular imaging. *Annu. Rev. Biomed. Eng.* **8**, 1–33.

13. Graves, E. E., Weissleder, R., and Ntziachristos, V. (2004) Fluorescence molecular imaging of small animal tumor models. *Curr. Mol. Med.* **4**, 419–430.
14. Tait, J. F., Brown, D. S., Gibson, D. F., Blankenberg, F. G., and Strauss, H. W. (2000) Development and characterization of annexin V mutants with endogenous chelation sites for 99mTc. *Bioconjug. Chem.* **11**, 918–925.
15. Levashova, Z., Backer, M.V., Backer, J.M., Blankenberg, F.G. (2008) Direct labeling of Cys-tag in scVEGF with technetium 99m. *Bioconjugate Chem.* **19**, 1049–1054.

Index

A

Aβ peptides
    aggregation and interaction with lipids, 74–75
    capture of, 75–83
    concentration of, 80
    detection of, 75
    in pathogenesis of AD, 71–73
    ProteinChip technology analysis of, 71–83
    quantitation of, 74
    toxicity introduced by, 73
ACE inhibitors. *See* Angiotensin-converting enzyme
        inhibitors
Acetylation, technique of, 3
AD. *See* Alzheimer's disease
Alkylation
    of disulfide bonds, 34–35, 38–39
    of peptides, 15
Alzheimer's disease (AD)
    derived monoclonal antibodies, 5
    pathogenesis of, Aβ peptides role in, 71–73
    peptides in, 5, 71–73
Amine, 233, 239
    free, 230
    reactivity of, 240
Amino acids, 267. *See also* Glycosylated amino
        acids; Phosphoamino acids
    coupling, 268
    deprotection of, 268
    derivatives of, 138, 220
    isobaric, 32
    naturally occurring, 87, 95
    nonnatural, 1, 3, 267
    position of, defining, 60
    sequencing, 21, 134
    solutions, preparation of activated, 51
    unusual, 267
Aminoalkylphosphonate, activation, 232
AMPs. *See* Antimicrobial peptides
Amyloid precursor protein, cleavage of, 71–73
Angiotensin-converting enzyme (ACE) inhibitors, 4
Animals, peptide extraction from, 12
ANN. *See* Artificial neural networks
Antibacterial peptides
    modeling of, 149
        QSAR, 135–136, 150–154

Antibiotics
    bacteria producing peptide, 10
    bacteria's resistance against, 25–26,
        31–32, 127
    therapy, 31
Antibodies. *See also* Monoclonal antibodies
    arrays coated with, 76–80
    production, 263
Antigens
    binding, 78
    epitopes, 260
    peptide, 66, 264
    viral, 67, 156
Antimicrobial assays, 16
    liquid, 23–24
Antimicrobial peptides (AMPs). *See also* Cationic
        antimicrobial peptides
    bacterial infections and, 3
    circulation of, 31
    derivatives of, 3–4
    extraction methods for, 36
    as host-defense peptides, 10
    identifying, 31–33
    immune effectors as members of
        armamentarium of, 31
    in innate community, 32
    in invertebrates, 10
    isolation of, 3, 32
    naturally occurring, 32
    from noninsect arthropods, 11
    QSAR descriptors used for AMPs studies,
        137–143, 150–151
    QSAR studies of, 137–146
    QSAR used to predict activity of,
        135–154
    tandem mass spectrometry of, 31–46
Arrays. *See also* Peptide arrays
    antibody-coated, 76–80
    ProteinChip technology, 73–74, 76–80
Arthropods
    AMPs from noninsect, 11
    venom of, 10
Artificial neural network, 151
Artificial neural networks (ANN), 127, 150
    high-activity peptides prediction and, 154

phosphopeptides influencing changes specific
  to, 209
Disulfide bonds
  formation of native, 278–279
  reduction/alkylation of, 34–35, 38–39
Drug design
  carbohydrate and glycoconjugate chemistry
    in, 187
  examples of NMR applications in peptide-based,
    103–108
  lead components for novel, 177
  NOE-based methods applied in, 99–101
  peptide-based, 1–6, 103–108
  peptides in process of, 87–88
  rational, 88
  screening based approach to, 88–89
  synthetic peptides in, 178
Drugs. *See also* Peptide drugs
  antiinfectious, 25
  antitumoral, 25
  biopolymers targeted by designer, 1–2
  designer, 1–2
  NMR in development of peptide, 87–108
  peptide, 87–108, 223
  peptide arrays as tool in development of, 47
  peptides v. small molecule, 178
  targeted, 276
DSSP. *See* Dictionary of Protein Secondary
    Structure

E

E. coli
  Cys-tag and, 278
  PRP resistant, 165–166, 169
Edman sequencing, 13
Electron microscopy, 74
Electrospray ionization (ESI), 13, 32
  mass spectrometry, 14, 22–23, 34, 38, 44
ELISA. *See* Enzyme-linked immunosorbent assay
Enzyme-linked immunosorbent assay (ELISA), 255
Epithelial tissue, 12
Epitopes. *See also* Peptide epitopes
  antigen, 260
  T-helper cell, 263, 264
    as built, 268
  vaccines and, 248, 263–267
ESI. *See* Electrospray ionization
Evolution, 9

F

Fertility, of mice, 256
9-fluorenylmethoxycarbonyl (Fmoc)

based phosphopeptide synthesis on solid
    phase, 210
  deprotection, 54–55
  glycopeptide synthesis based on, 190–198
  peptide synthesis by chemistry, 214
  phosphoamino acid derivatives' direct
    incorporation, 211
  solid phase methodology, 210, 226
Fmoc. *See* 9-fluorenylmethoxycarbonyl
Force fields, 116–117

G

Gangliosides, 106
Genes
  PtrB, 172
  resistance-associated, identification
    of, 163
  SbmA, 172–173
  sequencing, 72
Glycans
  in O-linked glycoproteins as attached, 188
  structure of, 188
Glycoamino acids, 3
Glycoconjugate, 187
Glycopeptides
  assembly of, 187
  chemical synthesis of, 191
  family of molecules, 187
  genetically synthetic methodology
    for, 187
  isolation from natural sources of, 189
  synthesis based on Fmoc protective group
    therapy, 190–198
  synthesis of, 187, 189–198
  synthetic, 190
Glycoproteins. *See also* N-linked glycoproteins;
    O-linked glycoproteins
  isolation from natural sources of, 189
  mucin type, 188
  role of, 188
  synthesis of, 187, 189
Glycosides
  formation of, 193
  preparation of, 196
  synthesis of, 193, 195, 200, 205
Glycosidic bonds, formation of, 187, 191–188
Glycosylamine, synthesis from glycosyl azides,
    200–197
Glycosylated amino acids
  N-, synthesis of, 199
  O-, 192–195
    synthesis of, 193
  synthesis, 187–198

.

Printed in the United States
126361LV00003B/1-93/P